WIND ENERGY
THEORY AND PRACTICE
Third Edition

SIRAJ AHMED

Professor and Former Head
Department of Mechanical Engineering
Maulana Azad National Institute of Technology
Bhopal, India

PHI Learning Private Limited

Delhi-110092
2016

₹ 395.00

WIND ENERGY: Theory and Practice, Third Edition
Siraj Ahmed

© 2016 by PHI Learning Private Limited, Delhi. All rights reserved. No part of this book may be reproduced in any form, by mimeograph or any other means, without permission in writing from the publisher.

ISBN-978-81-203-5163-9

The export rights of this book are vested solely with the publisher.

Fourth Printing (Third Edition) ... ... **October, 2015**

Published by Asoke K. Ghosh, PHI Learning Private Limited, Rimjhim House, 111, Patparganj Industrial Estate, Delhi-110092 and Printed by Rajkamal Electric Press, Plot No. 2, Phase IV, HSIDC, Kundli-131028, Sonepat, Haryana.

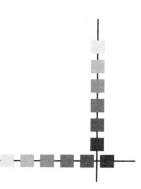

Contents

Preface ix
Acknowledgements xi
List of Symbols xiii
List of Abbreviations xv

1 Background   1–10

    1.1 Introduction *1*
    1.2 Modern Wind Turbines *2*
    1.3 Wind Resource *3*
    1.4 Wind Vs. Traditional Electricity Generation *4*
    1.5 Technology Advancements *5*
    1.6 Material Usage *6*
    1.7 Wind Energy Penetration Levels *6*
    1.8 Applications *6*
    Review Questions 9

2 Wind Resource Assessment   11–62

    2.1 Introduction *11*
        2.1.1 Spatial Variation *14*
        2.1.2 Time Variations *15*

2.2 Characteristics of Steady Wind 19
    2.2.1 Turbulence 20
    2.2.2 Types of Turbulence Models 21
    2.2.3 Turbulence Intensity 22
    2.2.4 Wind Power Density 23
2.3 Weibull Wind Speed Distribution Function 24
    2.3.1 Estimating Weibull Distribution Factor 26
2.4 Vertical Profiles of The Steady Wind 30
2.5 Wind Rose 32
2.6 Energy Pattern Factor 32
2.7 Energy Content of the Wind 32
2.8 Resource Assessment 33
    2.8.1 Measurement Plan 34
    2.8.2 Monitoring Duration 34
    2.8.3 Siting of Monitoring Systems 34
    2.8.4 Site-Specific Wind Data 36
    2.8.5 Topographic Indicators 37
    2.8.6 Field Surveys and Site Ranking 38
    2.8.7 Tower Placement 38
    2.8.8 Measurement Parameters 39
    2.8.9 Monitoring Station Instrumentation 40
    2.8.10 Remote Sensing Technique 47
Review Questions 56

3 Rotor Aerodynamic 63–109

3.1 Introduction 63
3.2 Aerofoil 64
    3.2.1 Two-dimensional Airfoil Theory 66
    3.2.2 Relative Wind Velocity 73
    3.2.3 Stall Control 77
3.3 Wind Flow Models 77
    3.3.1 Wind Flow Pattern 79
3.4 Axial Momentum Theory 80
3.5 Momentum Theory for a Rotating Wake 87
3.6 Blade Element Theory 89
3.7 Strip Theory 90
3.8 Tip Losses 91
3.9 Tip Loss Correction 93
    3.9.1 Condition for Maximum Power from Blade Element 93
    3.9.2 Optimum Design and Peak Performance 95
    3.9.3 Analytical Method for Optimum Design and Peak Performance Prediction* 96
3.10 Drag Translator Device 99

3.11 Wind Machine Characteristics *101*
      3.11.1 Fixed Speed and Variable Speed Machines *103*
      3.11.2 Load Matching *103*
Solved Examples *104*
Review Questions *106*

4. Wind Turbine   110–124

4.1 Introduction *110*
      4.1.1 Historical Aspects *111*
      4.1.2 Modern Wind Turbine *111*
      4.1.3 Current Wind Turbine Size *114*
4.2 Classification of Wind Turbines *115*
      4.2.1 Horizontal Axis Machines *115*
      4.2.2 Vertical Axis Machines *115*
4.3 Turbine Components *118*
      4.3.1 Power Train Subsystem *120*
Review Questions *124*

5. Wind Turbine Design   125–184

5.1 Introduction *125*
5.2 Rotor Torque and Power *126*
      5.2.1 Glauert Momentum Vortex Theory *130*
      5.2.2 Optimal Rotors *134*
      5.2.3 Dual Optimum *138*
5.3 Optimum Design for Variable Operation *139*
5.4 Influence of Reynolds Number *148*
5.5 Cambered Aerofoils *148*
      5.5.1 Rotor Sizing *150*
5.6 Load Calculation *151*
      5.6.1 Blade Loads *156*
      5.6.2 Rotor Loads *157*
      5.6.3 Codes and Standards *159*
      5.6.4 Design Loads *159*
5.7 Cost Modelling *161*
      5.7.1 Simplified Cost Model *161*
5.8 Power Control *163*
      5.8.1 Passive Stall Control *163*
      5.8.2 Active Pitch Control *164*
      5.8.3 Passive Pitch Control *164*
      5.8.4 Active Stall Control *164*
      5.8.5 Yaw Control *165*
      5.8.6 Teetering *165*

    5.9   Braking Systems   *165*
           5.9.1   Active Pitch Control Braking   *165*
           5.9.2   Braking by Pitching Blade Tips   *165*
    5.10   Turbine Blade Design   *166*
           5.10.1   Form of Blade Structure   *167*
           5.10.2   Blade Materials   *168*
           5.10.3   SERI Blade Sections   *171*
    5.11   Rotor Hub   *173*
    Review Questions   *175*

6 Siting, Wind Farm Design   185–238

    6.1   Introduction   *185*
    6.2   Wind Flow Modelling   *187*
           6.2.1   Physical Modelling   *187*
           6.2.2   Numerical Modelling   *188*
           6.2.3   Modelling Goals   *188*
    6.3   Capacity Factor   *189*
           6.3.1   Estimation of Capacity Factor   *190*
    6.4   Planning of Wind Farm   *194*
           6.4.1   Environmental Issues   *198*
           6.4.2   Constraints Map   *198*
    6.5   Siting of Wind Turbines   *199*
           6.5.1   Siting in Flat Terrain   *199*
           6.5.2   Uniform Roughness   *200*
           6.5.3   Changes in Roughness   *202*
           6.5.4   Shelterbelts   *204*
           6.5.5   Trees and Scattered Barriers   *204*
           6.5.6   Siting in Non-flat Terrain   *204*
           6.5.7   Ridges   *205*
           6.5.8   Isolated Hills and Mountains   *209*
           6.5.9   Passes and Saddles   *209*
           6.5.10   Gaps and Gorges   *210*
           6.5.11   Basins   *212*
           6.5.12   Cliffs   *212*
           6.5.13   Mesas and Buttes   *214*
    6.6   Ecological Indicators   *214*
    6.7   Site Analysis Methodology   *216*
    6.8   Layout of Wind Farm   *217*
    6.9   Initial Site Selection   *220*
    6.10   Measure-Correlate-Predict (MCP) Technique   *222*
    6.11   Micrositing   *222*
    6.12   Wake Models   *224*
           6.12.1   Park and Modified Park Model   *224*

    6.12.2 Eddy Viscosity Model *225*
    6.12.3 Initialization of the Model *226*
    6.12.4 Turbulence Intensity *227*
  6.13 Re-Powering *228*
  Review Questions *231*

7 Wind Energy Economics   239–273

  7.1 Introduction *239*
    7.1.1 Cost Calculation *240*
  7.2 Annual Energy Output (AEO) *241*
  7.3 Time Value of Money *243*
    7.3.1 Present Worth Approach *243*
    7.3.2 Inflation *246*
  7.4 Capital Recovery Factor *248*
  7.5 Depreciation *249*
  7.6 Life Cycle Costing *250*
    7.6.1 Cost of Wind Energy *251*
    7.6.2 Present Value of Annual Costs *253*
    7.6.3 Value of Wind Generated Electricity *255*
    7.6.4 Economic Merit *256*
    7.6.5 Net Present Value *256*
    7.6.6 Benefit Cost Ratio *257*
    7.6.7 Pay Back Period *257*
    7.6.8 Internal Rate of Return *258*
    7.6.9 Tax Deduction due to Investment Depreciation *261*
    7.6.10 Straight Line Depreciation *261*
    7.6.11 Declining Balance Depreciation *261*
    7.6.12 Sum of the Years' Digit Depreciation *262*
    7.6.13 Wind Energy Economics Worksheet *264*
  7.7 Project Appraisal *266*
    7.7.1 Project Finance *269*
  Solved Examples *270*
  Review Questions *273*

8 Environmental Impact   274–289

  8.1 Introduction *274*
  8.2 Biological Impact *275*
    8.2.1 Ecological Assessment *276*
    8.2.2 Avian Issue *277*
  8.3 Surface Water and Wetlands *278*
  8.4 Visual Impact *278*
    8.4.1 Shadow Flicker *279*
  8.5 Sound Impact *280*
    8.5.1 Measurement, Prediction and Assessment *285*

    8.6 Communication Impact   *286*
         8.6.1 Electromagnetic Interference (EMI)   *287*
    Review Questions   *288*

9 Electrical and Control Systems   290–319

    9.1 Introduction   *290*
    9.2 Classification of Generators   *296*
    9.3 Synchronous Generators   *297*
    9.4 Induction Generator   *298*
    9.5 Variable-Speed Generators   *299*
    9.6 Control Systems   *300*
         9.6.1 RPM Control (Mechanical)   *300*
         9.6.2 Electronic Controls   *301*
         9.6.3 Electrical Cut-in   *302*
    9.7 Power Collection System   *302*
    9.8 Earthing (Grounding) of Wind Farms   *303*
         9.8.1 Lightning Protection   *304*
    9.9 Wind Power Integration into Grid   *306*
         9.9.1 Power System Stability   *307*
         9.9.2 Economics of Grid Network   *307*
         9.9.3 Codes and Standards for Grid Integration   *307*
    9.10 Embedded (Dispersed) Wind Generation   *309*
         9.10.1 Electrical Distribution Networks   *309*
         9.10.2 Power Flows, Slow Voltage Variations and Network Losses   *311*
         9.10.3 Power Quality   *313*
         9.10.4 Wind Farm and Generation Protection   *314*
         9.10.5 Interface Protection   *315*
         9.10.6 Losses in Generation   *316*
    Review Questions   *316*

Appendix A *Units and Conversion*   *321–326*

Appendix B *Wind Characteristics of MANIT, Bhopal, India*   *327–333*

Appendix C *Newton–Raphson Method*   *334–335*

Glossary   *337–345*

Bibliography   *347–351*

Index   *353–360*

Preface

The book, now in its Third Edition, covers all the important aspects of, and advances in, wind energy. In this edition, substantial addition has been made in chapters on wind resource assessment (remote sensing application in wind energy), siting (re-powering) and electrical and control system (wind power integration into grid).

As wind energy technology grows in various stages nationally and internationally, the need for trained human resources continues to be a significant driver in future employment. This has resulted in universities and institutes offering various UG and PG programmes in Renewable Energy, and Wind Energy remains a core subject.

This book is intended for the undergraduate and postgraduate students of Mechanical/Electrical Engineering and students pursuing Renewable Energy Studies. It will also serve as a handbook and ready reference for practicing engineers and professionals in the field of Wind Energy.

Any suggestion for further improvement of the book is most welcome.

Siraj Ahmed

Acknowledgements

I gratefully acknowledge MANIT for providing me the initial opportunity and exposure to undertake training in wind energy technology for six months at Salford University under Indo-UK, REC Project of Energy Theme (1995–96). I gratefully acknowledge the following persons and sources of information for writing of the book:

- Central Institute of Agricultural Engineering, Bhopal.
- Central Library, IIT Delhi.
- Consolidated Energy Consultants Ltd. and M.P. Wind Farm, Bhopal, its website www.windpowerindia.com and popular Yearly Directory on Wind Power.
- ISTE for sponsoring and preparation of Lecture Notes in Vol. I and II for a Winter School on Wind Energy and Its Utilization, 2000.
- HEG Ltd. for a consultancy project on wind resource assessment, 2004.
- Ministry of Human Resource Development, GOI for sponsoring a project on Development of Wind Energy Laboratory, 2003–06.
- TEQIP for sponsoring and preparation of Lecture Notes for Short-Term Training Programme on Advances in Wind and Solar Energy Technology, 2006 and on Recent Development in Solar and Wind Energy System, 2008.
- Vipo Energy Resources Inc., Harte Ranch, Denver, Colorado, USA and Mr. Rohit Poddar for the International Consultancy Project on Wind Farm Development, 2007–10.

- Notes of Dr. S.B. Kedare of CEP Training Programme on Wind Energy Technology, IIT Mumbai, 2007 and 2008.
- MHRD and AICTE for sponsoring and preparation of Lecture Notes for Staff Development Programme on Solar and Wind Energy: Theory and Practice, 2009.
- National Institute of Wind Energy (NIWE) Chennai and its website www.niwe.res.in.
- American Wind Energy Association and its website www.awea.org.
- Ammonit Gmbh, Berlin-Germany and their website www.ammonit.com
- Seimens Wind Power A/S, Brande, Denmark and their website www.siemens.com/wind.
- Lecture Notes of Ph.D. Summer School: Remote Sensing for Wind Energy, 10–14 June 2013. DTU Wind Energy, Risoe Campus, Roskilde, Denmark.

I remain indebted to my teachers for their guidance and inspiration to take such work, particularly to Shri P.D. Tiwari, Dr. T.P. Sahay, Dr. Abdul Mubeen, Dr. S.M. Yahya, Dr. T.K. Kundra, Dr. P.N. Rao, Dr. A.R. Siddiqui, Dr. R.L. Gupta, Prof. P. Krishnamachar, Dr. R.A. Sawyer, Dr. M.C. Sharma, Dr. S.C. Soni, Dr. M.M. Pandey, Dr. A.R. Ansari, Dr. R.K. Dube, Dr. R.P. Singh and many more for showing me the light and direction on the path to way forward.

I am thankful to my students, friends and colleagues, especially to Dr. Subramanyam Ganesan, Dr. Nilesh Diwakar, Vilas Warudkar and many others for their support, encouragement and valuable suggestions from time to time.

I am thankful to my sons Moin and Shams, wife Seemi and mother Jahanara for their affection and support and being perpetual source of confidence and inspiration.

Finally, I wish to thank my publisher, PHI Learning for accepting to publish this book as well as their editorial and production team.

Siraj Ahmed

List of Symbols

P	Power	z	Vertical coordinate
P_e	Electrical power	U	Wind speed
p	Pressure	W	Relative wind velocity
p^+	Pressure upstream	U_∞ or V_∞	Unperturbed wind velocity
p^-	Pressure downstream	V_P	Particular wind speed
p_o	Standard sea level atmospheric pressure, 101325 Pa	$\bar{V}$	Average wind speed
g	Gravitational constant	U_d	Annual average wind speed
ρ	Density of air	U_C	Cut-in wind speed
A	Swept area of rotor	U_R	Rated wind speed
A_r	Swept area of rotor	U_F	Furling wind speed
C_P	Coefficient of power	$U(t)$	Instantaneous wind speed
kW	Kilowatt	$g(t)$	Fluctuating wind speed
t	Time	$U(y, z, t)$	Instantaneous horizontal free-stream wind velocity field
Δt	Time interval	$U(y, z)$	Steady horizontal free-stream wind velocity field
x	Axial coordinate	$g(y, z, t)$	Fluctuating wind velocity field; instantaneous deviation from $U(y, z)$
y	Lateral coordinate		

List of Symbols

σ	Standard deviation of wind speed
n	Number of records in the averaging interval
V_i	ith wind speed value
R	Specific gas constant for air, 287 J/kg °K
T	Air temperature in degree Kelvin
z	Elevation
f_U	Frequency distribution function of the steady wind speed, (h/y) (m/s)
$F(\ )$	Annual time that ( ); cumulative frequency distribution function, (h/y)
U_I	Arbitrary value of U
$e^{[\ ]}$	Exponent function of [ ]
C	Empirical Weibull scale factor, m/s
k	Empirical Weibull shape factor
$\Gamma(\ )$	Gamma function of ( )
α	Empirical wind shear component or angle of incident
C_l or C_L	Lift coefficient
C_d or C_D	Drag coefficient
C_m or C_M	Moment coefficient
C_t or C_T	Turbine thrust coefficient
c	Chord length
L	Lift force
D	Drag force
A_b	Blade area (chord × length)
Re	Reynolds number
Ω	Angular velocity of rotor
R	Radius of blade
r	Local radius of blade
λ	Tip speed ratio
λ_r	Local tip speed ratio
a	Axial interference factor
a'	Angular interference factor
ε	Ratio of drag coefficient to lift coefficient
μ	Viscosity
ν	Kinetic viscosity
$\varepsilon(x)$	Eddy viscosity
ε_{amb}	Ambient eddy viscosity
β	Blade setting angle
ϕ	Relative wind angle
θ	Blade angle
B	Number of blades
ϕ_T	Angle of helical surface with the slip stream boundary
$H(U_i)$	Number of hours in wind speed bin U_i
$P(U_i)$	Power output in wind speed bin U_i
K_K	von Karman Constant
$RMS[\ ]$	Root mean square average of [ ]

List of Abbreviations

AEO	Annual energy output
ALCC	Annualised life cycle cost
AWEA	American wind energy association
B/C	Benefit/Cost
BMP	Bit map picture
CF	Capacity factor
CFRP	Carbon filament-reinforced plastic
COE	Cost of energy
CRF	Capital recovery factor
dB	Decibels
dBA	Decibel (on an A-weighted scale)
DCF	Discount cash flow
DS	Danish standard
EC	Energy content
EMI	Electromagnetic interference
EPF	Energy pattern factor
EPROM	Extended programmable read only memory
GHG	Green house gases
GL	Germanischer Lloyd
GPS	Geographical positioning system
GRP	Glass fibre-reinforced plastic
HAWT	Horizontal axis wind turbine
Hz	Hertz
IEC	International electromagnetic council
ITCZ	Inter tropical convergence zone
KE	Kinetic energy
LCC	Life cycle costing
LIDAR	Light detection and ranging
MCP	Measure correlate predict
mph	Miles per hour
NACA	National Advisory Committee for Aeronautics

NPV	Net present value	SCADA	Supervisory control and data acquisition
NREL	National Renewable Energy Laboratory	SODAR	Sound detection and ranging system
Pa	Pascal	TN	True north
PDF	Probability density function	UTM	Universal transverse mercator
PFC	Power factor correction capacitors	VAWT	Vertical axis wind turbine
PPA	Power purchase agreement	WECS	Wind energy conversion system
RPM	Revolution per minute	WPD	Wind power density
SAR	Synthetic aperture radar	WTG	Wind turbine generator
		ZVI	Zone of visual influence

CHAPTER 1

Background

1.1 INTRODUCTION

The extraction of power from wind is an ancient endeavour, beginning with wind-powered ships, grain mills and threshing machines. The fact that the mechanical insight and technology of sailing ships was carried over into wind mills is evident from the canvas 'sails' of the Cretian and Dutch windmills. Only towards the beginning of the last century has the development of 'high speed' wind turbines for generation of electrical power been taken seriously.

Engineers are uncomfortable calling a machine a wind mill when it is in fact not a mill, and so have developed special terms such as wind turbine, wind turbine generator (WTG), wind energy conversion system (WECS), aerogenerator, horizontal-axis wind turbine (HAWT), vertical-axis wind turbine (VAWT), and others. Even the term 'wind generating system' was used but later discarded because it seemed to imply that the machine is generating the wind, not electricity. In this book, wind turbine will be used for a machine with rotating blades that converts the kinetic energy of a flow of air (wind) into useful power.

Wind energy has been in use since the earliest civilization to grind grains, pump water from deep wells, and power sail boats. Wind mills in pre-industrial Europe were used for many things, including irrigation or drainage pumping, sawmilling of timber, and the processing of other commodities such as spices, cocoa, paints and dyes, and tobacco. The stimulus for the development of wind energy in 1973 was due to the rising price of oil and concern over limited fossil-fuel resources. The main driver for use of wind turbines to generate electrical power is the very low CO_2 emissions over the entire life cycle of manufacture, installation, operation and decommissioning, and the potential of wind energy to help limit climate change.

The wind turbine to convert the power of the wind into electricity has many advantages, especially in parts of the world where the transmission infrastructure is not fully developed. It is modular and can be installed relatively quickly, so it is easy to match electricity supply and demand. The fuel, the wind, is free and plentiful, which eliminates or reduces the need to purchase, ship, and store expensive fuels. It is flexible, with the power generated, households can use appliances, such as lighting; schools can use computers; and industries can access reliable power source. Small wind turbines are becoming economical for power production for homes, villages and industries that may be remote from an established grid or wish to operate without burning fossil fuels. Large wind turbines can be price competitive with any other form of generating technology in good wind resource areas.

The reliable forecasting of wind variability on time-scale of 24 hours to 72 hours for operation of a wind power station is one of the research goals the reliability of electricity generation from wind energy.

1.2 MODERN WIND TURBINES

Consider a cylinder in space of cross-section A and wind is passing through it at a speed V, then rate of kinetic energy passing at any section will be

$$\text{Power} = \frac{1}{2} \dot{m} V^2 = \frac{1}{2} (\rho A V) V^2$$

where $\dot{m}$ is mass flow rate, ρ is the density of air (1.225 kg/m³ at sea level). If a wind turbine is placed inside it and part of this power is transferred to this wind turbine then the power output P, from wind turbine is given by the expression

$$P = \frac{1}{2} C_p \rho A V^3 \qquad (1.1)$$

C_P is the power coefficient, that is part of kinetic energy from wind is transferred to turbine, A is the rotor swept area.

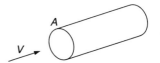

FIGURE 1.1 Wind front passing through a cylinder.

The density of air is rather low, 800 times less than that of water which powers hydro plant, and this leads directly to the large size of a wind turbine. The power coefficient describes that fraction of the power in the wind that may be converted by the turbine into mechanical work. Not all of the energy present in a stream of moving air can be extracted; some air must remain in motion after extraction. Otherwise, no fresh, more energetic air can enter the device. Building a wall would stop the air at the wall, but the free stream of energetic air would just flow around the wall. On the other end of spectrum, a device that does not slow the air is not extracting any energy either. The maximum energy that can be extracted from a fluid stream by a device with the same working areas as the stream cross section is only 16/27 part of the energy in the stream. As it was first derived by wind turbine pioneer Albert Betz, this maximum is known as the *Betz limit*. It has a theoretical maximum value of 0.593 (the Betz limit) and rather lower peak values are achieved in practice. The power coefficient of a rotor varies with the tip speed ratio, that is, the ratio of rotor tip speed to free wind speed and is only a maximum for a unique tip speed ratio. Incremental improvements in the power coefficient are continually being sought by detailed design changes of the rotor and, by operating at variable speed, it is possible to maintain the maximum power coefficient over a range of wind speeds. However, these measures will give only a modest increase in the power output. Major increase in power output can only be achieved by increasing the swept area of the rotor or by locating the wind turbines on sites with higher wind speeds. A doubling of the rotor diameter leads to four-times increase in power output. The influence of wind speed is, of course, more pronounced with a doubling of wind speed leading to an eight-fold increase in power. Thus, sincere efforts have been made to ensure that wind farms are developed in the areas of the highest wind speeds and the turbines optimally located within wind farms.

All modern electricity-generating wind turbines use the lift force derived from the blades to drive the rotor. A high rotational speed of the rotor is desirable in order to reduce the gear box ratio required, and this leads to low solidity rotor. The 'solidity' is defined as the ratio of blade area to rotor swept area.

Now with the advent of variable speed generators using permanent magnets with power electronics the wind turbines are designed as gear-less machines.

1.3 WIND RESOURCE

Winds result due to the earth's equatorial regions receiving more solar energy than the polar regions and this sets up large-scale convection currents in the atmosphere. Meteorologists estimate that about 1 per cent of the incoming solar radiation is converted to wind energy. Solar energy received by the earth in just ten days has an energy content equal to the world's entire fossil fuel reserves (coal, oil and gas). This means that the wind resource is extremely large. One per cent of the daily wind energy input, i.e. 0.01 per cent of the incoming solar energy is equivalent to the present world daily energy consumption. No one would suggest that wind energy is panacea, for energy crisis but as progress is made, technically and economically it can contribute substantially. It is encouraging to know that the global wind resource is so large and is very widely distributed.

Although global resource estimates have relevance, more detailed assessments are required to quantify the resource in a particular area. Winds are only an intermittent source of energy, day-by-day, though they represent a reliable energy source year-by-year. The many Grid System Integration studies for wind generated electricity completed in recent years show that the intermittency of wind energy is no barrier to its large scale use. It is now clear that the major applications of wind energy will involve electricity generation, with the wind turbines operating in parallel with utility grid systems or—in more remote locations. The wind turbines role is to reduce fossil fuel consumption and to reduce overall electricity generation costs.

Analysts use two measures to describe the role of wind in energy mix: units of capacity (MW), the amount of power wind turbines are capable of producing; and units of energy in kilowatt-hour(s) (kWh), the amount of electricity they actually generate. Capacity is an effective shorthand for describing the relative position of wind energy in nations around the globe. Yet, it is just that, a short hand. Capacity is never a substitute for the kWh of electricity generated by wind turbine, the bottom line in any discussion of wind energy. The goal of those who consider wind energy a timely, economical, and environmentally sensible way to meet society's energy needs, after all, is the production of wind-generated electricity, it is useful to keep that perspective in view. It is not the number of wind turbines, or even their size, that is important, it is the amount of energy they generate.

1.4 WIND VS. TRADITIONAL ELECTRICITY GENERATION

Wind power avoids several of the negative effects of traditional electricity generation from fossil fuels:

- Emissions of mercury or other heavy metals into the air.
- Emissions associated with extracting and transporting fuels.
- Lake and streambed acidification from acid rain or mining.
- Water consumption associated with mining and electricity generation.
- Production of toxic solid wastes, ash or slurry.
- Greenhouse gas (GHG) emissions.
- Each Megawatt-hour generated by wind could save as much as 600 gallons of water that would otherwise be lost for cooling of power plant based on fossil fuel.
- A single 1.5 MW wind turbine at its rated power displaces 2700 tons of CO_2 per year as studies show.
- An important factor to note is that wind energy projects use the same land area each year, coal and uranium must be mined from successive areas, with total distributed area increasing each year.
- Wind projects in agricultural areas, land use for wind generation projects has the potential to be compatible with some uses because only a few hectares are taken out of production, and no mining or drilling is needed to extract the fuel.

For the feasibility of a wind energy project, the following points are needed to be examined:

- Wind resource potential
- Access to the site for man, material and equipment
- Logistics
- Detailed analysis of a site
    - Meteorological station
    - Wind map atlas
    - Wind regime
    - Flat or complex terrain
    - Wind rose
    - Any other important influencing point
- Planning permission
- Power purchase agreement (PPA)
- Financing of project

- Planning of construction and actual construction
- Commissioning
- Sale of wind farm
- Operation and management
- Repowering or dismantling of wind farm after its life period.

1.5 TECHNOLOGY ADVANCEMENTS

The improvements due to technological advancements in various areas of wind turbine system are summarised in Table 1.1.

TABLE 1.1 Areas of potential technology improvement

Technical area	Potential advances
Advanced tower concept	• Taller tower in difficult locations • New materials and/or processes • Advanced structures/foundations • Self-erecting, initial, or for service
Advanced (enlarged) rotors	• Advanced materials • Improved structural aero-design • Active controls • Passive controls • Higher tip speed/lower accoustics
Reduced energy losses and improved availability	• Reduced blade soiling losses • Damage-tolerant sensors • Robust control systems • Prognostic maintenance
Drive train (Gear box, generators and power electronics)	• Fewer gear stages or direct-drive • Medium/low speed generators • Distributed gear box topologies • Permanent–magnet generators • Medium–voltage equipment • Advanced gear tooth profiles • New circuit topologies • New semi conductor devices • New materials
Manufacturing	• Sustained, incremental design and process improvements • Large-scale manufacturing • Reduced design loads

1.6 MATERIAL USAGE

The availability of critical resources is crucial for large-scale manufacturing of wind turbines. The most important resources are steel, fiberglass, resins for composites and adhesives, blade core materials, permanent magnets and copper.

- Turbine material usage is, and will continue to be, dominated by steel.
- Opportunities exist for introducing aluminium or other light weight composites, provided that cost, strength and fatigue requirements can be met.
- Glass-fibre-reinforced plastic (GRP) is expected to continue to be used for blades.
- The use of various composites and carbon filament-reinforced plastic (CFRP) might help reduce weight and cost.
- Low cost of materials and high reliability remain the primary drivers.
- Variable-speed generators will become more common.
- Permanent-magnet generators on larger turbines will increase the need for magnetic materials.
- Simplification of the nacelle machinery might reduce raw materials costs and also increase reliability.

1.7 WIND ENERGY PENETRATION LEVELS

Three different measures are used to describe wind energy penetration levels: energy penetration, capacity penetration, and instantaneous penetration. They are defined and related as follows:

Energy penetration: It is the ratio of the amount of energy delivered from the wind electricity generation to the total energy delivered. For example, if 200 MWh of wind energy is supplied and 1,000 MWh is consumed during the same period, wind's penetration is 20 per cent.

Capacity penetration: It is the ratio of the nameplate rating of the wind plant capacity to the peak load. For example, if 300 MW wind plant is operating in a zone with a 1,000 MW peak load, the capacity penetration is 30 per cent. The capacity penetration is related to energy penetration by the ratio of the system load factor to the wind plant capacity factor. Say the system load factor is 60 per cent and the wind plant capacity factor is 40 per cent. In this case, and with an energy penetration of 20 per cent, the capacity penetration would be 20 per cent × 0.6/0.4 or 30 per cent.

Instantaneous penetration: It is the ratio of the wind plant output to load at a specific point in time, or over a short period of time.

1.8 APPLICATIONS

Grid connected power

The cost of utility-scale wind power has been steadily declining. In a good wind regime, wind power can be least-cost resource. Wind power can help diversify a country's energy resources and can bring construction and maintenance jobs. In large-scale wind power applications, there are two keys to developing the most cost-effective projects: wind speed and project size. Since the power output is so highly dependent on the wind speed, differences in one metre per second can mean differences of several kWh in a year. Wind projects are also subject to scale economies. In general, given the same wind speed, a large project will be more cost-effective than a small one.

Industrial applications

The number of dedicated industrial applications for wind power continues to grow. Small wind power systems are ideal for applications where storing and shipping of fuel is uneconomical. Wind power can be used for:

- Telecommunications
- Radar
- Pipeline control
- Navigational aids
- Cathodic protection
- Weather stations/seismic monitoring
- Air-traffic control

Stand-alone system

In many places, wind power is the least-cost option for providing power to homes and businesses that are remote from an established grid. It is estimated that wind produces more power at less cost than diesel generators at any remote site with a wind power density above 200 W/m^2 at 50 m elevation.

Table 1.2 gives a representative idea of the power requirements of some household appliances. Wind turbine performance depends primarily on rotor diameter and wind speed. Table 1.3 gives an estimate of a wind turbines output, based on wind speed rotor diameter, in Whr/day.

TABLE 1.2 Household appliances details

Item description	Power, Watt	Daily energy Whr/Day
Incandescent light	60	1,800–720
Fluorescent light	15	45–180
Refrigerator	80–500	2,000–10,000
Television	15–100	30–600
Village household	60–300	300–1,200

TABLE 1.3 Output of wind turbine

Rotor diameter, (m)	Wind speed, m/s			
	4	5	6	7
1.5	35	70	94	117
3	152	269	386	468
5	421	761	1,053	1,287
7	831	1,522	2,107	2,575

Water pumping

The mechanical water pumping and/or lifting is the best option in some circumstances. However, because it must be placed close to the water source, it is often unable to capture the best wind resources. A wind electric pumping system overcomes some of the problems with the simple wind water pump. The system generates electricity, which, in turn, runs an electric pump. Wind electric pumping system allows greater siting flexibility, higher efficiency of wind energy conversion, increased water output, increased versatility in use of output power, and decreased maintenance and life-cycle costs.

Table 1.4 gives an idea about the water requirement. The size of the wind turbine required for water delivery depends on the average daily volume of water required, the total pumping head, the average wind speed, and the system efficiency. The total pumping head is the sum of the static head (the distance from water level below ground to water outlet at the water storage container) plus well drawdown (the level to which the water drops during pumping) and pipe friction.

TABLE 1.4 Water requirements

Type of use	*Consumption* (litres/day)
Household	
Drinking, cooking, sanitation and other uses	40–100 per person
Garden	25 per m² of land
Farm	
Cattle	20–40 per head
Sheep	2–6 per head
Chicken	20–90 per 100 birds
Sanitation	100–500

During pump operation, additional pumping head will be required due to well drawdown and frictional losses within the pipe. The additional head can be estimated at 10–15 per cent of the static head if more accurate information is not available.

A wind electric pumping system should be sized according to the months of highest water demand. Table 1.5 provides a rough guide to how much water can be delivered with different rotors in various wind regimes.

TABLE 1.5 Rotor sizing vs wind speed, m/s

Diameter, m	3	4	5	6
1.5	60	150	290	500
3.0	250	590	1,160	2,000
5.0	700	1,650	3,220	5,560
7.0	1,360	3,230	6,310	10,900

Offshore prospects

Offshore wind energy is not cheap. In fact, it can cost twice as much as land-based wind power. Offshore has great wind capacity factor, it blows at the right times, and the transmission challege is not difficult. Offshore wind energy project development requires a more complex design to bear up under storms, waves and tides. Thus, while turbines represent the greatest expense for land-based wind energy project, the foundation, towers, transmission lines and installation tend to account for offshore projects. Foundations, in particular, are costly for offshore facilities because they require more steel, concrete and specialised construction techniques. Water depth plays a big role as each metre of tower height adds more money in the project.

To offset the cost, off-shore wind energy needs larger scale with turbine of at least 5 MW for better economics than smaller projects. Offshore wind energy project capital cost must be put into perspective. First, wind velocity is higher over the unobstructed ocean surface, which, in turn, leads to greater capacity factor, more energy production, and greater revenue for off-shore wind farms than their onshore counterparts.

REVIEW QUESTIONS

1. A car with a mass of 1,000 kg moving at 15 m/s has kinetic energy
    $$KE = 0.5 \, mV^2$$
    $$= 0.5 \times 1,000 \times (15)^2 = 1,12,500 \text{ J}$$
2. A 5 kW electric motor that runs for 2 h consumes 10 kWh of energy.
3. Ten 100-watt light bulbs that are left on all day will consume 2.4 kWh of energy.
4. An electric generating plant uses steam at 700°C (973 K) and on the downside the steam is cooled by water to 300°C (573 K). The maximum efficiency possible is around 0.41 or 41 per cent.

    Since efficiency is always less than 1, for a system to continue to operate, energy must be obtained from outside the system. For a series of energy transformations there is a total efficiency which is the product (multiply) of the individual efficiencies.
5. Efficiency of incandescent lights in your home from a coal-fired plant:

Transformation	*Efficiency, (%)*
Mining of coal	96
Transportation of coal	97
Generation of electricity	38
Transmission of electricity	93
Incandescent bulb (electricity to light)	5
Overall efficiency	1.6

6. Present consumption of energy is 100 units/year and growth rate is 7 per cent per year. What would be the consumption per year after 100 years?

    [**Hint:** Value of future consumption, r, can be calculated from the present rate, r_o, and the fractional growth per time, k:

    The exponential growth, where e is the base of the natural log and t is the time,
    $$r = r_o e^{kt}$$
    $r_o = 100$ units/year, $k = 0.07$/year, $t = 100$ years

$$r = 100e^{0.07 \times 100} = 100e^7$$
$$= 100 \times 1{,}097$$
$$= 10^5 \text{ per year}$$

The consumption per year after 100 years is 1,000 times larger than the present rate of consumption.]

Note: Exponents never have any units associated with them.

7. List three ways in which you are going to save energy this year.
8. What is doubling time if the growth rate is 0.5 per cent? The world population according to Population Reference Bureau's in 2008 was around 6.7×10^9.
9. Assume electricity demand increases by 10 per cent per year over the next 30 years for the world. To meet all the increased demand, how many 1,000 MW wind farm would have to be installed by the end of 30 years? What is the total cost for those wind farms at $2200/kW?
10. For your house/hostel, estimate the power installed for lighting. Then, estimate the energy used for lighting for 1 year.
11. Estimate the wind installed capacity in coming 5 years for the world and for India.
12. Estimate the capacity of offshore wind farms in India and the world today.
13. What factors contributed to the initial wind farm boom in India?

Chapter 2

Wind Resource Assessment

2.1 INTRODUCTION

Wind is air in motion relative to the surface of the earth. The wind vector is considered to be composed of a steady wind plus fluctuations about the steady wind. The primary cause of air motion is uneven heating of the earth by solar radiation. By and large, the air is heated directly, but solar radiation is first absorbed by the earth's surface and is then transferred in various forms back into the overlying atmosphere. Since the surface of the earth is not homogeneous, i.e. land, water, desert, and forest, the amount of energy that is absorbed varies both in space and time. This creates difference in atmospheric temperature, density and pressure, which in turn create forces that move air from one place to another. For example, land and water along a coastline absorb radiation differently, as do valleys and mountains, and this creates breezes.

During the year, tropical regions receive more solar energy than they radiate to space, and polar regions receive less. Since the tropics do not get hotter continuously from year-to-year nor do the poles get colder, there is an exchange of thermal energy across latitudes with wind as the medium of convection. In addition to uneven heating, a second important factor in large-scale air movement is the earth's rotation, which gives rise to two effects. First, the rotation results in *Coriolis forces* which accelerate each moving particle of air. This acceleration moves an air particle to the right of its direction of motion in the Northern hemisphere, and to the left in the Southern hemisphere. When air movements reach a steady state, Coriolis forces balance pressure gradient forces, leaving a resultant motion approximately along isobars, i.e. lines of equal pressure, which is called the *geostrophic wind*.

The second effect of the earth's rotation on wind flow becomes apparent in the mid-latitudes. By virtue of the rotation, each air particle in the atmosphere has an angular momentum directed from west-to-east. As a particle moves towards the poles, remaining at approximately the same altitude, it draws closer to the axis of rotation. The conservation of angular momentum requires an increase in the component of its velocity in the west-to-east direction. This effect is small near the equator, but in the temperate zones, it accounts for the Westerlies which are in a direction opposite to the general flow, in both hemispheres.

Solar heating and rotation of the earth establish certain semi-permanent patterns of circulation in the atmosphere, as illustrated in Figure 2.1. From the foregoing discussion, it is evident that regions of the earth in the Westerlies are preferable to others for extracting power from the wind in the boundary layer of the atmosphere. In addition to these major wind creating factors, local topographical features may alter the energy distribution considerably, so many exceptions to this general picture exist.

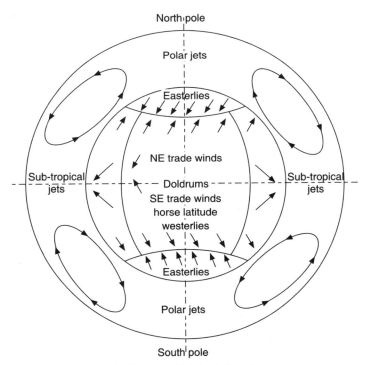

FIGURE 2.1 Global wind pattern.

The atmosphere operates on many time and space scales, ranging from seconds and fraction of a metre to years and thousands of kilometres. Time and space scales of the atmospheric motions and their importance to wind energy utilization are summarized in Figure 2.2. The very large or climatic scale includes seasonal and annual fluctuations in the wind, which are useful for assessing regional wind resources. On a scale comparable with the weather maps are large-scale synoptic fluctuations identified by the patterns of isobars moving across a country. These large-scale fluctuations influence the output of wind power stations, so the selection of sites depends on the

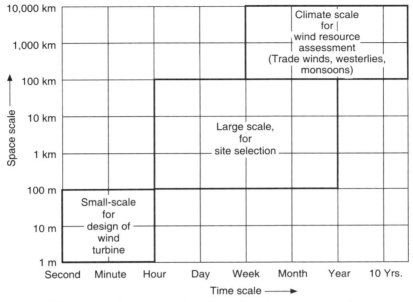

FIGURE 2.2 Temporal and spatial scales of atmospheric motion.

knowledge of regional atmospheric motions. Small-scale fluctuations, which are even more local in size and higher in frequency, are best observed on local anemometre records, providing data for wind turbine design and micro siting (i.e. siting of individual turbines).

Given the record of the meteorological variable, such as the horizontal wind speed, we can always decompose it into components representing a time or spatial average and fluctuations superimposed on the average. After doing this, we view the average value of the variable as a deterministic quantity and assign statistical characteristics to only the fluctuations. Wind loadings on the turbine can also be classified into two parallel groupings: those associated with the mean wind speed (which are conveniently described as quasi-steady or time averaged loadings) and those associated with the gustiness or turbulence of the wind (which are predominantly dynamic in character).

Figure 2.3 is a wind energy spectrum which shows that the majority of the fluctuating energy is contained at the macro and micro meteorological scale, and that a region of low energy exists between them. The spectral gap for periods between 0.1 and 5 hours defines a convenient range of averaging periods, Δt, to which a steady wind speed can be referenced. The period should be long enough to minimize non-stationary, and short enough to reflect short-term stormy activity. Periods from 10 minutes to one hour have been found to be suitable for defining the steady wind speed, giving it only a weak dependence on the averaging period.

Steady wind speeds required for wind resource assessments may vary throughout each day and from day-to-day, but they are essentially constant, relative to the dynamic response frequencies of typical wind turbines. Much of the wind data used in the assessment methodologies, however, is derived from one-minute averages centred on the hour and sampled once an hour or once in three hours (one-hour and three-hour readings, respectively). Figure 2.3 clearly shows that a

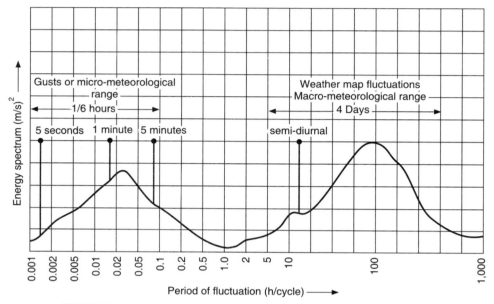

FIGURE 2.3 Energy spectrum of wind speed fluctuation in the atmosphere.

one-minute average wind (i.e. $\Delta t \approx 0.02$ h) contains a high level of fluctuation energy to which a wind turbine might be sensitive. However, since the most extensive sources of wind data provide information in this format, one-minute averages are commonly used to formulate statistical data on the frequency distribution (cumulative duration within a given wind speed range per year) and the persistence (duration within a wind speed range per occurrence) of steady wind.

2.1.1 Spatial Variation

On a planetary scale, semi-permanent features are part of the general circulation of the earth's atmosphere and are found at every latitude. Some of these features are associated with useful wind energy potential and lend themselves to energy extraction, while others do not. Turbulence, for example, cannot be considered as a potential wind energy resource that can be converted into useful energy.

The *trade wind*, that emerge from subtropical, anticyclone cells in both hemispheres are known for being the steadiest wind systems in the lower atmosphere. They blow throughout the year and, on an average, are best developed and strongest in the winter hemisphere. *Travelling-wave* perturbations are superimposed on the trades, causing them to vary in time and space. Most parts of the trade regions, however, have good wind energy potential throughout the year.

Monsoons are seasonal winds that last for a number of months. Monsoons are caused by the larger annual range of air temperatures over large land masses compared with those over neighbouring ocean and sea surfaces. These temperature differentials create pressure gradients which move the monsoon winds. Many monsoons are strong enough for wind energy use, especially where topography and other features such as upwelling of cool water in coastal areas further increase temperature differential between land and water areas.

Broad belts of the *Westerlies* extend across the mid-latitudes in both hemispheres. They blow throughout the year, but are quite variable in space and time, depending on many factors, such as season, topography, and configuration of land masses. The Westerlies and *sub-polar flows* are regions of very disturbed weather. Mid-latitude cyclonic storms frequently come in series and have a periodicity of 4–6 days. When a wave passes over a location, large variations are experienced in both wind speed and wind direction. Despite this variability, however, most regions of the middle-latitude belts have Westerlies that can be used for energy conversion, with peaks occurring most commonly in spring and winter and lulls in summer.

Synoptic-scale motions (i.e. correlated over a wide geographical region) are associated with periodic systems, such as travelling waves in the tropical *Easterlies* or the temperate Westerlies, or at temperate latitudes. Some parts of these waves have very good wind potential (Class 5 or better see Table 2.5). Typically, the area of influence of a travelling wave is of the order of 1,000 to 1,500 km, with a time scale of about 2–4 days.

Mesoscale wind systems can be associated either with travelling disturbances (such as squall lines) or with topographical features (such as valleys and coastal areas). Squall lines are generally convective systems that consist of several convective cells of the cumulo-nimbus type. Squall-line winds can be very violent and destructive, and may not always be of value for wind energy conversion. Mesoscale winds caused by differential heating of topographical features are generally referred to as *breezes*. A breeze is similar to a monsoon, but it operates on much smaller scales, typically a few hundred kilometres and a few hours. In many areas, breezes are a regular daily occurrence and, therefore, are of great value as a wind energy resource, especially when they enhance the existing basic wind.

Convective-scale motion is associated with vertical activity in the lower atmosphere, especially in connection with cumulus clouds. Since the scales of convective flow are a few kilometres and minutes to a couple of hours, this motion of the air does not contribute significantly to the wind energy resource. An exception to this is the condition where topographic lifting occurs on the windward side of a mountain. The convective activity that may result could keep local circulations going for several hours. In regions where the winds are prevailing from one direction (e.g. the trades), this phenomenon may repeat itself from day-to-day, and the enhanced low-level winds may contain sufficient energy to be extracted and used.

2.1.2 Time Variations

It has been pointed out that the motions of the atmosphere vary over a wide range of time scales (seconds to months) and space scales (metres to thousands of kilometres), and that these time and space scales are related. In this section, time variation of the wind is discussed in general terms. Statistical methods and data are described later.

Long-term variability

The first concern about a site that is under consideration for a wind power station is with the long-term mean wind speed. Can the winds be counted onto 'fuel' cost-effective power production over many years? What is the year-to-year variability at this site? What period of wind speed measurement at the site is adequate to establish a reliable estimate of the long-term mean wind speed? Wind power station operators have indicated that the ability to estimate the interannual variability at a site is almost as important as estimating its annual mean wind speed. The complexity of the interactions of the meteorological and topographical factors that cause the mean wind to vary from one year to the next have hampered the development of a reliable prediction method.

One statistically-developed rule of thumb is that one year of record is generally sufficient to predict long-term seasonal mean wind speeds within an accuracy of 10 per cent with a confidence level of 90 per cent. A study that compared seven different methods of estimating long-term average wind speeds from short-term data samples showed that the accuracy of more-sophisticated methods (including *principal component analysis* and *weather pattern classification*) was not significantly higher than that of simpler, linear statistical methods. Accuracy was measured by the degree of correlation between each of the estimates and actual long-term wind speed data.

Seasonal and monthly variability

Significant variations in wind speeds from season-to-season are common in most parts of the world. The degree of seasonal variation in the wind at a given site depends on latitude and position with respect to specific topographic features such as land masses and water. In general, mid-latitude continental locations that are well-exposed will experience higher winds in winter and spring, primarily because of large-scale storm activity. However, mountain passes in coastal areas may experience strong winds in the summer when cool maritime air moves into a hot interior valley. Within a given season, time variations in the wind over periods of one to several days can be caused by disturbances in the overall flow pattern such as *cyclonic storms* (in temperate latitudes) and *travelling wave systems* (in the tropics). These disturbances are quite capable of causing the output of a wind power station to cycle between zero and rated power several times a month. This type of wind variability is illustrated in Figure 2.4 for t, a mid-latitude site. Here, the fluctuations in wind speed have several different time periods, but three storms approximately 10 days apart are dominant.

In tropical latitudes, pronounced wind changes from season to season are well recognized. A belt of feeble winds named the *Inter Tropical Convergence Zone* or ITCZ (commonly referred to as the *doldrums*) moves north and south within the tropics, following the annual march of the sun. The trades or monsoons on either side of the ITCZ are also affected. Therefore, large seasonal time variations of the wind regime occur at most latitudes in the tropics, and they are most pronounced over continents. During a year in the tropics, it is reasonable to expect three or four months of low-to-mediocre winds and eight or nine months of good-to-excellent wind energy conditions.

The skill in reliable forecasting of wind variability on the time scale of a day or two has been shown to be valuable in the operation of a wind power station. However, there is little evidence that the appropriate level of reliability in wind forecasting is available through public or private organizations.

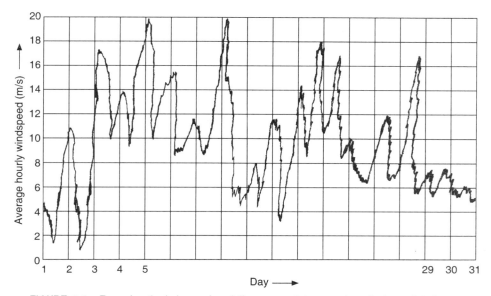

FIGURE 2.4 Example of wind speed variations caused by a series of storm disturbances.

Diurnal variations

In both tropical and temperate latitudes, large wind variations can occur on the *diurnal* (daily) time scale. In the tropics, these variations are most pronounced over land areas and during dry seasons, when the humidity content of the air is very low and the skies are often cloudless or almost so. Variations in radiation flux during the day enhance momentum transfer in the vertical direction during daytime and inhibit such transfer during night. As a result, wind speeds are maximum during the afternoon and minimum during the early morning. In extreme cases, this diurnal range of wind speed may approach 10 m/s. Such wind variations cannot be detected from monthly and seasonal averages, they require sampling with time resolutions of one hour or less.

Daily variations in solar radiation are also responsible for diurnal wind variations in temperate latitudes over relatively flat land areas. The largest diurnal changes generally occur in spring and summer, and the smallest in winter. In some cases, they extend downward to low enough elevations (30 m) to benefit large-scale HAWTs, but they disappear during the day. Finally, wind speed and direction variations with periods on the order of a few minutes are important to turbines with active pitch and/or yaw controls. Wind variability on this time scale is the evidence of turbulence in the flow that may have been generated by upstream topographic features, surface roughness elements, or thermal stratification. Modelling studies have shown that operational strategies involving wind measurements and the logic of turbine start-up, shut-down, and yaw control actions can play an important role in optimizing performance and reducing control-generated fatigue loads.

The atmospheric boundary layer (ABL) is lower part of atmosphere, where the atmosphere variables change from free atmospheric characteristics to surface values such as wind speed free in the atmosphere becomes almost same as at ground level. Similarly, temperature and humidity approach to surface values as illustrated in Figure 2.5.

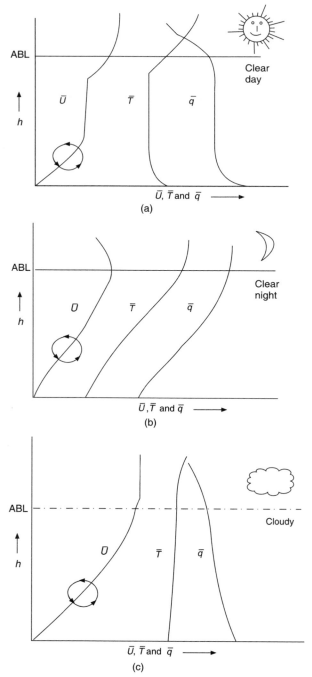

FIGURE 2.5 Variation of mean values of wind speed $\bar{U}$ temperature $\bar{T}$ and humidity $\bar{q}$ with height h above ground. Above atmospheric boundary layer (ABL), it is free atmosphere (a) Clear day (b) Clear night and (c) Cloudy conditions.

When the ground is hotter than air, then air will rise in the atmosphere. The shear produced eddies are enhanced by thermal structure, and it is seen as thermally unstable situation. When air is hotter than ground, then shear produced eddies will diminish and situation is seemed as stable. Note that above ABL, the temperature in general will increase with height. Marine boundary layer is distinct as compared to atmospheric boundary layer. The water is semi-transparent due to which heating and cooling is distributed. It has high mixing properties and circulation. When water is heated, the surface water is evaporated. If surface is cooled, then it shrinks. All these situations enhance the mixing of upper 10 m of the ocean. The roughness element over water is made by small steep waves of wavelength of about 50 mm. Roughness normally follows the images, the larger and sharper the protruding elements, the larger the roughness. In nature the roughness of sea is one of the smallest as compared to other natural forms like plains, deserts, forests, waste land etc. Consequent to this the turbulence in the wind over water is lower than land. The differential heating of land and water surface in coastal areas influence the atmospheric dynamics on a diurnal scale. During the day time, the temperature over the land increases quickly as compared to water surface, therefore, air will tend to rise over the land and sink over the water due to cooler surface. The air circulation is called land–sea breeze system as shown in Figure 2.6. In the night time, land cools faster than water body and air flow will reverse. The land–sea breeze system is more marked near major cities due to *urban heat island effect.* When such land–sea breeze system is developed on continental scale, it takes the form of monsoons which is typically seasonal. However, breeze system can develop on many scales spatially at the same location due to differential heating and cooling rates of land and water bodies.

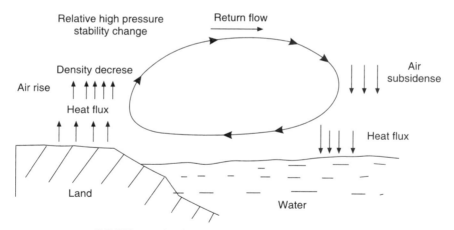

FIGURE 2.6 Land–sea breeze system during day-time.

2.2 CHARACTERISTICS OF STEADY WIND

The wind vector is considered to be composed of a *steady value* of wind speed plus a *fluctuation* of speed about the *steady wind*. For the horizontal component of free stream wind velocity,

$$u(t) = U + g(t) \tag{2.1}$$

$$\int_0^{\Delta t} g(t)\, dt = 0 \tag{2.2}$$

$$RMS[u(t) - U] = \sqrt{\frac{1}{\Delta t} \int_0^{\Delta t} g^2(t)\, dt} = \sigma_a \tag{2.3}$$

Moreover, the steady value is assumed to be quasi-static, so that its time variation is negligible. It is generally the steady wind speed which is referred to when discussing wind energy resources and wind turbine siting. Two parameters commonly used to characterize the steady wind speed at a given elevation are frequency distribution of wind speed on an annual basis and persistence. *Frequency* indicates the cumulative time the wind blows at prescribed values of speed as distinct from *persistence* which provides statistics on the continuous time the wind maintains that speed. Frequency distribution and persistence are important factors in both the design and siting of a wind turbine generator. The energy input to a wind turbine can be calculated from the frequency distribution of the wind speed. Turbulence usually increases directly with steady wind speed.

2.2.1 Turbulence

Referring once again to the energy spectrum of the wind shown in Figure 2.3 all of the time variations in wind speed and direction with periods less than about 1/10 hour are generally considered to be turbulence. These turbulent fluctuations in the *micro-meteorological range* create unsteady, non-uniform aerodynamic forces on the wind turbine that must be taken into account in the design of its structure and controls. The wind loads on the turbine can be classified into two groups: those associated with the steady wind speed, which are described as *quasi-static* or *time-averaged;* and those associated with the gustiness or turbulence of the wind, which are predominantly *dynamic*. Expanding Eqs. (2.1), (2.2) and (2.3) to represent a wind field instead of a streamtube,

$$u(y, z, t) = U(y, z) + g(y, z, t) \tag{2.1a}$$

$$\int_{A_r} \int_0^{\Delta t} g(y, z, t)\, dt\, dA = 0 \tag{2.2a}$$

$$RMS[u(y, z, t) - U(y, z)] = \sqrt{\frac{1}{A_r \Delta t} \int_{A_r} \int_0^{\Delta t} g^2(y, z, t)\, dt\, dA} = \sigma_0 \tag{2.3a}$$

where
y, z = lateral and vertical coordinates, respectively (m)
$u(y, z, t)$ = instantaneous horizontal free-stream wind velocity field (m/s)
$U(y, z)$ = steady horizontal free-stream wind velocity field (m/s)
$g(y, z, t)$ = fluctuating wind velocity field; instantaneous deviation from $U(y,z)$ (m/s)
A_r = swept area of the rotor (m^2)
Δt = averaging time interval (h)
$RMS[\ ]$ = root-mean-square average of [ ]
σ_0 = ambient or natural turbulence (m/s)

Equation (2.3a) states the common assumption that the turbulence is *homogeneous* (i.e. has the same structure over the swept area of the rotor). Wind models which describe various fluctuating velocity field functions $g(y,z,t)$ that create dynamic wind forces are discussed in this section.

2.2.2 Types of Turbulence Models

Mathematical models of the fluctuating or turbulent wind field at the rotor of a wind turbine can be grouped into four categories of increasing complexity and realism as follows:

1. Non-uniform in space, steady in time $g = g(y, z, 0)$
2. Uniform in space, unsteady in time $g = g(0, 0, t)$
3. Non-uniform in space, unsteady in time $g = g(y, z, t)$
4. Stochastic $g = g$(power spectrum vs. discrete gust parameters)

The earliest models of wind inflow to a rotor were of first type and were developed primarily to represent *tower shadow*, i.e. flow distortion to a downwind rotor caused by the tower wake. However, the free-stream wind field itself can be highly non-uniform, as illustrated in Figure 2.7. Rotational sampling is a procedure that uses an array of anemometers distributed along the hypothetical path of a section of a turbine blade. These anemometers record the time history of wind speed that would be experienced by that blade section.

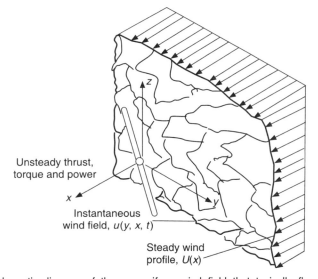

FIGURE 2.7 Schematic diagram of the non-uniform wind field that typically flows into the swept.

In order to use a non-uniform, steady wind model to represent rotationally-sampled turbulence, the assumption is made that spatial variations in wind speed are *quasi-static*, changing slowly in comparison with the rotational period of the rotor. Any mathematical function with periodic frequency might be used to convert the time series of wind speed at each rotor station to (y, z) coordinates. The swept area of the rotor is then simply divided into sectors, and a wind speed is specified for each sector. Presently, most turbulence inflow modelling uses this procedure.

Models uniform in space and unsteady in time: This type of model may also be called a *planar gust front* or a *discrete gust model*. The gust profile (in the coordinates y and z) is constant and envelopes the rotor uniformly, but its level changes with time. However, a given turbulence eddy may appear uniform to a small-scale rotor, but non-uniform to a larger one. This illustrates the need to model gust *amplitude,* duration, number of occurrences, and dimensions together. Generally, the larger the scale of the rotor, the larger the amplitude, the longer the duration, and the lower the frequency of occurrence of the planar gust fronts that envelop it.

The *ultimate strength* design of a wind turbine and the design of shutdown and safety controls are typically driven by extreme, somewhat isolated, gusts which are embedded in the turbulent winds. For the analysis of extreme values, one generally resorts to a *probability of maximum value* or *external gust* model.

Models non-uniform in space and unsteady in time: The actual inflow to a turbine is both unsteady and non-uniform, so this type of model is the most realistic and, of course, the most complex. It considers a non-uniform flow front, that is perhaps two-dimensional, which varies with time. The need for this level of complexity will differ for different rotor sizes. For example, both spatially and temporally varying inflow models may be required for large-scale rotors, whereas temporal variations alone may suffice for small rotors.

Stochastic models: The final turbulence model type is stochastic (sometimes called *probabilistic*). The three types of inflow models described previously contain *deterministic* descriptions of the wind fluctuations. Stochastic models, on the other hand, are based on the concept that turbulence is made up of sinusoidal waves or eddies with many periods and random amplitudes. Stochastic models may also use probability distribution or other statistical parameters. The stochastic analysis can be invaluable in developing models of the wind inflow from measurements of the response of an operating wind turbine, because it facilitates the following critical tasks:

- Representation of many data points taken in field tests
- Rapid evaluation of fatigue loads
- Comparison of model predictions with historical field data.

The *method of bins* is a simple application of stochastic methods to power performance testing of wind turbines. Another application of a stochastic approach is to represent the time wise variation of a wind 'front' with probabilistic formulations. This might be the probability of experiencing a given shape and magnitude of non-uniformity in the inflow. In this case, the spatial variation is still deterministic. The number of times the wind speed exceeds a prescribed value (i.e. *exceedance statistics*) is another example of a stochastic model. Such a model is applicable to fatigue analysis. There are many other applications of stochastic modelling in wind engineering.

2.2.3 Turbulence Intensity

Wind turbulence is the rapid disturbances or irregularities in the wind speed, direction, and vertical component. It is an important site characteristic, because high turbulence levels may decrease power output and cause extreme loading on wind turbine components. The most common indicator of turbulence for siting purposes is the standard deviation σ of wind speed. Normalizing this value with the mean wind speed gives the turbulence intensity (*TI*). This value allows for an overall assessment of a site's turbulence. *TI* is a relative indicator of turbulence with low levels indicated by values less than or equal to 0.10, moderate levels to 0.25, and high levels greater than 0.25.

$$TI = \frac{\sigma}{V} \qquad (2.4)$$

where
σ = standard deviation of wind speed
V = mean wind speed

2.2.4 Wind Power Density

Wind power density (WPD) is a truer indication of a site's wind energy potential than wind speed alone. Its value combines the effect of a site's wind speed distribution and its dependence on air density and wind speed. WPD is defined as the wind power available per unit area swept by the turbine blades and is given by the following equation:

$$WPD = \frac{1}{2n} \sum_{i=1}^{n} \rho(v_i^3) \qquad (2.5)$$

where
n = number of records in the averaging interval
ρ = air density
v_i^3 = cube of the ith wind speed value

EXAMPLE 2.1 Assume that over a two-hour period, the average wind speed at a site is 6.7 m/s (15 mph); 4.5 m/s (10 mph) the first hour and 8.9 m/s (20 mph) the next. Calculate the WPD using the combined two-hour average ($n = 1$) and then with the two distinct hourly average values ($n = 2$).

Solution: Applying standard temperature and pressure (101,325 Pa and 288 K) to the above equation, the calculated WPD using the overall average value is 184 W/m² while that using the two average values is 246 W/m². The latter value represents the average of that calculated for 4.5 m/s (55 W/m²) and 8.9 m/s (438 W/m²). There is actually 34 per cent more power available at the site than would have been realized if the equation was used incorrectly.

The air density term in the WPD must be calculated. It depends on temperature and pressure (thus altitude) and can vary 10 per cent to 15 per cent seasonally. If the site pressure is known, e.g. measured as an optional parameter, the hourly air density values with respect to air temperature can be calculated from the following equation:

$$\rho = \frac{P}{RT} \qquad (2.6)$$

where
P = air pressure (Pa or N/m²)
R = specific gas constant for air (287 J/kg K)
T = air temperature in degrees Kelvin (°C + 273)

If site pressure is not available, air density can be estimated as a function of site elevation (z) and temperature (T) as follows:

$$\rho = \left(\frac{P_o}{RT}\right) e^{\left(\frac{-gz}{RT}\right)} \qquad (2.7)$$

where

P_o = standard sea level atmospheric pressure (101,325 Pa), or the actual sea level adjusted pressure reading from a local airport
g = gravitational constant (9.8 m/s²)
z = site elevation above sea level (m)

This air density equation can be substituted into the WPD equation for the determination of each hourly average value.

How much power is contained in the wind? In other way, it is calculated as follows:

$$p_w = \frac{d}{dt}(q_d x) = \frac{1}{2}\rho U^2 \frac{dx}{dt} = \frac{1}{2}\rho U^3 \tag{2.8}$$

where U = average wind speed.

Since for capturing energy from the wind, the best indication of the size of local wind energy resource is its annual average wind power density, or

$$p_{w,d} = \frac{1}{2}\frac{\rho}{8760}\int_{year} U^3 \, dt \tag{2.9}$$

From Eq. (2.9), for annual average,

$$w_d = \frac{1}{2}\frac{\rho}{8760}\int_{year} U^3 \, dt \tag{2.10}$$

It can also be written as,

$$w_d = \frac{1}{2}\frac{\rho}{8760}\int_{year}^{\infty} U^3 f_u \, dU \tag{2.11}$$

where

$$f_U = \left[\frac{dt}{dU}\right]_{year} = \frac{d}{dU}[F(U_1 \geq U)] \tag{2.12}$$

where

f_U = frequency distribution function of the steady wind speed [(h/y) (m/s)]
$F(\ )$ = annual time that ( ); cumulative frequency distribution function (h/y)
U_1 = arbitrary value of U (m/s)

2.3 WEIBULL WIND SPEED DISTRIBUTION FUNCTION

In Eq. (2.12), F is the cumulative distribution function which defines the so-called wind duration curve. The Weibull equation for the duration curve is:

$$F(U_1 \geq U) = 8760 e^{[-(U/C)^k]} \tag{2.13}$$

where

$e^{[\ ]}$ = exponential function of [ ]
C = empirical Weibull scale factor (m/s)
k = empirical Weibull shape factor

Duration curves for several values of the shape factor k are shown in Figure 2.8. The range from 1.5 to 3.0 for k includes most site wind conditions. The Rayleigh distribution is a special case of the Weibull distribution in which the shape factor is 2.0. Substituting Eq. (2.13) into Eq. (2.12), the Weibull frequency distribution function is:

$$f_U = \left(\frac{8760}{C}\right) k \left(\frac{U}{C}\right)^{k-1} e^{\left[-\left(\frac{U}{C}\right)^k\right]} \qquad (k>0, U>0, C>1) \qquad (2.14)$$

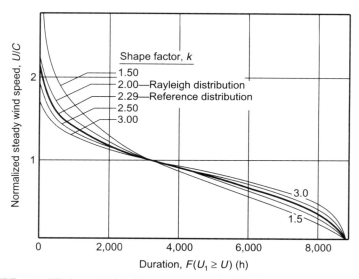

FIGURE 2.8 Wind speed duration curves according to Weibull distribution model.

From Eq. (2.14), the average annual wind speed is to be calculated:

$$U_d = C\Gamma\left(1 + \frac{1}{k}\right) \approx (0.90 \pm 0.01)C \qquad (2.15)$$

where

U_d = annual average wind speed
$\Gamma(\ )$ = gamma function of ( )

Figure 2.9 shows the Weibull frequency distribution curves associated with the duration curves in Figure 2.8 and annual average wind speed for each value of k. Note that the annual average wind speed is higher than the most frequent wind speed, but these two parameters approach equality at higher values of k.

A reference annual wind speed distribution is intended to serve as a uniform basis for research and development projects. The annual average wind speed was selected to be 6.24 m/s (14 mph) at an elevation of 9.1 m (30 ft) above level terrain. Weibull factors for this reference distribution are C = 7.07 m/s and k = 2.29. The resulting duration and frequency curves are shown in Figure 2.8 and Figure 2.9.

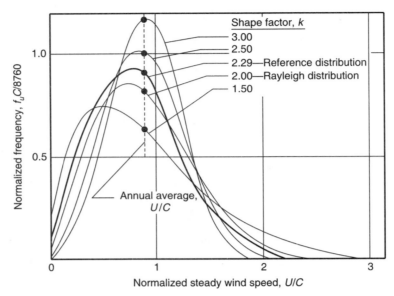

FIGURE 2.9 Wind speed frequency distribution in dimensionless form, based on the Weibull duration curves.

2.3.1 Estimating Weibull Distribution Factor

There are several methods which can be used to estimate the Weibull factor C and k, depending on the available wind statistics and the desired level of sophistication in data analysis. The *least squares method* requires an observed wind speed histogram, which is frequency data for n speed intervals or bins. A duration curve is then constructed from the histogram and the results plotted in the linearized form.

$$y = y_0 + mx \tag{2.16}$$

$$x = \ln(U) \tag{2.17}$$

$$y = \ln\left[-\ln\left(\frac{F}{8760}\right)\right] \tag{2.18}$$

The data are least squares curve-fit with a line of slope m and intercept y_0, from which

$$k = m \tag{2.19}$$

$$C = e^{\left(\frac{y_0}{m}\right)} \tag{2.20}$$

The other methods are *median and quartile, annual average and standard deviation*, and *variance vs. annual average trend*. The *median and quartile* method is useful if a complete histogram is not available, since it requires only the wind speed at $F = 2190$, 4380 and 6579 hours. The *annual average and standard deviation* requires the annual average wind speed, U_a, and the annual standard deviation from this average, σ_a. The *variance* vs. *annual average trend* method uses qualitative estimate of the average wind speed to make a rough estimate of Weibull factors.

It must be borne in mind that wind speed duration curves are quite sensitive to location because of surface roughness conditions, and are valid only for relatively flat terrain. Steady wind speed data include the influence of atmospheric stability and terrain features peculiar to the site at which they were measured but the influence of these factors is normally small enough under higher wind conditions that the available data are a good representation of the wind for design purpose. The factors in the Weibull distribution are elevation-specific and must be adjusted to account for wind shear. One method of adjusting the C and k values for changes in elevation uses power law equations.

Two methods are discussed in detail for calculating the parameters of the Weibull wind speed distribution for wind energy analysis: maximum likelihood method for use with time series wind data, and the modified maximum likelihood method for use with wind data in frequency distribution format.

Maximum likelihood method: In wind energy analysis, it is used to represent wind speed probability density function by Weibull distribution function as given by

$$p(V < V_i < V + dV) = p(V > 0) \left(\frac{k}{C}\right)\left(\frac{V_i}{C}\right)^{k-1} \exp\left[-\left(\frac{V_i}{C}\right)^k\right] dV \quad (2.21)$$

dV is an incremental wind speed, $p(V < V_i < V + dV)$ is the probability that the wind speed is between V and $V + dV$, and $p(V > 0)$ is the probability that wind speed exceeds zero. Equation (2.21) can be interpreted as the fractional probability that the wind speed exceeds zero or the number of hours per year that the wind speed exceeds zero.

The cumulative distribution fnction is given by

$$p(V < V_i) = p(V \geq 0)\left[1 - \exp\left\{-\left(\frac{V_i}{C}\right)^k\right\}\right] \quad (2.22)$$

where $p(V < V_i)$ is the probability that the wind speed is less than V_i and $p(V \geq 0)$ is the probability that the wind speed equals or exceeds zero.

The two Weibull parameters and the average wind speed are related by

$$\bar{V} = C\Gamma\left(1 + \frac{1}{k}\right) \quad (2.15)$$

where $\bar{V}$ is the average wind speed and $\Gamma(\ )$ is the gamma function.

EXAMPLE 2.2 Measured wind data is available in time-series format in Table 2.1. Estimate the Weibull parameters by maximum likelihood method.

TABLE 2.1

Hour	speed (m/s)	Hour	speed (m/s)	Hour	speed (m/s)
1	3.3	9	5.2	17	6.5
2	3.8	10	6.7	18	4.2
3	4.2	11	6.8	19	4.3
4	3.3	12	6.8	20	3.7
5	2.8	13	5.7	21	4.0
6	3.0	14	8.3	22	2.8
7	4.0	15	9.2	23	3.7
8	2.7	16	9.3	24	3.3

Solution: The Weibull distribution can be fitted to time-series wind data using maximum likelihood method. The shape factor k and the scale factor c are estimated using the following two equations:

$$k = \left(\frac{\sum_{i=1}^{n} V_i^k \ln(V_i)}{\sum_{i=1}^{n} V_i^k} - \frac{\sum_{i=1}^{n} \ln(V_i)}{n} \right)^{-1} \quad (2.23)$$

$$C = \left(\frac{1}{n} \sum_{i=1}^{n} V_i^k \right)^{1/k} \quad (2.24)$$

where V_i is the wind speed in time step i and n is the number of nonzero wind speed data points. Equation (2.23) must be solved using an iterative procedure ($k = 2$ is a suitable initial guess), after which Eq. (2.24) can be solved explicitly. Care must be taken to apply Eq. (2.23) only to the nonzero wind speed data points.

Modified maximum likelihood method: When wind speed data are available in frequency distribution format as given in Table 2.2, the modified maximum likelihood method can be applied to estimate the Weibull parameters using the following equations:

TABLE 2.2 Wind speed data in frequency distribution format

Wind speed (m/s)	0–1	1–2	2–3	3–4	4–5	5–6	6–7	7–8	8–9	9–10
Frequency (%)	2	7	9	15	20	17	8	1	0	0

$$k = \left(\frac{\sum_{i=1}^{n} V_i^k \ln(V_i) p(V_i)}{\sum_{i=1}^{n} V_i^k p(V_i)} - \frac{\sum_{i=1}^{n} \ln(V_i) p(V_i)}{p(V \geq 0)} \right)^{-1} \quad (2.25)$$

$$C = \left(\frac{1}{p(V \geq 0)} \sum_{i=1}^{n} V_i^k p(V_i) \right)^{1/k} \quad (2.26)$$

where V_i is the wind speed central to bin i, n is the number of bins, $p(V_i)$ is the frequency with which the wind speed falls within bin i, $p(V \geq 0)$ is the probability that the wind speed equals or exceeds zero.

Equation (2.25) must be solved iteratively, after which Eq. (2.26) can be solved explicitly.

It is sometimes necessary to estimate the Weibull parameters in the absence of any information about the distribution of wind speeds. For example, only annual or monthly averages may be available. In such a situation, the value of k must be estimated. The value of k is usually between 1.5 and 3, depending on the variability of the wind. Smaller k values correspond to more variable (more gusty) winds, with $k = 2$ for moderately gusty winds. If no information about the variability of wind is available, a k value 2 is often assumed.

With the estimated value of k and the average wind speed, the value of C can be obtained using Eq. 2.26.

EXAMPLE 2.3 For the sample data set as shown in Table 2.3, Table 2.4 shows the frequency distribution and the cumulative frequency distribution of same data set. The sample data set consists of three days of hourly wind speed data.

TABLE 2.3 Sample time series data set.

Hour	Wind speed (m/s)			Hour	Wind speed (m/s)		
	Day 1	Day 2	Day 3		Day 1	Day 2	Day 3
1	3.3	4.0	4.7	13	5.7	4.5	6.3
2	3.8	4.0	4.5	14	8.5	5.8	9.4
3	4.2	2.0	4.2	15	8.9	4.8	7.7
4	3.3	2.7	5.7	16	9.3	4.8	6.0
5	2.8	2.7	2.7	17	6.5	5.5	8.9
6	3.0	3.3	4.3	18	4.2	5.7	7.7
7	4.0	2.7	4.3	19	4.3	5.0	6.2
8	2.7	2.7	4.5	20	3.7	4.3	5.7
9	5.2	5.8	4.5	21	4.0	4.0	5.7
10	6.7	5.7	6.0	22	2.8	3.5	7.5
11	6.8	6.2	10.4	23	3.7	5.0	7.5
12	6.8	6.5	6.7	24	3.3	3.7	5.3

Solution: By maximum likelihood method using $n = 72$ and an initial guess of $k = 2$, successive applications of Eq. (2.23) to the data in Table 2.3 give $k = 4.07$, 2.30 and 3.60 converging after several iterations to 2.93. Equation (2.24) then gives $C = 5.75$ m/s.

By modified maximum likelihood method using $n = 12$ and an initial guess of $k = 2$, successive applications of Eq. (2.25) to the data in Table 2.23 give $k = 4.20$, 2.31 and 3.70, converging after several iterations to 2.99. Equation (2.26) then gives $C = 5.77$ m/s.

TABLE 2.4 Sample data frequency distribution and cumulative frequency distribution

Speed (m/s)	Frequency (%)	Cumulative frequency (%)
2–3	12.50	12.50
3–4	13.89	26.39
4–5	26.39	52.78
5–6	18.06	70.83
6–7	15.28	86.11
7–8	5.56	91.67
8–9	4.17	95.83
9–10	2.78	98.61
10–11	1.39	100.00

2.4 VERTICAL PROFILES OF THE STEADY WIND

The mean horizontal wind speed is zero at the earth's surface and increases with altitude in the atmospheric boundary layer. The variation of wind speed with elevation is referred to as the vertical profile of the wind speed or the wind shear as shown in Figure 2.10. The variation of wind speed with elevation above ground has important influence on both the assessment of wind energy resources and the design of wind turbines. A power law for vertical profiles of steady wind is commonly used in wind energy for defining vertical wind profiles. If heights at which measurements have been performed do not match the hub height of a wind turbine, then it is necessary to extrapolate the wind speeds to the hub height of the wind turbine. The computational steps required to extrapolate the available wind speeds to the turbine hub height are given below. The basic equation of the wind shear power law is,

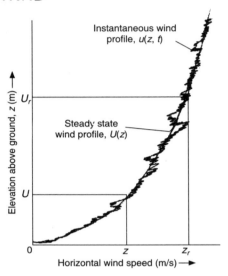

FIGURE 2.10 Typical vertical profile of the wind speed, steady and instantaneous both are shown.

$$U = U_r \left(\frac{z}{z_r} \right)^\alpha \qquad (2.27)$$

where
U = wind velocity at elevation z
U_r = wind velocity at higher elevation z_r
α = empirical wind shear exponent

Therefore, the value of α can be determined as:

$$\alpha = \frac{\log \dfrac{U_r}{U}}{\log \dfrac{z_r}{z}} \qquad (2.28)$$

If the hub height of a wind turbine is z_t, then the extrapolated wind speed U_t corresponding to z_t is given by,

$$U_t = U_r \left(\frac{z}{z_r}\right)^\alpha \quad (2.29)$$

For majority of sites the value of α is taken as 1/7 or it is called *one seventh exponent*.

Since wind speed varies with height, Weibull parameters C and k must be corrected for height at the wind-turbine hub. 'Frost' proposed Weibull parameters are as follows.

$$k = k_r = \frac{\left[1 - 0.088 \ln\left(\frac{z_r}{10}\right)\right]}{\left[1 - 0.088 \ln\left(\frac{z}{10}\right)\right]} \quad (2.30)$$

$$C = C_r \left(\frac{z}{z_r}\right)^\alpha \quad (2.31)$$

and

$$\alpha = \frac{[0.37 - 0.088 \ln(C_r)]}{\left[1 - 0.088 \ln\left(\frac{z_r}{10}\right)\right]} \quad (2.32)$$

where
k = shape factor at elevation z
k_r = shape factor at higher elevation z_r

If wind speed measurement data are available for more than two heights, the shear exponent α is determined by large eddy simulation (LES) technique as:

$$\alpha_i = \frac{M \sum_{j=1}^{M} l_n(v_{ij}) l_n(z_i) - \sum_{j=1}^{M} l_n(z_j) \cdot \sum_{j=1}^{M} l_n(v_{ij})}{M \sum_{j=1}^{M} l_n(z_i)^2 - \left(\sum_{j=1}^{M} l_n z_i\right)^2} \quad (2.32a)$$

where v_{ij} = wind speed in i_{th} 10-min measurement interval ($i = 1, 2, 3, \ldots, N$) at height z_j ($j = 1, 2, 3, \ldots, M$);

M = total number of heights on the measurement mast where the measurement have been taken

N = total number of observed 10-min measurement intervals.

Now,

$$U_i = U_{Mi} \left(\frac{z}{z_M}\right)^{\alpha_i} \quad (2.32b)$$

where U_{Mi} = wind speed over i_{th} 10-min interval at the reference height
z_M = highest measurement point

2.5 WIND ROSE

The wind rose is a diagram showing the percentage of winds blowing from each of the leading 12 points of the compass. A typical wind rose is shown in Figure 2.11 with twelve sectors shows percentage of time distribution of one year for which wind blows from that particular direction. The prevailing wind direction is from west. Wind rose also show *prevailing wind direction* i.e. the direction from which the wind blow most of the time.

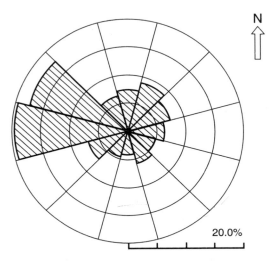

FIGURE 2.11 A typical wind rose diagram.

2.6 ENERGY PATTERN FACTOR

The energy pattern factor (EPF) for a particular site and month is defined as the ratio:

$$\text{EPF} = \frac{\text{Actual energy available in the wind for the month}}{\text{Energy calculated from mean monthly wind speed}}$$

EPF is calculated from the series of hourly wind speed values; the mean monthly and annual values. Higher values of EPF are associated with lower wind speeds and lower values with higher wind speeds. It is a useful parameter for calculating the available energy in the wind from the knowledge of the mean annual or monthly wind speed. It is also useful while choosing locations with limited wind data, because long-term data from neighbouring stations can be correlated with on-site short-term measurement.

2.7 ENERGY CONTENT OF THE WIND

The energy content (EC) of the wind per square metre normal to the wind flow for any specified period is given by:

$$\text{EC} = 0.01073 \rho \sum_{i=1}^{N_m} V_i^3 \ \text{Whm}^{-2} \tag{2.33}$$

and
$$EC = 0.00001073 \sum_{i=1}^{N_m} V_i^3 \quad \text{kWhm}^{-2} \tag{2.34}$$

where V_i is the hourly wind speed in km/h, for hour i, and N_m is the total number of hours in the specified period. The value of i ranges from 1 to 744 for a month of 31 days. The energy content of the wind for a month in each of the years is obtained by summing V_i^3 over the total number of hours in the month. The energy content of the wind for the year is the sum of the values for the twelve months of the year. From these values for several years, the mean energy content is obtained for each month and for the year as a whole.

2.8 RESOURCE ASSESSMENT

The first step for wind resource assessment is the preliminary area identification. In this process, a relatively large region is screened for suitable wind resource areas based on information such as airport wind data, topography, flagged trees, and other indicators.

The second step is wind resource evaluation and this applies to wind measurement programmes to characterize the wind resource in a defined area or set of areas where wind power development is being considered. The most common objectives of this scale of wind measurement are to:

- Determine or verify whether sufficient wind resources exist within the area to justify further site-specific investigations.
- Compare areas to distinguish relative development potential. Wind resources are represented by wind power classes and each class represents a range of annual average wind power densities and equivalent mean wind speed as illustrated in Table 2.5.
- Obtain representative data for estimating the performance and/or the economic viability of selected wind turbines.
- Screen for potential wind turbine installation sites by matrix analysis of wind, site and wind turbine.

The third step is of micrositing at the smallest scale of wind resource assessment. Its main objective is to quantify the small-scale variability of the wind resource over the terrain of interest. Ultimately, micrositing is used to position one or more wind turbines on a parcel of land to maximize the overall energy output of the wind plant.

TABLE 2.5 Classes of wind power density

Wind power class	30 m (98 ft)		50 m (164 ft)		Remark
	Wind power density (W/m²)	Wind speed m/s (mph)	Wind power density (W/m²)	Wind speed m/s (mph)	
1	≤ 160	≤ 5.1 (11.4)	≤ 200	≤ 5.6 (12.5)	Poor
2	≤ 240	≤ 5.9 (13.2)	≤ 300	≤ 6.4 (14.3)	Marginal
3	≤ 320	≤ 6.5 (11.6)	≤ 400	≤ 7.0 (15.7)	Fair
4	≤ 400	≤ 7.0 (15.70	≤ 500	≤ 7.5 (16.8)	Good
5	≤ 480	≤ 6.4 (16.6)	≤ 600	≤ 8.0 (17.9)	Excellent
6	≤ 640	≤ 8.2 (18.3)	≤ 800	≤ 8.8 (19.7)	Outstanding
7	≤ 1600	≤ 11.0 (24.7)	≤ 2000	≤ 11.9 (26.6)	Superb

2.8.1 Measurement Plan

The plan should specify the following features:
- Measurement parameters
- Equipment type, quality and cost
- Number and location of monitoring stations
- Sensor measurement heights
- Minimum measurement accuracy, duration and data recovery
- Data sampling and recording intervals
- Data storage format
- Data handling and processing procedures.

2.8.2 Monitoring Duration

The minimum monitoring duration should be one year, but two or more years will produce more reliable results. One year is usually sufficient to determine the diurnal and seasonal variability of the wind. With the aid of a well-correlated, long-term reference station such as an airport, the inter-annual variability of the wind can also be estimated. The data recovery for all measured parameters should be at least 90 per cent over the programmes duration, with any data gaps kept to a minimum (less than a week).

2.8.3 Siting of Monitoring Systems

The main objective of a siting programme is to identify potentially windy areas that also possess other desirable qualities of a wind energy development site. There are three steps in the siting effort:
- Identification of potential wind development areas
- Inspection and ranking of candidate sites
- Selection of actual tower location(s) within the candidate sites.

Earlier recorded wind data are useful in the early stage of the siting process. These data represent records of actual wind conditions, so they must be evaluated before the windiest areas of a particular region are sought. Unfortunately, most historical wind data were not collected for wind energy assessment purposes. Thus, the results often represent the mean conditions near population centres in relatively flat terrain or low elevation areas. Their primary benefit to the analyst, therefore, is to provide a general description of the wind resource within the analysis area, not to pinpoint the windiest locales. Common sources of wind information include the Weather Stations/Office, Airport and meteorological offices for regional wind resource estimates. Certainty ratings were also generated for each grid cell to depict a confidence level in the wind resource estimate. The degree of certainty depends on the following three factors:
- Abundance and quality of wind data
- Complexity of the terrain
- Geographical variability of the wind resource.

The highest degree of confidence (rating 4) was assigned to grid cells containing abundant historical data and relatively simple terrain; the lowest certainty (rating 1) was assigned to data sparse regions or those within complex terrain.

The wind resource atlas of India is shown in Figure 2.12 with varying wind power density in different regions.

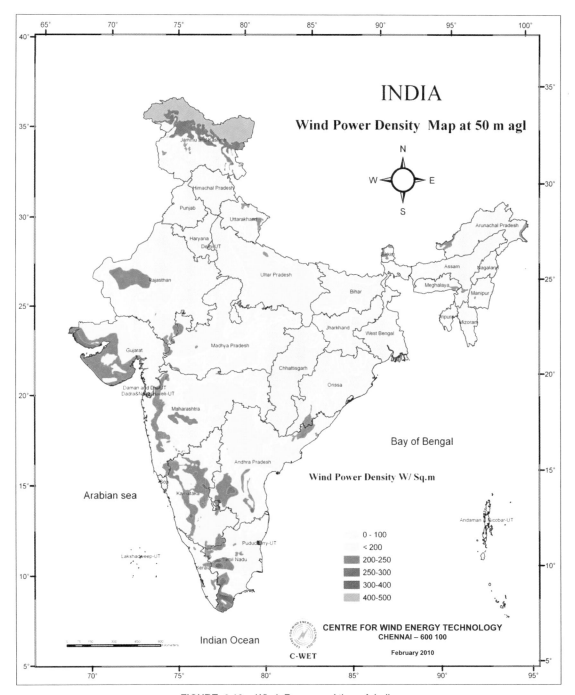

FIGURE 2.12 Wind Resource Atlas of India.
(*Source:* Centre for Wind Energy Technology, Chennai, India)

A basic hypothesis of wind atlas is that wind data are distributed according to the Weibull distribution. The Weibull distribution usually gives a good fit to the observed wind speed data. Numerical wind atlas methodologies have been devised to solve the issue of insufficient wind measurements.

The Centre for Wind Energy Technology (C-WET) published the Indian Wind Atlas, showing large areas, with annual average wind power density of more than 200 W/m^2 at 50 meter above ground level. This is considered to be a benchmark criterion for establishing wind farms in India as per C-WET. The potential sites have been classified according to annual mean wind power density ranging from 200 W/m^2 to 500 W/m^2. The wind Atlas has projected Indian wind power installable potential (name plate rating) as 49,130 MW at 2 per cent land availability*. This is seen as a conservative estimate of wind power potential in India. With the improvement in technology and increase in the hub height of the wind turbine, it has become possible to generate more electricity than assumed in earlier estimates. Further, India's as yet un-assessed offshore wind potential was not included in the C-WET study. A long coastline and relatively low construction cost could make India a favoured destination for offshore wind power.

2.8.4 Site-Specific Wind Data

If you wish to closely examine wind data from selected stations, several attributes about the data should be determined including:
- Station location
- Local topography
- Anemometer height and exposure
- Type of observation (instantaneous or average)
- Duration of record.

Data are more representative of the surrounding area where the terrain is relatively flat. In complex terrain, the ability to reliably extrapolate the information beyond a station's immediate vicinity is limited. In recent decades, most airport measurements have been taken adjacent to the runways where the surrounding area is open and unobstructed. Measurements taken from rooftops may be unreliable due to the building's influence on the wind flow and should be used with caution. Typical airport anemometer heights are in the 6 m to 15 m range. When comparing data with other stations, all wind speed data should be extrapolated to a common reference height (e.g. 50 m or above).

As a first approximation, the wind shear exponent is often assigned a value, known as the 1/7th power law, to predict wind profiles in a well-mixed atmosphere over flat, open terrain. However, higher exponent values are normally observed over vegetated surfaces and when wind speeds are light to moderate (i.e. under 7 m/s). Referenced data sets should be at least one year in duration and possess consistent data for at least 90 per cent of that period. A useful format is a time series

---

* The assessment in the Indian Wind Atlas is assumed at 2 per cent land availability for all states except the Himalayan states, North-East states and the Andaman and Nicobar Island, where it is assumed as 0.5 per cent. However, the potential would change as per the real land availability in each state. Further, the installable wind power potential is calculated for each wind power density range by assuming 9 MW (average of 7D × 5D, 8D × 4D and 7D × 4D spacing is the rotor diameter of the turbine) could be installed per square km.

of hourly wind speed and wind direction measurements, which can be analyzed for a number of user-specified wind characteristics. In many instances, wind data summaries will already be available, which eliminates the need to process the data.

2.8.5 Topographic Indicators

The analysis of topographic maps is an effective means of streamlining the siting process. Maps on a 1:24,000 scale are available from the Geological Survey of India for identifying suitable terrain features. The topographic screening should attempt to identify features that are likely to experience a greater mean wind speed than the general surroundings. This process is especially important for areas containing little or no relevant historical wind speed data. Features that are likely to be windier are given in Figure 2.13 (please see for further detailed discussion on geographical features of sites in Chapter 6). Like the crest of a hill with shape feature as convex will give accelerating effect to wind flow at the top of hill and concave shape give rise to reverse effect.

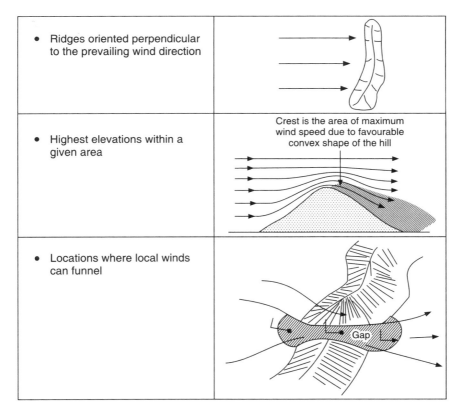

FIGURE 2.13 Topographical indicators.

Features to be avoided include areas immediately upwind and downwind of higher terrain, the lee side of ridges, and excessively sloped terrain. In each of these situations, increased turbulence may occur.

Topographic maps also provide the analyst with a preliminary look at other site attributes, including:

- Available land area
- Positions of existing roads and dwellings
- Land cover (e.g. forests)
- Political boundaries
- Parks
- Proximity to transmission lines.

Following the topographic screening, a preliminary ranking can be assigned to the list of candidate sites based on their estimated wind resource and overall development potential.

2.8.6 Field Surveys and Site Ranking

Visits should be conducted to all potentially suitable areas with the main goal of verifying site conditions. Items of importance include:

- Available land area
- Land use
- Location of obstructions
- Trees deformed by persistent strong winds (flagged trees)
- Accessibility into the site
- Potential impact on local aesthetics
- Cellular phone service reliability for data transfers
- Possible wind monitoring locations.

The evaluator should use a topographic map of the area to note the presence or absence of the above site characteristics. A Global Positioning System (GPS) receiver should be used to record the location coordinates (latitude, longitude, elevation) of the sites. A video or still camera record is useful for future reference and presentation purposes. While at the site, the evaluator should determine the soil conditions so that the proper anchor type can be chosen if a guyed meteorological tower is to be installed. An updated ranking of all candidate sites should be developed following the site visits. This can be obtained by constructing a matrix that assigns a score to each siting criterion. For example, suppose the siting criteria are similar to the features listed above. The evaluator assigns a numerical score (e.g. 1–10) to every criterion for each site that was visited. If some criteria are more important than others, their scores can be weighted accordingly. The weighted scores are then summed up and sorted by magnitude to reach a composite ranking.

2.8.7 Tower Placement

Two important guidelines should be followed when choosing an exact location for the monitoring tower:

- Place the tower as far away as possible from local obstructions to the wind.
- Select a location that is representative of the majority of the site.

Siting a tower near obstructions such as trees or buildings can adversely affect the analysis of the site's wind characteristics. Figure 2.14 illustrates the effects of an undisturbed airflow that

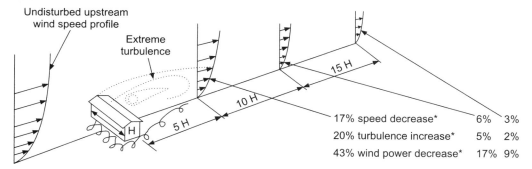

FIGURE 2.14 Obstruction effects on airflow.

encounters an obstruction. The presence of these features can alter the perceived magnitude of the site's overall wind resource, wind shear, and turbulence levels. As a rule, if sensors must be near an obstruction, they should be located at a horizontal distance not closer than 10 times the height of the obstruction in the prevailing wind direction.

Formal lease agreements should be negotiated between the utility and the landowner to protect the parties involved. They should include:

- Tower location
- Total area required for monitoring
- Duration of monitoring period
- Liability
- Insurance
- Title to data and equipment
- Access to premises
- Payment schedule.

2.8.8 Measurement Parameters

The core of the monitoring programme is the collection of wind speed, wind direction, and air temperature data. Wind speed data are the most important indicator of a site's wind energy resource. Multiple measurement heights are encouraged for determining a site's wind shear characteristics, and conducting turbine performance simulations at several turbine hub heights typical at 50 m and above and also at hub height of proposed wind turbine at the site. To define the prevailing wind direction(s), wind vanes should be installed at all significant monitoring levels. Wind direction frequency information is important for identifying preferred terrain shapes and orientations and for optimizing the layout of wind turbines within a wind farm. Air temperature is an important descriptor of a wind farm's operating environment and is normally measured either near ground level (2 to 3 m), or near hub height. In most locations, the average near ground level air temperature will be within 1°C of the average at hub height. It is also used to calculate air density, a variable required to estimate the wind power density and a wind turbine's power output. The vertical wind speed provides more details about a site's turbulence and can be a good predictor of wind turbine loads. Historically, this parameter has been a research measurement, but as wind energy development spreads into new regions of the country, regional information on vertical wind velocity may become important. To measure the vertical wind component W as an indicator of wind turbulence, a w anemometer (see Figure 2.14) should be located near the upper basic wind speed monitoring level but not exactly at that level to avoid instrument clutter. All parameters should be sampled once every one or two seconds and recorded as averages, standard deviations, and maximum and minimum values. Data recording should be serial in nature and designated by a corresponding time and date stamp.

The *average* value should be calculated for all parameters on a ten-minute basis, which is now the international standard period for wind measurement. Except for wind direction, the average is defined as the mean of all samples. For wind direction, the average should be a unit vector (resultant) value. Average data are used in reporting wind speed variability, as well as wind speed and direction frequency distributions.

The *standard deviation* should be determined for both wind speed and wind direction and is defined as the true population standard deviation for all one or two second samples within each averaging interval. The standard deviations of wind speed and wind direction are indicators of the turbulence level and atmospheric stability. Standard deviation is also useful in detecting suspect or erroneous data when validating average values.

Maximum and minimum values should be determined for wind speed and temperature at least daily. The maximum (minimum) value is defined as the greatest (lowest) one or two second reading observed within the preferred period. The coincident direction corresponding to the maximum (minimum) wind speed should also be recorded.

2.8.9 Monitoring Station Instrumentation

Cup or propeller anemometers (Figure 2.15) are the sensor types most commonly used for the measurement of near-horizontal wind speed. This instrument consists of a cup assembly (three or four cups) centrally connected to a vertical shaft for rotation. At least one cup always faces the oncoming wind. The aerodynamic shape of the cups converts wind pressure force to rotational torque. The cup rotation is nearly linearly proportional to the wind speed over a specified range. A transducer in the anemometer converts this rotational movement into an electrical signal, which is sent through a wire to a data logger.

Cup anemometers are standard type of anemometer as shown in Figure 2.15. They are robust and resistant to the turbulence and skew winds caused by masts and traverses. Each anemometer should be individually calibrated and equipped with a certified calibration report. The number of anemometers applied at one mast can vary from a minimum of 3 sensors. Most anemometers can be equipped with electronically regulated heating.

A propeller anemometer measures the airflow parallel to its axis of rotation as shown in Figure 2.16. They are usually applied in wind farm, monitoring by showing how the turbine react to airflow. A propeller anemometer utilises a fast-response helicoid propeller and high-quality tach-generator transducer to produce a DC voltage that is linearly proportional to air velocity. Airflow from any direction may be measured, but the propeller responds only to the component of the airflow that is parallel to its axis of rotation.

FIGURE 2.15 Cup anemometer.
(*Source:* Published with permission from Ammonit.)

FIGURE 2.16 Propeller anemometer.
(*Soruce:* Published with permission from Ammonit.)

Ultrasonic anemometers measure horizontal wind speed and wind direction as well as the speed of sound and virtual temperature. The typical ultrasonic anemometers are shown in Figure 2.17 (a), (b) and (c), which can measure wind speed, direction and temperature. The anemometer shown in Figure 2.17(b) can measure wind speed in x, y and z direction.

However, because of their high power consumption, a connection to mains power supply is required. Ultrasonics are applied for wind farm monitoring on turbines and on offshore projects. The measurement is based on the following concept:

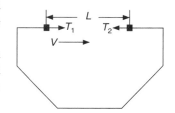

$$T_2 = \frac{L}{C - V} \text{ and } T_1 = \frac{L}{C + V} \qquad (2.35)$$

$$V = \frac{L}{2}\left(\frac{1}{T_1} - \frac{1}{T_2}\right)$$

Therefore,
$$C = \frac{L}{2}\left(\frac{1}{T_1} + \frac{1}{T_2}\right)$$

where
L = distance between transducer faces
C = speed of sound
V = velocity of wind
T_1 = transit time of ultrasound
T_2 = transit time of ultrasound

(a) Ultrasound Anemometer for wind speed and direction with temperature measurement.

(b) Ultrasound anemometer for wind speed in x, y and z direction, wind direction and temperature measurement.

(c) Compact ultrasound anemometer for wind speed direction and temperature measurement.

FIGURE 2.17 Ultrasound anemometers.

(*Source:* Published with permission from Ammonit.)

A wind vane is used to measure wind direction shown in Figure 2.18. The most familiar type uses a fin connected to a vertical shaft. The vane constantly seeks a position of force equilibrium by aligning itself into the wind. Most wind vanes use a potentiometer type transducer that outputs an electrical signal relative to the position of the vane. This electrical signal is transmitted via wire to a data logger and relates the vane's position to a known reference point (usually true north). Therefore, the alignment (or orientation) of the wind vane to a specified reference point is important.

Typical wind vanes are shown in Figure 2.18 (a) and (b). The evaluation of the wind direction enables the best possible positioning of wind turbines. A three-level wind monitoring station is shown in Figure 2.19 with direction sensors, anemometer and data logger.

(a) Wind vane.

(b) Compart wind vane.

FIGURE 2.18 Two configuration of typical wind vanes.

(*Source:* Published with permission from Ammonit.)

In Figure 2.19, a typical wind monitoring station with usual instruments is shown on a lattice meterological mast, with data logger, solar module for continuous power supply, anemometers, wind vanes, barometric pressure sensor, etc.

A typical ambient air temperature sensor is composed of three parts: the transducer, an interface device, and a radiation shield. The transducer contains a material element (usually nickel or platinum) with a relationship between its resistance and temperature. Thermistors, resistance thermal detectors (RTDs), and semiconductors are common element types recommended for use. The propeller anemometer is especially suited for measuring the vertical wind component. It consists of a propeller mounted on a fixed vertical arm.

Data loggers (or data recorders) come in a variety of types and have evolved from simple strip chart recorders to integrated electronic on-board cards for personal computers. Many manufacturers offer complete data logging systems that include peripheral storage and data transfer devices. The data logger should be electronic and compatible with the sensor types, number of sensors, measurement parameters, and desired sampling and recording intervals. It should be mounted in a non-corrosive, water-tight, lockable electrical enclosure to protect itself and peripheral equipment from the environment and vandalism. It should also:

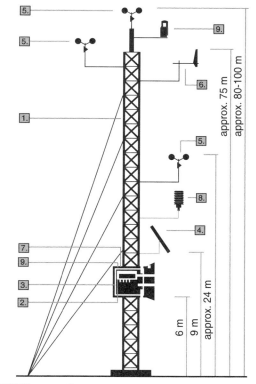

FIGURE 2.19 Typical wind monitoring station.
1. Lattice met-mast, 2. Data logger, 3. Logger accessories, standard steel cabinet, data transfer system and solar module, 4. Solar module, 5. Anemometers, 6. Wind vane, 7. Barometric pressure sensor, 8. Hygro- thermal sensor and 9. Additional components: precipitation sensor pyranometer, obstacle lights, surge protection and lightening protection.

(*Source*: Published with permission from Ammonit).

- Be capable of storing data values in a serial format with corresponding time and date stamps.
- Contribute negligible errors to the signals received from the sensors.
- Have an internal data storage capacity.
- Operate in the same environmental extremes.
- Offer retrievable data storage media.
- Operate on battery power, which is rechargeable by solar panel.

The manual *data transfer* involves two steps: (i) remove and replace the current storage device (e.g. data card) or transfer data directly to a laptop computer; and (ii) upload the data to a central computer in an office. The advantage of the manual method is that it promotes a visual on-site inspection of the equipment. The disadvantages include additional data handling steps (thus increasing potential data loss) and frequent site visits. Remote transfer requires a telecommunication system to link the in-field data logger to the central computer. The communications system may incorporate direct wire cabling, modems, phone lines, cellular phone equipment, or Radio Frequency (RF) telemetry equipment, or some combination thereof. An advantage of the manual method is that you can retrieve and inspect data more frequently than you can conduct site visits. Cellular phone data loggers are gaining popularity today for their ease of use and reasonable cost. You need to determine the minimum signal strength requirements of the data logger and relate that to actual field testing when researching this type of system. A portable phone can be used at the proposed site to determine the signal strength and the cellular company. Locations that experience weak signal strengths may be improved by selecting an antenna with higher gain. Guidelines for establishing a cellular account are usually provided by the data logger supplier.

There are two basic *tower* types for sensor mounting: tubular and lattice. For both, tilt-up, telescoping, and fixed versions are available. In addition, these versions may be either guyed or self-supporting. For new sites, tubular, tilt-up guyed types (with or without Gin-pole) are recommended for their ease of installation (the tower can be assembled and sensors mounted and serviced at ground level), minimal ground preparation, and relative low cost. Another typical wind monitoring station with tubular or lattice tower and guyed wires support is shown in Figure 2.20 with standard mountings and instruments. Towers should:

- Have an erected height sufficient to attain the highest measurement level.
- Be able to withstand wind and ice loading extremes expected for the location.
- Be structurally stable to minimize wind-induced vibration.
- Have guy wires secured with the proper anchor type, which must match the site's soil conditions.
- Be equipped with lightning protection measures including lightning rod, cable, and grounding rod.
- Be secured against vandalism and unauthorized tower climbing.
- Have all ground-level components clearly marked to avoid collision hazards.
- Have blinking light at the top as per aviation standard norms.
- Be protected against corrosion from environmental effects, including those found in marine environments.
- Be protected from cattle or other grazing animals.

The detailed description of location in terms of geographical positioning system (GPS) of latitude and longitude and position of sensors in terms of elevation and their configuration in terms of angle in degrees measured clockwise with respect to true north (TN) are shown in Figure 2.21 on a typical 60 m tall tower.

How high a tower should be?

The modern turbine blades extend upto 120 m high. The results of the forecast of lower height data collected (say at 80 m) from higher height (say 100 m or 120 m) were incredibly varied. A high lattice tower costs higher which includes:

- Tower itself
- Construction cost

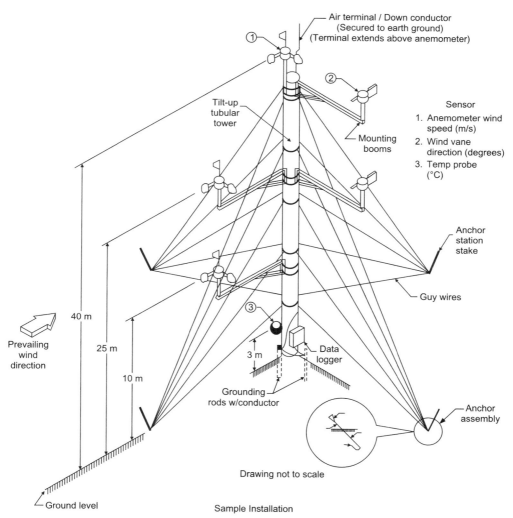

FIGURE 2.20 Typical wind monitoring station.
(*Source:* Wind Monitoring Station, Harte Ranch, Denver, Colorado, USA)

- Erection of tower
- Permission/clearance from aviation authority
- Site acquisition
- Local building and other permits
- Removing the tower at the end.

Technically, shorter towers are inferior as wind mast covers less blade sweep of a typical turbine and wind speed data from a shorter towers needs to be projected up to hub height of the proposed turbine. The coefficients used to project short tower data up to the hub height and beyond are not immutable laws of nature as atmospheric winds are non-linear and chaotic. Full blade sweep data is a simple solution but rarely practically possible.

SENSOR CONFIGURATION
Typical Boom Length: 95.0″

POSITION

#1/109976 — 316°, 136° — #2/109978

#7 — Offset 46°, 226°

#3/109980 — 316°, 136° — #13/109981

#8 — Offset 46°, 226°

#14/082628 — 316°, 136° — #15/109983

HEIGHTS
Tower: 197′–6″
L.R.: 202′–0″

194′–0″
(59.13 m)

190′–0″
(57.91 m)

148′–9″
(45.34 m)

144′–9″
(44.12 m)

103′–10″
(31.65 m)

Magnetic Declination: **8.6°E**

Tower: *NRG 60 m XHD Tilt-up*

Data Logger Type: *NRG Symphonie*

Time Zone: *-7 (MST)*

GPS *(Garmin)*
Latitude: *N 37° 43.004′*
Longitude: *W 104° 23.924′*
Datum: **WGS84**
Elevation: *1819 m (5968 ft)*
UTM: Easting: *0552992*
       Northing: *4174557*
       Zone: 13S

Instruments/Equipment
Temperature: *#12 @ 6′*
Barometric Press: *#11 @ 5′, SN/1805 8278*
Voltmeter: *#10*
Obstruction Light: *Carmanah A601 solar*
Types: *Anemometer: NRG #40C*
       Direction Vane: NRG #200P

Ground Voltage: *0.0 V*
Ground Resistance: *5.08 Ω*

Notes: *The tower was set on the ground toward the NNE at true 30°*

TN   True North
MN   Magnetic North
GPS Geographical Positioning System
UTM Universal Transverse Mercator
WGS World Geodic System

FIGURE 2.21 Typical description of location and sensor configuration. (*Source*: Harte Ranch, Wind Resource Monitoring Tower, 2010)

2.8.10 Remote Sensing Technique

In recent times, the wind turbines are being installed at an increasing rate in hilly, forested and complex moutaneous terrain as well as in off-shore sites with ever increasing hub heights. The conventional method of accredited wind resource assessment is to mount cup anemometers and direction sensors on tall meteorological (met) mast. Due to increasing hub height of modern wind turbine the cost of met towers are increasing approximately to the third power of height. For example, the measurement of wind speed at 100 m height for a modern 5 MW wind turbine hub level at a single point by cup anemometer is no longer a representative speed of wind for a rotor diameter spanning more than 120 m. The rotor blade tip from lower most point to upper most point for such a wind turbine will be at a height from 50 m to 200 m. Cup anemometer is a point measurement device, whereas using a sodar (sound detection and ranging) or lidar (light detection and ranging) at 100 m height probes a sampling volume in order of 1000 m^3. Remote sensing application in wind energy is used to supplement as well as to replace tall met mast measurement for the following purposes:

- Wind resource assessment: global wind resource are mapped on-shore, plains, over hilly and mountaneous terrain, coastal area, off-shore et cetera.
- Evaluating various wind flow models and developing wind atlases of micro, macro and global scale.
- Power performance verification of wind turbines.
- Wind turbine online and forward-feed controls.

IEC 61400-12-1 Standard for power curve measurement specify that a calibrated cup-anemometer at hub height and two to four rotor diameter up-wind in front of wind turbine is to be mounted. With this arrangement the wind speed measurement is based which is only at a single point measurement. For a modern wind turbine of rotor diameter spanning 120 m, the wind field over the entire plane can not be represented by a single point measurement that too especially in mountaneous and complex terrain. Wind turbines operate in atmospheric boundary layer which is having high turbulence levels due to wind shear caused by roughness of the earth surface. To characterise the wind flow over the entire rotor plane, multi-point and multi-height wind measurement will be required. The remote sensing application is now available commercially to fulfill above requirement.

The remote sensing technique based on sound and light waves propagation and backscatter detection such as sound detection and ranging (sodar), light detection and ranging (lidar), satellite based sea surface micro-wave scatterometry and synthetic aperture radar (SAR) find useful replacement of expensive met mast based sensors. It also replaces the time-consuming erection and installation of high met mast. Sodar and lidar are direct measurement of wind speed based on *Doppler shift,* and SAR is based on proxy-empirical calibration methods.

Sodar

Sodar are based on audio-frequent technology. In sodar technique, a ground based instrument transmits sequence of short bursts of sound waves of audible range frequencies from 2000 Hz to 4000 Hz at various heights in the atmosphere in three directions. A small fraction of transmitted sound wave propagation is scattered and reflected in all directions from the temperature differences and turbulence in the atmosphere. A very small fraction of scattered waves received back into the receiver or detector of sodar instrument. The height at which the wind speed is measured, will be determined by the delay in time in the back scatter of the transmitted pulse. In standard atmospheric conditions, the speed of sound propagation is about 340 m/s. For example, the back scatter of sound wave from a height of 170 m above the ground will be received back into the detector after delay of 1 s. The wind speed is determined as a function of *Doppler shift* observed from the frequency difference between the transmitted sound wave or light beam and frequency of back scattered received by the sodar instrument. Sodar are typically applicable for measuring 10 min averaged vertical profile from 20 m to 200 m height for the following parameters: mean wind speed, mean wind direction and turbulence in three directions.

There are two types of sodar systems available: mono-static and bi-static. In mono-static system, transmitter and receiver are co-located on the ground. In bi-static system, the transmitter and receiver are separated 100 m to 200 m apart on the ground. The *carrier to noise ratio* (C/N) is higher for bi-static system as compared to mono-static system. Bi-static *continuous wave* (CW) sodar system is preferred due to higher accuracy considered for wind energy applications. In number of applications, it is shown that mean wind speed and direction measurement by sodar is within the accuracy of ± 3 per cent as compared to cup-anemometer. It is to be noted that 1 per cent uncertainty in mean wind speed results in 3 per cent uncertainty in mean wind power. Sodar is relatively cheaper option compared to lidar with low power consumption of about 10 W per unit. Some of the maufacturers of sodar system are Remtech, Atmospheric System Corporation, Metek, Scientec, Second Wind and AQ System.

Lidar

It is remote sensing of wind parameter like speed, direction and turbulence by using laser light. A lidar is used to measure wind speed and direction by probing the wind flow in the atmosphere from a ground based instrument using laser beams. The wind measurement is achieved by detecting the *Doppler shift* in the back scattered light. The Doppler shift is directly proportional to wind speed in the direction of laser beam in the adjustable measurement volume in the atmosphere. The lidar can be used at various ranges, angles and different heights. In lidar system, electromagnetic radiation of light in the form of laser of wavelength of about 1.5 µm near infra-red is transmitted in the atmosphere at the point of interest. The back scattering of light from many small aerosols suspended and moving with the wind will be detected by receiver of lidar in the form of small frequency shifts or *Doppler shifts*. The time period of round trip of laser light beam introgating the atmosphere at a target height is called *time of flight*. For a height of 100 m, it is less than 1 µs. *Aerosol* may be consisted of dust, organic matter (e.g. pollen), soot or water droplets. For such particles, viscous force is dominant. A monostatic lidar system is shown in Figure 2.22, the fraction of trasmitted light is received back by detector as scattered light from aerosol in the target atmosphere to determine aerosol velocity.

Wind Parameter Determination

The function of transmitter is to provide a focussed laser beam at a particular location with arrangement of altering the focus range. The coherent lidar system measures the Doppler shift of scattered light from the target (aerosol) along the laser beam or in the direction of line-of-sight. In the transverse direction of the beam, the Doppler shift is zero. If lidar is placed at (0, 0, 0) and

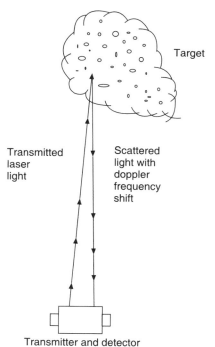

FIGURE 2.22 Monostatic lidar system.

measuring at target location of (x, y, z), where the wind vector components are (u, v, w) then the line-of-site velocity V detected by lidar system will be given by dot product of wind vector and unit vector along the beam direction:

$$V = \left| (u + v + w) \cdot \left(\frac{x + y + z}{\sqrt{x^2 + y^2 + z^2}} \right) \right| \tag{2.36}$$

Scattered light shows Doppler shift in frequency df is given as:

$$\delta f = \frac{2Vf}{c} = \frac{2V}{\lambda} \tag{2.37}$$

where f is laser frequency, λ is wavelength and c is speed of light (2.998×10^8 m/s). In wind industry standards, it is required to calculate 10 min averaged wind data for compatibility. The values of line-of-site velocity V and azimuth angle φ for uniform wind flow is represented by cosine wave of the form:

$$V = |a\cos(\varphi - b) + c| \tag{2.38}$$

a, b and c are floating parameters and can be estimated by standard least-square fitting method. The angle θ is scan cone half angle of order of 30°.

Horizontal speed $V_H = \dfrac{a}{\sin \theta}$

Vertical speed $W = -\dfrac{c}{\cos\theta}$

Bearing $B = b$ or $b \pm 180°$

If lidar probes the atmosphere with three beams in three directions, North, East and Zenith, the three measured line of sight (LOS) velocities V_{ri} are given as follows:

$$V_{rN} = u\sin\theta + w\cos\theta$$

$$V_{rE} = v\sin\theta + w\cos\theta$$

$$V_{rZ} = w$$

where θ is the angle between Zenith and North and East or the cone-angle. The wind speed components u, v and w are given as:

$$u = \frac{V_{rN} - V_{rZ}\cos\theta}{\sin\theta}$$

$$v = \frac{V_{rE} - V_{rZ}\cos\theta}{\sin\theta}$$

$$w = V_{rZ}$$

Horizontal wind speed V_h and wind direction D are given as:

$$V_h = \sqrt{u^2 + v^2}$$

$$D = \mathrm{mod}(360 + a\tan 2(v, u)360)$$

Wind lidars are working on two-measurement principles, namely continuous wave (CW) lidar and pulsed lidar.

Continuous Wave (CW) LIDAR

A continuous laser beam is transmitted in the atmosphere from lidar system and backscattering is received continuously at the receiver to record *Doppler shift* at a particular height. It focuses beam at different heights for measuring wind speeds. The measurement heights or range of measurement and spatial resolution of a continuous wave lidar is controlled by focal properties of the telescope. It works like a photographic camera to control the aperture, more for shorter distances and less for longer distances.

Continuous wave Doppler lidar system combines simplicity with high sensitivity for measurement of wind profiling. It is a ground based or wind turbine nacelle mounted instrument. There are two broad categories: coherent lidar and direct detection lidar. Former measures doppler shift by comparing backscattered radiation with a reference light beam. In latter system, the measurement of frequency shift is performed by passing the light through an optical filter. In coherent lidar system, a beam of coherent radiation illuminates the target as shown in Figure 2.21. A small fraction of light is backscattered and detected by a receiver. The motion of the target in the beam direction leads to a change in frequency of light as doppler shift. The frequency shift is measured as difference between original and scattered light beam. In monostatic lidar system, transmitted and received scattered light follow the same optical arrangement. In bistatic lidar system, the transmitter of laser light and detector are distinct and separate.

Pulsed LIDAR

The principle of pulsed lidar measurement of wind speed is based on aerosol movement at which laser pulses are directed and scattered off. The frequency difference of transmitted and reflected beam provide the data for determination of wind speed based on Doppler effect. Pulsed lidar emits regularly time spaced emissions of highly collimated light for a specified period of time of pulse length. It transmits a sequence of short pulses of laser beam in the order of 30 m wavelength and detects back scattered light. The measurement of wind speed in the direction of beam is related to *Doppler shift*. Comparatively CW lidar showed higher spatial resolution and faster data acquisition than pulsed lidar, and therefore, former is preferred for turbulence measurements. Pulsed lidar is preferred for wind speed measurements at multiple heights simultaneously with higher ranges.

A comparison between continuous wave lidar and pulsed lidars is given in Table 2.6.

TABLE 2.6 Comparision between CW and Pulsed Lidar

Parameter	*CW Lidar*	*Pulsed Lidar*
Velocity accuracy	Limited by coherence time of the atmosphere	Limited by pulse duration
Sensitivity to target out of focus	High	Low
Maximum range	Few hundred metres	Few kilometres
Laser source	1000 mW	10 mW
Linear polarisation	Not required	Required

Wind Profiler

A wind profiler measures average parameters of vertical wind speed profiles, vertical direction profiles, vertical turbulence profiles through a series of radial direction measurement of several components of wind speed normally based on 10 min data sampling in the atmosphere in beam direction. It is a ground based system transmitting a continuous or pulsed laser beam. It works on the same principle of *Doppler shift* detection in backscattered radiation. A wind profiler has transmitter and receiving antennas, and both are combined in a single optical telescope.

Lidar system of measurement shows higher accuracy due to following features:

- Nature of light which is 10^6 times faster than a sound wave.
- Antenna aperture size of lidar is based on the wavelength of light. The ratio of its lense diameter to wavelength is more than 1000 which results in better beam control and higher data sampling rate.
- Both types of lidars, continuous and pulsed wave, are based on the assumption that measured wind speed is proportional to *Doppler shift,* and also that the aerosols are homogenouly distributed and follow the mean wind flow.

- Lidars can be mounted on hub at the spinner or in the blades. The lidar can see the wind front 300 m to 400 m ahead of wind turbine and by the time wind strikes the blades it can activate the pitch of the blade for appropriate angle of attack for optimum aerodynamic performance. It works on active control of blade pitch and nacelle yaw movement as forward-feed control strategy. It also reduces the fatigue damage from extreme wind speeds and wind shear due to forward-feed control which results in prolonging the life of wind turbine. Improve power performance results due to enhanced yaw control and lead-time control of blade pitching for approaching wind front. Figure 2.23 shows lidar mounted on the hub and pointing into the on-coming wind front. It is used for wind profiling in conical scan pattern with the half cone angle θ is typically in the order of 30°.

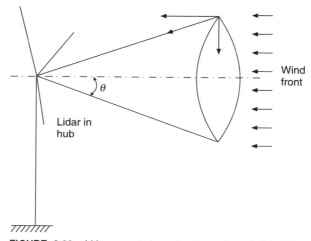

FIGURE 2.23 Lidar mounted on the hub of a wind turbine.

- Turbine spinner mounted continuous wave lidar is always pointing into the upcoming wind field and can be used for power curve measurement, feed forward data for turbine control system, warning of approaching gusts and optimizing rotor pitch control, hub height wind speed, shear exponent determination, rotor equivalent wind speed (REWS), etc. The lidar can be mounted on nacelle roof, on axis of spinner or can be built into the rotor blades.
- The lidar system is based on volume-averaged wind measurement, and therefore, more representative than point based cup-anemometer.
- Wind lidars are mobile instruments as compared to fixed location met mast mounted cup anemometres and better suited for onshore as well as offshore applications.

Scatterometry

Scatterometry is technique used for remote sensing the wind characteristics like speed, direction and turbulence aloft the sea surface. Scatterometers are active radars that send pulses towards the surface of sea and measure the backscattered signal due to small scale waves in the order of 20 mm. Post processing of backscattered signals relate to wind speed. An instrument known as synthetic aperture radar (SAR) is based on similar working principle and measures the backscattered signals related to wind speed. Sun synchronous satellite over the ocean fitted with radar scatterometers operating at different sub-bands of micro-wave are widely used to measure near-surface wind speed and direction. Radars operate at different sub-bands within the microwave range of electromagnetic radiation spectrum. As per IEEE standard, radar band spectrum of scatterometers are typically based on C and Ku sub-bands as follows:

Designation	Frequency	Wavelength
C-band	4–8 GHz	75–37.5 mm
Ku-band*	12–18 GHz	25–16.7 mm

Working principle

A microwave radar pulse is transmitted towards the ocean surface and the reflected signal is received and measured. Small scale ripples generated on the surface of water by the flow of wind satisfy the wavelength requirement for *Brag Scattering* of the incident pulse. From the energy reflected back to the receiver of the total measured signal the normalised radar cross-section is determined. The noise signal is sum of instrument noise and natural emmisivity of the atmosphere–earth system at the frequency of the radar pulse. The noise signal is to be subtracted from total signal of the scattering to get the NRCS σ_0. One of the empirical relation between wind velocity and normalised radar cross-section σ_0 is called geophysical model function (GMF). A general form of GMF as proposed by Naderi et al (1991) is given here, and geometry is shown in Figure 2.24.

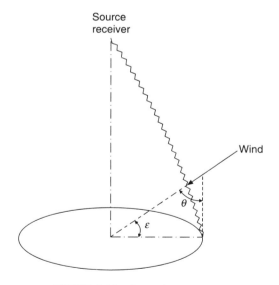

FIGURE 2.24 Scatterring geometry.

---

* Indian Space Research Organisation (ISRO) launched satellite Oceansat-2 in 2009 with Ku band scatterometer.

$$\sigma_0 = f(|u|, \xi, ..., \theta, f, \text{pol}) \tag{2.39}$$

where $|u|$ is wind speed, ξ is azimuth angle between the wind vector and the incident radar pulse, θ is incident angle of the radar signal measured in vertical plane, f is frequency of the radar signal and 'pol' is polarization.

The number of measurements are required for different azimuth angles and different polarizations to determine wind speed and direction.

Equivalent neutral wind

As per convention σ_0 values are referred to the wind at 10 metres above sea surface for neutral atmospheric stratification. The equivalent neutral wind (ENW) according to Liu and Tang (1996) is given as modified logarithmic wind profile for the atmospheric stratification as:

$$u = \frac{u_*}{K} \left(l_n \left(\frac{z}{z_o} \right) - \phi_M \right) \tag{2.40}$$

where u is wind speed at height z; z_0 is sea roughness length; u_* is friction velocity; K is von Karman constant and φ_M is stability correction. For validation, the observed σ_0 and computed wind speed is compared. The backscattered signal from water waves is affected by rain, sea surface temperature, currents and stratification of the atmosphere which also affect the ocean surface roughness. The sea surface temperature also influences the viscosity of surface layer of ocean. Higher temperature causes lower viscosity and more backscaterring and vice-versa, therefore, correction factors are used in the scatterometer.

EXAMPLE 2.4 Determine the expected number of hours per year of wind speed between 7.9 m/s and 8.8 m/s at site 1 in Figure 2.25 at an elevation of 20 m.

Solution: For site 1, at 10 m, $C = 5.4$ m/s, $K = 2.01$ and $\overline{V} = 4.8$ m/s. Correcting $z = 20$ m, we have

{using $K = K_{\text{ref}} [1 - 0.088 \ln (z_r/10)]/[1 - 0.088 \ln (z/10)]$}

$K = [1 - 0.88 \ln (3)]/[1 - 0.088(6)]$

$= 2.156$

{using $\alpha = [0.37 - 0.088 \ln (C_r)]/[1 - 0.088 \ln (z_r/10)]$}

$\alpha = [0.37 - 0.088 \ln (5.4)]/[1 - 0.088 \ln (3)]$

$= 0.22/0.90$

$= 0.241$

and

$$C = C_r \left(\frac{z}{z_r} \right)^\alpha$$

$= 5.4(2)^{0.241} = 6.38$ m/s

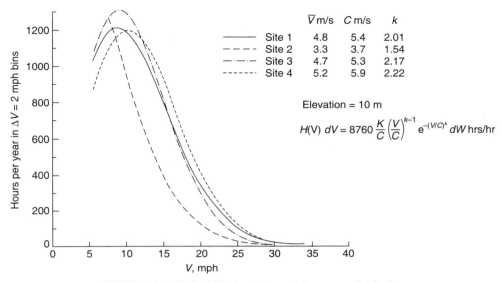

FIGURE 2.25 Weibull-fitted wind speed frequency distribution.

Then

$$H(7.9 < V < 8.8) = 8760(k/C)[(V/C)^{k-1}] \exp[-(V/C)^k] \, dV$$

$$= 8760 \left(\frac{2.156}{6.38}\right) \left[\left(\frac{8.35}{6.39}\right)^{1.56}\right] \exp\left[-\left(\frac{8.35}{6.38}\right)^{2.156}\right] 0.9$$

which yields estimate that $H = 806.62$ hr between 7.9 and 8.8 m/s at an elevation of 20 m at site 1.

EXAMPLE 2.5 A wind turbine at site 2 in Figure 2.25 will be mounted on a 33.3 m tower. If the turbine has a cut-out speed of 13.3 m/s, how many hours of wind speed greater than 13.3 m/s will probably be lost?

Solution: At site 2, $C = 3.7$ m/s, $k = 1.54$ and $\overline{V} = 3.3$ m/s for a reference height of 10 m. Correcting these values for $z_r = 33.3$ m, we have

$$k = k_r \frac{1 - 0.088 \ln\left(\frac{z_r}{10}\right)}{1 - 0.088 \ln\left(\frac{z}{10}\right)}$$

$$= 1.54 \frac{1 - 0.088 \ln(3)}{1 - 0.088 \ln\left(\frac{100}{10}\right)}$$

$$= 1.745$$

$$\alpha = \frac{0.37 - 0.088 \ln(C_r)}{1 - 0.088 \ln\left(\frac{z_r}{10}\right)}$$

$$= \frac{0.37 - 0.088 \ln(3.7)}{1 - 0.088 \ln\left(\frac{33.3}{10}\right)}$$

$$= 0.28$$

$$C = C_r \left(\frac{z}{z_r}\right)^\alpha = 3.7 \, (100)^{0.28} = 7.0 \text{ hr}$$

Thus, 7-hour operation per year will probably be sacrificed if a cut-off speed of 13.3 m/s is used.

REVIEW QUESTIONS

1. (a) Explain the causes and nature of atmospheric wind on the global level.
   (b) What is jet stream? How low-level jet stream may be utilised for very large turbines?
2. Explain the phenomenon of boundary layer.
3. What are the characteristics of the steady wind? How is it quantified?
4. Explain the vertical profile of the steady wind. What is power law index?
5. Explain how the wind data monitoring, recording and analysis is performed. How variations of velocity and directions are represented graphically?
6. How the wind energy potential is assessed at a site according to its wind characteristics?
7. Write a short essay on wind energy development in India and in the various leading countries in the world in the last decade with current capacities.
8. What are the future prospects for wind energy development in India in particular, and the world in general?
9. Explain the frequency distribution of mean hourly wind speed for a year as Weibull distribution and its scale and shape parameters.
10. How the topography and geographical features influence the wind regime of a region? Explain with suitable examples.
11. How Weibull distribution is used for the selection of an appropriate wind turbine for a specific site?
12. (a) What is atmospheric turbulence?
    (b) Explain the reasons of atmospheric turbulence.
13. Wind measurement data at a site for 20 m sensor height gives mean wind speed 6.66 m/s with $\sigma = 3.52$ m/s. Find Weibull parameters, C and k, for this site.
    (a) Find approximate annual values of Weibull parameters, C and k based on Gamma function (see Table 2.7).

TABLE 2.7 Gamma function values

No.	Gamma function
1.667	0.9027
1.625	0.8966
1.588	0.8922
1.556	0.8893
1.526	0.8874
1.500	0.8862
1.476	0.8857
1.455	0.8856
1.435	0.8859
1.417	0.8865
1.400	0.8873
1.385	0.8882
1.370	0.8893
1.357	0.8905
1.345	0.8917
1.333	0.8930

(b) For this site, the surface roughness length is 0.15 m. If it is found that k remains same for the wind measured at 60 m sensor height, find the mean wind speed and k at 60 m height.

(c) Find the power law index α for speeds near the mean wind speeds at 20 m and 60 m.

14. For Vijaydurg, the annual average wind speed and the standard deviation are 5.44 m/s and 2.88 m/s, respectively
    (a) Find the approximate annual values of Weibull parameters, C and k.
    (b) What is the energy pattern factor based on these C and k values?
    (c) Find the wind power density for Vijaydurg based on these C and k values.
    (d) What is the annual energy content of the wind based on these C and k values?

15. A wind machine mounted on a 20 m tower works only between wind speeds of 5 m/s and 15 m/s.
    (a) Find the number of hours it will work in a year at the following place with Weibull shape factor, $k = 2.0$, Weibull scale factor, $C = 8$ m/s at 20 m height and power law index, $\alpha = 0.2$.
    (b) What will be the number of hours it will work in a year if the tower's height is increased to 40 m?

16. A wind machine with rated power $P_{eR} = 1$ MW has the rated wind speed V_R of 7.5 m/s. It has cut-in speed V_C of 3 m/s and cut-off speed V_F of 20 m/s. It is installed at a place with annual Weibull parameters, $k = 2.1$ and $C = 6$ m/s. Find its capacity factor CF average annual power, and annual energy output.

$$CF = \frac{e^{-\left(\frac{V_C}{C}\right)^k} - e^{-\left(\frac{V_R}{C}\right)^k}}{\left\{\left(\frac{V_R}{C}\right)^k - \left(\frac{V_C}{C}\right)^k\right\}} - e^{-\left(\frac{V_F}{C}\right)^k}$$

17. The annual values of the scale parameter and the shape parameter for Weibull distribution at Tuticorin are given as 5.5 m/s and 2.3, respectively.

(a) Calculate the annual average wind speed for Tuticorin based on Weibull distribution.
(b) Find the probability of wind speed lying between 6 m/s to 12 m/s and 12 m/s to 20 m/s.
(c) Find the energy pattern factor for Tuticorin.

18. The wind measurement data at a site for 20 m sensor height gives the mean wind speed 6.66 m/s with $\sigma = 3.52$ m/s. Find Weibull parameters, C and k, for this site. Use exact analytical equations involving Gamma function.

(a) For this site, the surface roughness length is 0.15 m. If it is found that k remains same for wind measured at 60 m sensor height, write an expression for Weibull distribution for wind speed at 60 m sensor height.
(b) What is the mean wind speed at 60 m sensor height? Find the power law index, α for speeds near mean wind speeds at 20 m and 60 m.

19. Vijaydurg has Weibull parameters at 20 m height, $C = 5.83$ m/s and $k = 1.9$, annual power law index $\alpha = 0.13$, energy pattern factor = 1.6 and annual wind power density = 192 W/sq.m. If a 2-bladed propeller wind machine has diameter = 97 m and tower height = 60 m. The machine is designed for $Cp = 0.38$ at the rated wind speed and mechanical and electrical efficiencies are 0.96 and 0.95, respectively.

(a) What is the design rated wind speed for this machine such that its annual energy output is maximized? What are the values of cut-in and furling wind velocities? Make and state suitable assumptions.
(b) Find capacity factor, its rated power and the annual energy output.

20. A wind machine with rated power $P_{eR} = 1$ MW has rated wind speed of 7.5 m/s. It has cut-in speed of 3 m/s and cut-off speed of 20 m/s. It is installed at a place with annual Weibull parameters, $k = 2.1$ and $C = 6$ m/s.

(a) Find its capacity factor, average annual power and annual energy output. For selecting ideal rated speed for a wind machine at a given site, refer Figure 2.26, normalized power (P_N) maximized with respect to $U_{rated/C}$.

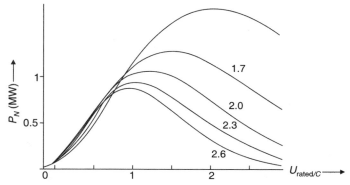

FIGURE 2.26 P_N vs U_{rated}/C for different values of k for $U_{cut-in}/U_{rated} = 0.5$ and $U_{cut-off}/U_{rated} = 2$.

(b) Select an ideal rated wind speed for a wind machine that has $U_{cut-in}/U_{rated} = 0.5$ and $U_{cut-off}/U_{rated} = 2$ to be installed at the place in the above question (a).
(c) Find its capacity factor.

21. The wind speed frequency distribution at Panchgani is given in Table 2.8.
    (a) Find the mean wind speed, U_{mean} and the standard deviation for the same.
    (b) If Weibull shape parameter, k can be expressed as $k = (\sigma/U_{mean})^{-1.07}$, then find the value of k. Based on this value of k, find Weibull scale parameter, C for Panchgani using Gamma function.
    (c) Derive an expression for the most frequent wind speed for a Weibull distribution.
    (d) Calculate the most frequent wind speed for Panchgani. How does it compare with the data?
    (e) Calculate the energy pattern factor, power density and annual energy content for Panchgani.

TABLE 2.8 Wind speed frequency distribution

Bin limits (kmph)	Representative wind speed (kmph)	Annual average (% frequency)
0	0	0.5
1–2	2	1.2
3–4	4	2.5
5–6	6	3.9
7–8	8	5.6
9–10	10	7.2
11–12	12	8.2
13–14	14	8.5
15–16	16	8.4
17–18	18	6.7
19–20	20	7.9
21–22	22	7.9
23–24	24	7.5
25–26	26	6.1
27–28	28	5.3
29–30	30	4.2
31–32	32	3.2
33–34	34	2.2
35–36	36	1.3
37–38	38	0.7
39–40	40	0.4
41–42	42	0.3
43–44	44	0.1
45–46	46	0.1
47–48	48	0.1
49–50	50	0
51–52	52	0

22. Okha in Gujarat has annual Weibull parameters at 30 m height, C is 5.72 m/s, k is 2.3, annual power law index is 0.21, energy pattern factor is 1.5, annual wind power density is 137.7 W/sq.m., and mean air density at the surface is 1162 gm/cu.m.

    (a) What is the ideal range for the rated wind speed for a propeller type machine with hub height 30 m for this site such that its annual energy output is maximum? What are the values of cut-in and furling wind velocities? Make and state suitable assumptions.

    (b) Find the capacity factor for such machine at this site.

23. For Saputara in Gujarat the annual average wind velocity is 3.23 m/s and the standard deviation is 1.56 m/s. Air density at the surface is 1156 gm/cu.m.

    (a) Find the approximate annual values of Weibull parameters, C and k.

    (b) What is the most frequently occurring wind speed calculated on the basis of these C and k values?

    (c) What is the energy pattern factor based on these C and k values?

    (d) Find the wind power density for Saputara based on these C and k values.

    (e) What is the annual energy content of the wind based on these C and k values?

24. What is the wind power class for Laddak area? Find out by internet searches.

25. Calculate the power, in kW, across the following areas for wind speeds of 5, 10 and 25 m/s. Use diameters of 5, 10, 50 and 100 m for the area. Air density = 1.2 kg/m³.

26. Solar power potential is around 1 kW/m². What wind speed gives the same power potential?

27. Calculate the factor for increase in wind speed if the original wind speed was taken at a height of 10 m. New heights are at 20 m and 50 m. Use the power law with an exponent $\alpha = 0.4$.

28. Calculate the factor for the increase in wind speed if the original wind speed was taken at a height of 10 m. New heights are 50 m and 100 m. Use the power law with an exponent $\alpha = 0.20$.

29. What is the air density difference between sea level and a height of 10,000 m?

30. At a place there is a wide temperature difference between summer (44°C) and winter (–2°C). What is the difference in air density? Assume you are at the same elevation and average pressure is the same.

    [Note: For problems with wind speed distributions, remember the wind speed has to be the number in the middle of the bin. If you use a bin width of 1 m/s, then the numbers have to be 0.5, 1.5, etc. In general, bin widths of 1 m/s are more than adequate. Smaller bin widths mean more calculations.]

31. Calculate the wind speed distribution using the Rayleigh distribution for an average wind speed of 8 m/s. Use 1 m/s bin widths.

32. Calculate the wind speed distribution for a Weibull distribution for $C = 8$ m/s and $k = 1.7$. Use 1 m/s bin widths.

33. Calculate the wind speed distribution for a Weibull distribution for $C = 8$ m/s and $k = 3$.

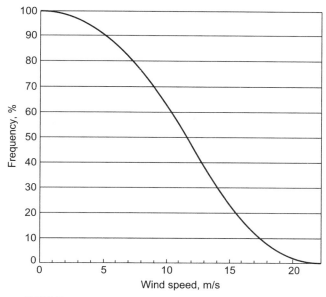

FIGURE 2.27 Wind speed-duration curve at 50 m height.

34. From Figure 2.27, what is the per cent of time the wind is 5 m/s or greater?
35. From Figure 2.27, what is the per cent of the time the wind is 12 m/s or greater?
36. For a 10 min period, the mean wind speed is 8 m/s and the standard deviation is 1.5 m/s. What is the turbulence intensity?
37. At a windy site, very high winds with gusts over 60 m/s were recorded. An average value for 15 min. was 40 m/s with a standard deviation of 8 m/s. What was the turbulence intensity?

Use the following table (Table 2.9) to calculate answers to Questions 39 to 41. The most convenient way is to use a spread sheet.

TABLE 2.9 Wind speed frequency distribution

Bin j	Speed m/s	No. observations	Bin j	Speed m/s	No. observations
1	1	20	6	11	150
2	3	30	7	13	120
3	5	50	8	15	80
4	7	100	9	17	40
5	9	180	10	19	10

38. Calculate the frequency for each class (bin). Remember sum of $f_j = 1$.
39. Calculate the power/area for $j = 5$ bin and $j = 10$ bin.
40. Calculate the average (mean) wind speed.
41. From wind speed histogram or wind speed distribution estimate annual energy production (Table 2.10), calculate annual energy production for 1 MW wind turbine.

TABLE 2.10 Wind energy frequency distribution

Wind speed (m/s)	Power (kW)	Bin hours (h)	Energy (kWh)
1	0	119	0
2	0	378	0
3	0	594	0
4	0	760	171
5	34	868	29,538
6	103	914	94,060
7	193	904	1,74,281
8	308	847	2,60,760
9	446	756	3,37,167
10	595	647	3,84,658
11	748	531	3,96,855
12	874	419	3,66,502
13	976	319	3,11,379
14	1,000	234	2,33,943
15	1,000	166	1,65,690
16	1,000	113	1,13,369
17	1,000	75	74,983
18	1,000	48	47,964
19	1,000	30	29,684
≥ 20	1,000	40	39,540
25	0	0	0
		Total 8,760	30,60,545

42. Calculate the wind speed frequency distribution for the data in Table 2.10.
43. Calculate the annual energy production for a mean wind speed of 8.2 m/s, average air density = 1.1 kg/m³. Use Rayleigh distribution to obtain a wind speed histogram. Use the power curve from Table 2.10.
44. In the preliminary data collection for a wind farm, for how long should data be collected if:
    (a) No regional data are available
    (b) Good regional data are available
    (c) There are other wind farms in the area.

Chapter 3

Rotor Aerodynamic

3.1 INTRODUCTION

What is aerodynamics? It comes from Greek word *Aerios* means air and dynamic concerns with forces and moments. Therefore, aerodynamics is the study of forces, moments and resulting motion of objects through the air.

To know how a modern wind turbine works, two terms from the field of aerodynamics need to be understood. They are 'drag' and 'lift'. An object in an airstream experiences a force that is imparted from the airstream to that object, [see Figure 3.1(a) and (b)]. We can consider this force to be equivalent to two component forces, acting in perpendicular direction, known as *drag* force and *lift* force. The magnitude of the drag and lift forces depends on the shape of the object, its orientation to the direction of the airstream, and the velocity of the airstream.

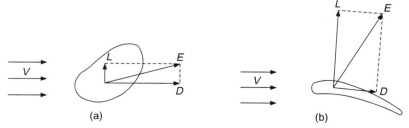

FIGURE 3.1(a) & (b) An object in an airstream is subjected to a force, *F*, from the airstream made of two component forces: the drag force *D*, acting in line with the direction of airflow; and the lift force, *L* acting at 90° in the direction of airflow.

Drag forces are those forces experienced by an object in an airstream that are in line with the direction of the airstream. A flat plate in an airstream, for example, experiences maximum drag forces when the direction of the airflow is perpendicular (that is, at right angles) to the flat side of the plate; when the direction of the airstream is in line with the flat side of the plate, the drag forces are at a minimum. A number of devices have been designed that make use of drag forces in their function. One example is the parachute, which relies on drag forces to slow down the rate of descent of a parachutist. Objects designed to minimize the drag forces experienced in an airstream are described as *streamlined*, because the lines of flow around them follow smooth, stream-like lines. Examples of streamlined shapes are teardrops, the shape of fish such as sharks and trout, and aeroplane wing sections (see Figure 3.2).

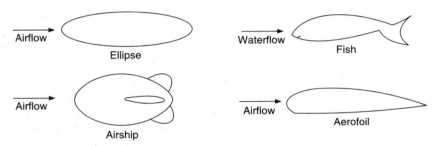

FIGURE 3.2   Some examples of streamlined shapes.

Lift forces are those forces experienced by an object in an airstream that are perpendicular to the direction of the airstream. They are termed *lift* forces because they are the forces that enable aeroplane to lift off the ground and fly. Lift forces acting on a flat plate are smallest when the direction of the airstream is at zero angle to the flat surface of the plate. At small angles relative to the direction of the airstream (that is, when the so-called *angle of attack* is small), a low pressure region is created on the 'downstream' or 'leeward' side of the plate as a result of an increase in the velocity on that side. In this situation, there is a direct relationship between air speed and pressure: the faster the airflow, the lower the pressure. This phenomenon is known as the *Bernoulli effect* after Daniel Bernoulli, the Swiss mathematician who first explained it. The lift force thus acts as a 'suction' or 'pulling' force on the object, in a direction at right angles to the airflow. Lift forces are also used to propel modern sailing yachts, and to support and propel helicopters. They are also the principal forces that cause a modern wind turbine to operate.

3.2   AEROFOIL

The angle that an object makes with the direction of an airflow, measured against a reference line in the object, is called the *angle of attack*, or angle of incident α. The reference line on an aerofoil section is usually referred to as the *chord line*. Arching or cambering a flat plate will cause it to induce higher lift forces for a given angle of attack, but the use of the so-called *aerofoil section* is even more effective. When employed as the profile of a wing, these sections accelerate the airflow over the upper surface. The high air speed induced results in a large reduction in pressure over the upper surface relative to the lower surface. This results in a 'suction' effect which lifts the aerofoil shaped wing. The strength of the lift forces induced by the aerofoil section is demonstrated most dramatically by their ability to support jumbo jets in the air.

There are two main types of aerofoil section: asymmetrical and symmetrical, as shown in Figure 3.3. Both of them have a markedly convex upper surface, a rounded end called the *leading edge* (which faces the direction from which the airstream is coming), and a pointed or sharp end called the *trailing edge*. It is the under surface of the sections that distinguishes the two types. The asymmetrical aerofoils are optimized to produce most lift when the underside of the aerofoil is closest to the direction from which the air is flowing, whereas the symmetrical aerofoils are able to induce lift equally well (although in opposite directions) when the airflow is coming from either side of them. Aerobatic display aeroplanes have wings with symmetrical aerofoil sections, which allow them to fly equally well when upside down. When airflow is directed towards the underside of the aerofoil, the angle of attack is usually referred to as positive.

(a)   (b)   (c)   (d)

FIGURE 3.3 Types of aerofoil section (a) is a symmetrical aerofoil section and (b), (c) and (d) are various forms of an asymmetrical aerofoil section.

The lift and drag characteristics of many different aerofoil shapes, for a range of angle of attacks, have been determined by measurements taken in wind tunnel tests, and catalogued. The lift and drag characteristics measured at each angle of attack can be described using non-dimensional lift and drag coefficients (C_l and C_d), or as lift to drag ratios $\left(\dfrac{C_l}{C_d}\right)$.

The 'length' (from the tip of its leading edge to the tip of its trailing edge) of an aerofoil section is known as the chord (see Figure 3.4 and Figure 3.5). The chord is also the width of the blade at a given position along the blade.

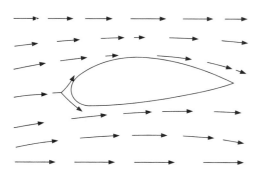

FIGURE 3.4 Streamlined flow around an aerofoil section.

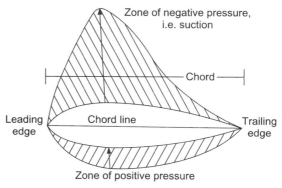

FIGURE 3.5 Zone of low and high pressure around an aerofoil section in an airstream.

3.2.1 Two-dimensional Airfoil Theory

Wind tunnel tests are usually necessary to validate a new airfoil design and explore performance in off-design condition. For airfoils (Figure 3.6), the drag force, D, is measured in the direction of the free stream relative wind felt by the foil, whereas the lift force, L, is measured perpendicular to this relative wind. The aerodynamic moment is usually measured about the 25 per cent chord point, i.e. one-fourth of the way back from the leading edge of the foil on a line joining the leading edge and trailing edge. The length, c, is called the chord of the foil. The lift force, L, and the drag force, D, together with the moment on an airfoil, M, are given in standard dimensionless ratios, C_l, C_d and C_m are called section data, because it applies to a certain particular airfoil section.

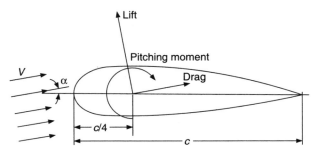

FIGURE 3.6 Aerodynamic forces and moments.

Wind turbine blades are long and slender structures where the spanwise velocity component is much lower than the streamwise component. In two-dimensional flow, the velocity component in the z-direction is zero and the flow is comprised of a plane as shown in Figure 3.7. To realize a flow of 2-dimensional plane, it is necessary to extrude an airfoil into a wing of infinite span. In Figure 3.7, the leading edge stagnation point is shown. The reaction force F from the flow is decomposed into two components, one parallel to flow as drag force and the other perpendicular to flow as lift force and the same is shown in Figure 3.8.

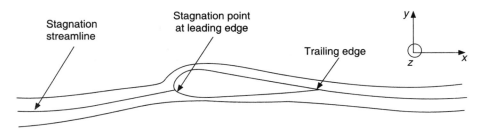

FIGURE 3.7 Schematic view of streamline past an airfoil.

Lift and drag coefficients C_l and C_d are defined as:

$$C_l = \frac{L}{\frac{1}{2}\rho V_\infty^2 c} \tag{3.1}$$

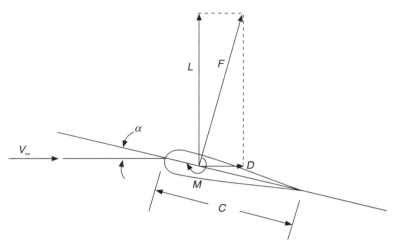

FIGURE 3.8 Definition of lift and drag forces.

and
$$C_d = \frac{D}{\frac{1}{2}\rho V_\infty^2 c} \quad (3.2)$$

Where ρ is the density and c the length of the chord, L, lift force, V_∞ the velocity of air approaching the airfoil and D, drag force. To know the moment M about a point in the aerofoil, it is normally located on the chord line at $\frac{c}{4}$ from the leading edge and considered positive when it turns the aerofoil in clockwise direction (nose up) and the coefficient of moment is defined as

$$C_m = \frac{M}{\frac{1}{2}\rho V_\infty^2 c^2} \quad (3.3)$$

The shape of the aerofoil forces the streamlines to curve around the geometry, as shown in Figure 3.9. The atmospheric pressure far from the aerofoil must be lower than atmospheric pressure on the upper side of the aerofoil and higher than atmospheric pressure on the lower side of the aerofoil. This pressure difference gives a lifting force on the aerofoil. When the aerofoil is almost aligned with the flow, the boundary layer stays attached and the associated drag is mainly caused by friction with the air.

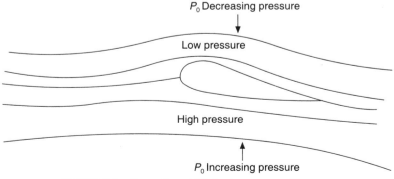

FIGURE 3.9 Generation of lift due to pressure difference.

The lift, drag and moment coefficients are usually given as a function of the angle of attack, α, which is the angle from the direction of flow to the airfoil chord line. The format for graphical presentation of airfoil data is shown in Figure 3.10. The lift and pitching moment coefficients are usually plotted versus angle of attack, whereas the drag coefficient is plotted versus the lift coefficient. Figure 3.11 shows the velocity diagram for a typical aerofoil.

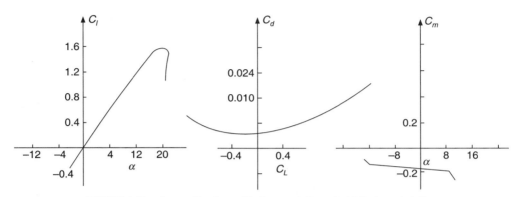

FIGURE 3.10 Conventional graphical presentation of airfoil characteristics.

In wind tunnel tests, the airfoils are constructed to exact shape and polished to a very smooth surface finish, since roughness on the surface, especially near the leading edge, can materially affect the boundary layer and change the airfoil characteristics. The characteristics of the flow depend upon the size of the foil and on the speed of the relative wind W. This dependence is quantified using a dimensional variable, the Reynolds number, Re:

Re = (velocity) (length) (density)/(viscocity)

$$Re = \frac{Wc\rho}{\mu} = \frac{Wc}{\upsilon} \quad (3.4)$$

and Mach number M_a:

$AO = r\Omega = U$

OF = Drag Force
OE = Lift Force
OD = Resultant Force
CD = Axial Force (Thrust)
OC = Tangential Force

$\alpha + \beta = \phi$

$V_{axial} \cong \frac{2}{3} V_\infty$

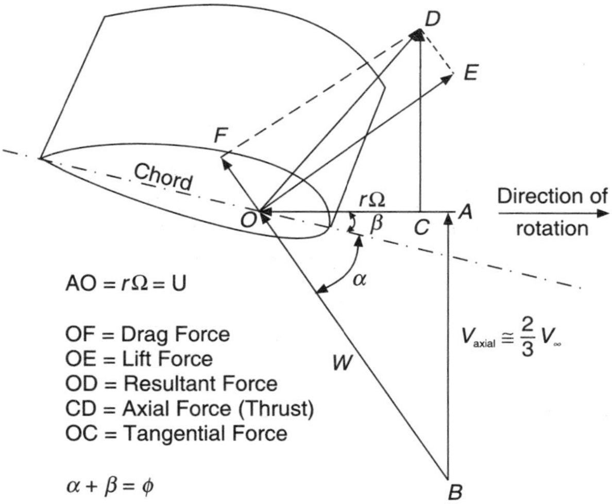

FIGURE 3.11 Flow velocity and aerodynamic forces at airfoil-section for a blade element.

$$M_a = \frac{V_\infty}{V_s} \quad (3.5)$$

where V_s is the speed of sound and W is the relative wind speed at a point on the blade and can be interpreted as the ratio of inertial to viscous forces on the fluid.

The geometrical shape of an airfoil is usually given in a tabular form (Table 3.1) in which the *x, y* coordinates of both upper and lower surfaces are given as measured with respect to the chord line, which is used as the reference. Both *x* and *y* coordinates are given as fractions of chord length *c* (Figure 3.12). Cambered (i.e. curved) airfoils are generated by defining a camber line and adding a thickness distribution equally above and below the camber line at each station along the foil. The exact shape of the leading edge is determined by leading edge radius, a circular arc centred on the camber line going through the leading edge. Cambered airfoils are capable of generating lift with a smaller drag penalty than symmetrical foils. Since the blades of a vertical axis rotor must generate both positive and negative lift during each rotation, they cannot effectively take advantage of the low drag capabilities of cambered airfoils. This gives an aerodynamic performance advantage to horizontal-axis wind turbines over vertical-axis machines.

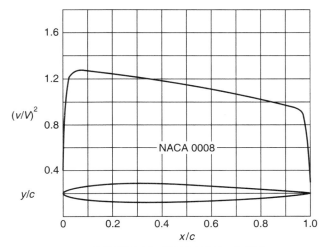

FIGURE 3.12 NACA 0008 basic thickness.

TABLE 3.1 NACA 0008 basic thickness form leading edge radius 0.70 per cent *c* (Ref. Abbott & Doenhoff)

x per cent c	y per cent c	$(v/V)^2$
0	0	0
0.5	—	0.792
1.25	1.263	1.103
2.5	1.743	1.221
5.0	2.369	1.272
7.5	2.800	1.284
10	3.121	1.277
15	3.564	1.272
20	3.825	1.259
25	3.961	1.241
30	4.001	1.223
40	3.869	1.186
50	3.529	1.149
60	3.043	1.111
70	2.443	1.080
80	1.749	1.034
90	0.965	0.968
95	0.537	0.939
100	0.084	—

In four-digit XXXX NACA airfoil designation, following scheme is adopted for different digits:

First digit indicates maximum separation between the mean camber line as a percentage of chord. Second digit indicates distance from the leading edge to the position of the maximum camber in tenths of the chord. Third and Fourth digits indicate maximum thickness of the airfoil as a per cent of chord.

Example NACA 2412 with a chord length of 200 mm,
Maximum camber = $0.02 \times 200 = 4$ mm
Position of maximum camber = $0.4 \times 200 = 80$ mm of leading edge
Maximum thickness of airfoil = $0.12 \times 200 = 24$ mm

There may be other schemes of calculations due to much modifications for five or more digits airfoil designations.

Airfoils differ in their lift, drag and moment coefficients as a result of their shape. Some shapes are structurally better than others, and some shapes are more easily manufactured. It is possible to keep a favourable boundary layer pressure gradient for longer distance along the foil by moving the point of maximum thickness away from leading edge. This facilitates additional structure to provide bending strength to the blade. Airfoils with a thin trailing edge may have superior aerodynamics but are more sensitive to mechanical damage in handling and shipping. An important parameter of an airfoil is its thickness. The thickness is given as a fraction of the chord length, i.e. a 12 per cent thick foil has a maximum thickness, that is, 12 per cent of the chord length. Since airfoil characteristics vary with thickness, it is important that the per cent thickness of the actual rotor corresponds with the data used in the design.

The lift and drag coefficients of an aerofoil can be measured in a wind tunnel at different angles of attack and wind velocities. The results of such measurements can be presented either in a tabular or graphical form as shown in Figure 3.13. Each aerofoil has an angle of attack at which the lift-to-drag ratio (C_l/C_d) is at a maximum and this angle of attack results in the highest efficiency of the blade of a HAWT.

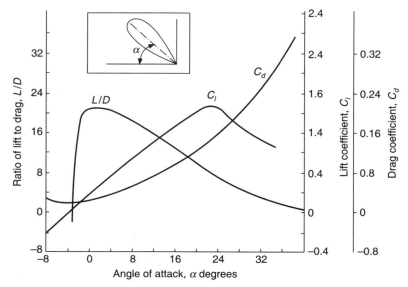

FIGURE 3.13 Lift coefficient, C_l, drag coefficient, C_d, and lift-to-drag ratio versus angle of attack α, for an aerofoil section. The region just to the right of the peak in the C_l curve corresponds to the angle of attack at which stall occurs.

Another important characteristic relationship of an aerofoil is its angle of stall. This is the angle of attack at which the aerofoil exhibits stall behaviour. Stall occurs when the flow suddenly leaves the suction side of the aerofoil (when the angle of attack becomes too large), resulting in dramatic loss in lift and an increase in drag (Figures 3.13 and 3.14). When this happens during an aeroplane flight, it can be extremely dangerous. Some wind turbine blades have been designed to take advantage of this phenomenon as a means of limiting the power extracted by the rotor in high winds. The knowledge of these coefficients is essential when selecting appropriate aerofoil sections for wind turbine blade design. Lift and drag forces are both proportional to the energy in the wind.

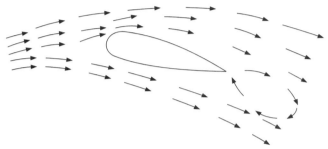

FIGURE 3.14 Aerofoil section in stall.

C_l increases linearly with α, until a certain value of α, where a maximum value of C_l is reached. Hereafter, the aerofoil is said to stall and C_l decreases in a very geometrically dependent manner. For small angle of attack, the drag coefficient C_d is almost constant but increases rapidly after stall. The way an aerofoil stalls is dependent on geometry. Thin aerofoils with a sharp nose, in other words with high curvature around the leading edge, tend to stall more abruptly than thick aerofoils.

Figure 3.15 shows the streamlines for a NACA 63–415 aerofoil for small angle of attack in the range of 0° to 14°. The forces on the aerofoil stem from the pressure distribution $p(x)$ and the skin friction with the air. The shear stress is expressed as

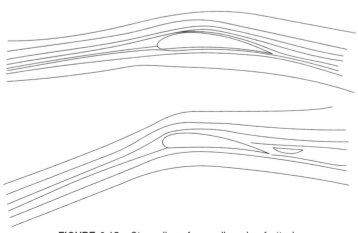

FIGURE 3.15 Streamlines for small angle of attack.

$$\tau = \mu \left(\frac{dU}{dy}\right)_{y=0} \tag{3.6}$$

(x, y) is the surface coordinate system as shown in Figure 3.16 and μ is the dynamic viscosity. The skin friction is mainly contributing to the drag, whereas the force found from integrating the pressure has lift and drag component. The drag component from the pressure distribution is known as the form drag and becomes very large when the aerofoil stalls.

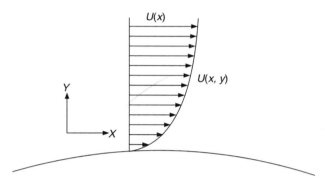

FIGURE 3.16 Viscous boundary layer at the wall of an airfoil.

A boundary layer thickness is often defined as the normal distance $\delta(x)$ from the wall where $u(x)/U(x) = 0.99$. Close to the aerofoil, there exists a viscous boundary layer due to no-slip condition on the velocity at the wall as shown in Figure 3.16. The fluid which flows over the aerofoil accelerates as it passes the leading edge and, since the leading edge is close to the stagnation point and the flow accelerates, the boundary layer is thin. It is known from viscous boundary layer theory that the pressure is approximately constant from the surface to the edge of the boundary layer, i.e. $\frac{\partial p}{\partial y} = 0$. At the trailing edge, the pressure must be same at the upper and lower sides and, therefore, the pressure must rise, $\frac{\partial p}{\partial x} > 0$, from a minimum value somewhere on the upper side to a higher value at the trailing edge. An adverse pressure gradient, $\frac{\partial p}{\partial x} > 0$, may lead to separation.

The u velocity profile in an adverse pressure gradient, $\frac{\partial p}{\partial x} > 0$, is s-shaped and separation may occur, whereas the curvature of the velocity profile for $\frac{\partial p}{\partial x} < 0$ is negative throughout the entire boundary layer and the boundary layer stays attached. A schematic picture showing the different shapes of the boundary layer is given in Figure 3.17. Since the form drag increases dramatically when the boundary layer separates, it is of utmost importance to the performance of an aerofoil to control the pressure gradient.

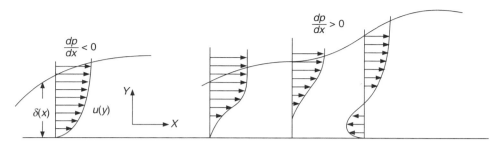

FIGURE 3.17   Schematic view of the shape of the boundary layer for a favourable and an adverse pressure gradient.

For small values of x, the flow is laminar, but for a certain x, the laminar boundary layer becomes unstable and a transition from laminar to turbulent flow occurs. In Figure 3.18 transition from a laminar to turbulent boundary layer is shown.

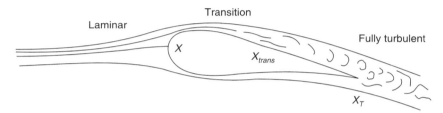

FIGURE 3.18   Schematic view of the transitional process.

3.2.2   Relative Wind Velocity

When wind turbine is stationary, the direction of the wind as 'seen' from a wind turbine blade is same as the undisturbed wind direction. However, once the blade is moving, it has velocity, and this can be graphically represented as a two-dimensional vector. The velocity of the wind as seen from a point on a moving blade is known as the *relative wind velocity* (usually symbolized by W). This is a vector which is the resultant of the undisturbed wind velocity vector, V_0 and the tangential velocity vector of that point on the blade, U (Figure 3.11). Now, let us

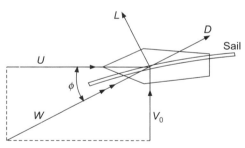

FIGURE 3.19   Velocity and force vector diagram for a sailing boat.

consider the example of a sailing boat travelling at a velocity, U, 10 ms^{-1} across (that is, perpendicular to) a wind stream with an undisturbed wind velocity, V_0, of 5 ms^{-1} (Figure 3.19). The relative or apparent wind velocity, W, is the resultant of the velocity vectors of the boat and the undisturbed wind, U and V_0, respectively. The relative wind velocity is the actual wind, both in speed and direction, that is always on board the boat and used by the sails to propel a sailing boat. When a boat is moving, the relative wind is always different from the undisturbed wind. The angle that the relative wind velocity vector makes with the boat's velocity vector is known as the *relative wind angle* (usually represented by ϕ), and given by the formula:

$$\tan \phi = \frac{V_0}{U} \tag{3.7}$$

In the above example,

$$\tan \phi = \frac{5}{10} = 0.5$$

$$\phi = 26.5°$$

Now, determine the magnitude of the relative wind velocity, W, using the relationship:

$$W = \frac{V_0}{\sin \phi} = \frac{5}{\sin 26.5} = 11.2 \text{ ms}^{-1}$$

How do wind turbines work?

Both horizontal and vertical axis wind turbines make use of the aerodynamic forces generated by aerofoil in order to extract power from the wind, but each harnesses these forces in a different way. In a fixed-pitch HAWT, assuming the rotor axis is in constant alignment with the (undisturbed) wind direction, for a given wind speed and constant rotation speed, the angle of attack at a given position on the rotor blade stays constant throughout its rotational cycle. During the normal operation of a horizontal axis rotor, the direction from which the aerofoil 'sees' the wind is such that the angle of attack remains positive throughout. In the case of vertical axis rotor, however, the angle of attack changes from positive to negative and back again over each rotation cycle. This means that the 'suction' side reverses during each cycle, so a symmetrical aerofoil has to be employed to ensure that power can be produced irrespective of whether the angle of attack is positive or negative.

Horizontal axis wind turbines operate with their rotation axes in line with the wind direction and are therefore called *axial flow devices*. The rotation axis is maintained in line with wind direction by a yawing mechanism that constantly realigns the wind turbine rotor in response to changes in wind direction. The performance of a horizontal axis wind turbine rotor is dependent on the number and shape of its blades and the choice of aerofoil section, together with the blade chord, relative wind angle and blade pitch angle at positions along the blade, and the amount of twist between the hub and the tip.

Modern VAWTs, unlike HAWTs, are 'cross flow devices'. This means that the direction from which the undisturbed airflow comes at right angles to the axis of rotation, that is, airflow across the axis. As the rotor blades turn, they sweep a three-dimensional surface, as distinct from the single circular plane swept by an HAWT's rotor blades. In contrast to traditional vertical axis windmills, the blades of modern vertical axis wind turbines extract most of the power from the wind as they pass across the front and rear (relative to the undisturbed wind direction) of the swept volume. When the blade is moving at a velocity several times greater than the undisturbed wind velocity, the angle of attack from which it sees the relative wind velocity, though varying, remains small enough to enable it to absorb aerodynamics forces which impart a tangential driving force and torque to the rotor.

Figure 3.20 shows a section through a moving rotor blade of an HAWT. Also shown is a vector diagram of the forces and velocities at a position along the blade at an instant in time. Because the blade is moving, the direction from which the blade 'sees' the relative wind velocity, W, is the resultant of the tangential velocity, U, of the blade at that position and the wind velocity,

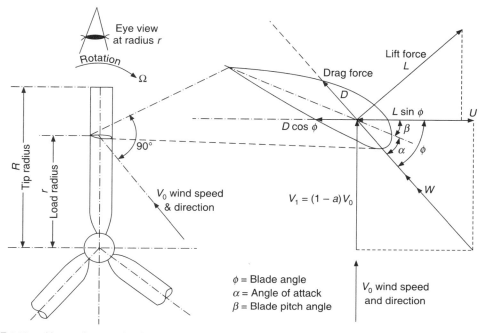

FIGURE 3.20 Vector diagram showing a section through a moving HAWT rotor blade. Notice that the drag force D, at the point shown in line with the relative wind direction, W, and the lift force, L is acting at 90° to it.

V_1 at the rotor. The tangential velocity, U at a point along the blade is the product of the angular velocity, Ω (in radians per second) of the rotor and the local radius, r, at that point, that is:

$$U = \Omega r \qquad (3.8)$$

The wind velocity at the rotor, V_1, is the undisturbed wind velocity upstream of the rotor, V_0, reduced by a factor that takes account of the wind being slowed down as a result of power extraction. This factor is often referred to as the axial interference factor, and is represented by a. The relative wind angle, ϕ, is the angle that the relative wind makes with the blade (at a particular point with local radius r along the blade) and is measured from the plane of rotation. The angle of attack, α, at this point on the blade can be measured against the relative wind angle, ϕ. The blade pitch angle (usually represented by β) is then equal to the relative wind angle of attack.

Because the rotor is constrained to rotate in a plane at right angles to the undisturbed wind, the driving force at a given point on the blade is that component of the aerofoil lift force that acts in the plane of rotation. This is given by the product of the lift force, L, and the sine of the relative wind angle, ϕ (that is, $L \sin \phi$). The component of the drag force in the rotor plane at this point is the product of the drag force, D, and the cosine of the relative wind angle ϕ (that is, $D \cos \phi$).

The torque, q (that is, the moment about the centre of rotation), in Newton-metres (Nm) at this point on the blade is equal to the product of the net driving force in the plane of rotation (that is, the component of the lift force in the plane of rotation minus the component of drag force in the rotor plane) and the local radius r. The total torque Q, acting on the rotor can be calculated by summing the torque at all points along the length of each blade and multiplying by the number of blades. The power from the rotor is the product of the total torque and the rotor's angular velocity, Ω.

The vertical axis wind turbine will function with wind blowing from any direction, but let us assume initially that it is blowing from one particular direction and also that the setting angle of the blade is such that its chord is in line with the tangent of the circular path of rotation (that is, it has 'zero set pitch'). Clearly, the angle of the blade to the direction of the undisturbed wind changes from zero to 360° over each cycle of rotation. It might appear that the angle of attack of the wind to the blade would vary by the same amount, and so it might seem impossible for a VAWT to operate at all. However, we have to take into account the fact that when the blade is moving, the relative wind angle seen by the blade is the resultant of the wind velocity V_1 at the rotor and the blade velocity U. Provided that the blade is moving sufficiently fast, relative to the wind velocity (in practice, this means tip speed ratio of three or more), the angle of attack that the blade makes with the relative wind velocity W will vary within a small range (see Figure 3.21).

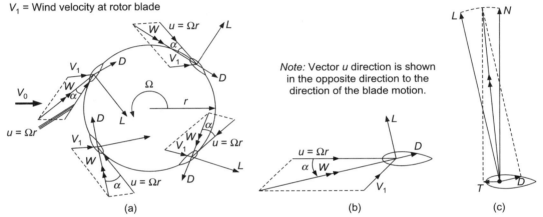

FIGURE 3.21 (a) Blade forces and relative velocities for a VAWT, showing angles of attack at different positions, (b) Details: aerodynamic forces on a blade element of a VAWT rotor blade, (c) Normal (radial) and tangential (chord-wise) components of force on a VAWT blade.

The lift and drag forces acting on the blade can be resolved into two components: 'normal' (N) (that is, in line with the radius) and 'tangential' (T) (that is, perpendicular to the radius). The magnitude of both components varies as the angle of attack varies during the rotation cycle and as a result, the torque output fluctuates as the turbine rotates (see Figure 3.21).

The magnitude and direction of the relative wind angle, ϕ, varies along the length of the blade according to the local radius, r. This is because the local tangential speed, u of a given blade element is the fuction of the rotor's angular velocity (Ω) times the local radius, r, of the blade element. As the tangential speed decreases towards the hub, the relative wind angle ϕ progressively increases. If a blade is designed to have a constant angle of attack along its length, it will have to have a built-in twist, the amount of which varies progressively from tip to root. Figure 3.22 demonstrates the progressive twist of a HAWT rotor blade. Most manufacturers of HAWT blades produce twisted blades, although it is possible to build adequate HAWT rotor blades that are not twisted. These are cheaper, but less efficient. The pressure and velocity relationships in the vicinity of the rotor are shown in Figure 3.23.

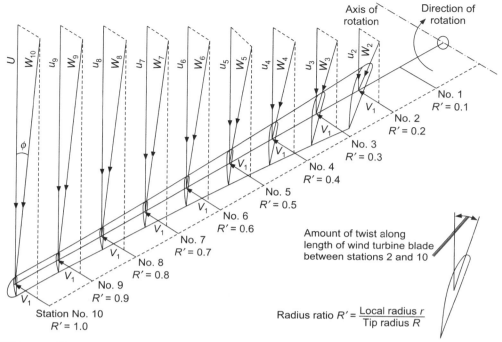

FIGURE 3.22 Three-dimensional view of an HAWT rotor blade design showing the changing angle of the relative wind velocity, ϕ, along the blade span.

3.2.3 Stall Control

Let us assume that a wind turbine is rotating at a constant rotation speed, regardless of wind speed, and that the blade pitch angle is fixed. As the wind speed increases, the tip speed ratio decreases. At the same time, the relative wind angle increases, causing an increase in the angle of attack. It is possible to take advantage of this characteristic to control a turbine in high winds, if the rotor blades are designed so that above the rated wind speed, they become less efficient because the angle of attack is approaching the stall angle. This results in loss of lift, and thus torque, on the regions of the blade that are in stall. This method of so-called stall-control has been employed successfully on numerous fixed pitch HAWT rotors. It is also employed on most VAWT rotors.

3.3 WIND FLOW MODELS

For wind turbine design, it is necessary to generate mathematical models of the aerodynamic forces and moments on a rotor working in natural winds. The simplest wind model is to think of wind as constant, homogenous velocity field of fixed direction. The actual wind field experienced by a given rotor may be inhomogeneous and unsteady. To get a concept of some of the problems involved, consider a rigid tower having a set of anemometer type axis centred at its top at hub height in such a way that the x-axis points into the wind, the y-axis is horizontal and points to the right, and the z-axis points downwards. Let the direction of the local wind be represented by the

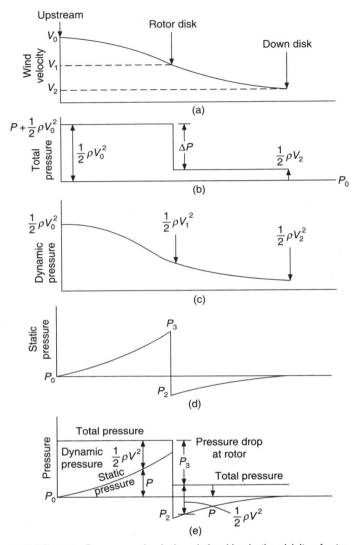

FIGURE 3.23 Pressure and velocity relationships in the vicinity of rotor.

angle $\phi(x, y, z, t)$ measured in a horizontal plane between the x-axis and true north, as shown in Figure 3.24(a). It can be expressed as a steady homogenous wind as $v(x, y, z, t) = $ constant, with $\phi(x, y, z, t)$ a constant of course, the direction, $\phi(x, y, z, t)$ in the preferred plane of rotor, $x = 0$, may not be constant. The rotor would nominally rotate in this plane, although a rotor free to yaw usually does not do so. If the rotor is not aligned with the instantaneous wind, a yaw condition exists. A rotor yawed from the wind experiences cross-flow components whose direction changes as the blades rotate. This results in unsteady flow at the blades and dynamic (periodic) variations in the angle between the relative wind and the blade at a frequency of one cycle per revolution. Horizontal axis wind turbines use either passive or active yaw control in an attempt to keep the rotor plane oriented perpendicular to the wind. Passive yaw control systems sometimes allow yaw angles of

30 to 40 degrees (sometimes even 180°) during operation, and even active yaw controls sometimes move so slowly that substantial yaw angles develop. Moreover, since yaw control system can fail, no one can guarantee that a given rotor will never have to operate in a yawed condition.

For a homogenous, constant direction, time varying wind, for a rotor plane at $x = 0$ and a wind $v(0, y, z, t) = v(t)$, with ϕ a constant. This might be an acceptable model for, say, control system analysis if $v(t)$ consisted of an average value plus small random variations defined by a particular spectrum. A steady wind with vertical shear could be defined as $v(0, y, z, t) = v(z)$, with ϕ a constant, as in Figure 3.24(b). The effect of wind shear should be smaller, the smaller the rotor diameter. Wind shear results in each section of the rotor blade (especially those sections near the tip) seeing unsteady (periodic) flow. The wind field through a real rotor probably looks more like that of Figure 3.24(c) and changes continually with time.

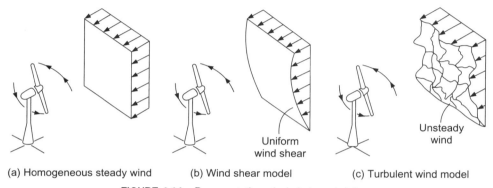

FIGURE 3.24 Representation of wind characteristics.

In general, then, the operational environment of a wind turbine rotor may be quite complex. We attempt, in analysis, to formulate conditions representative of certain modes of operation and wind regimes. In design, we try to build mathematical models of the most critical conditions for the machine and gain as much insight from these models as we can. We note that when a wind turbine rotor is tested, there are conditions under which well-defined flow is experienced, and other conditions under which the flow breaks away, becomes turbulent, or may even reverse. Aerodynamic modelling of states, where well-defined flows exist and turbulence is very small, is fairly well established. Modelling of unsteady flow processes for wind turbine rotors, however, is in its infancy.

3.3.1 Wind Flow Pattern

One must enter into wind turbine aerodynamics analysis with a proper feeling for the complexity of the subject. The real flows of interest are governed by the Navier-Stokes equations. For steady flow over a flat plate at zero angle of attack, the resulting flow pattern turns out to be of the complexity shown in Figure 3.25. Progressing aft from the leading edge, we start with a section having laminar flow. After the critical Reynolds number is reached, the effect of spanwise vorticity due to the low aspect ratio of the plate, three-dimensional vortex breakdown, turbulent spots, edge contamination and finally fully developed turbulent flow. One can imagine how this 'simple' real

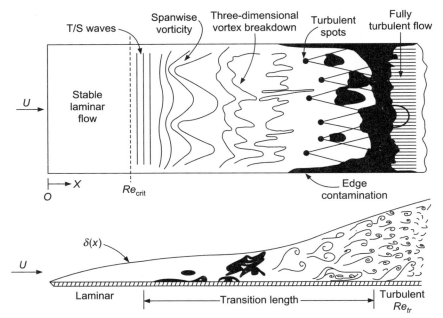

FIGURE 3.25 Idealized sketch of the transition process on a flat plate.

flow will be affected if the flat plate changes into an airfoil section of a rotor blade at an angle of attack. To this complication must be added the effects of surface imperfection from manufacturing, squashed bugs, surface water droplets, and turbulent eddies causing unsteady motions and aggravating natural flow instabilities. Then, we must consider flow separation and reattachments, as well as three-dimensional effects such as possible span wise flow and interactions with blade hub and roots. If that were not enough, blade flapping and torsional oscillations cause important dynamic effects in the fluid flow. This is clearly not a game for the faint hearted. The simplest model for propeller or wind turbine aerodynamic is the actuator disc model. This model requires some extremely simplifying assumption and yet yields a number of useful approximate results.

3.4 AXIAL MOMENTUM THEORY

The Rankine-Fraude momentum theory predicts forces acting on a rotor. It helps in predicting the ideal efficiency of the rotor. The function of a wind turbine is to convert kinetic energy of the moving air into mechanical energy. The rotational motion of the fluid that is imparted by the blades, and frictional drag affect the energy extraction. As a first approximation to determine the maximum possible output of a wind turbine, the following assumptions are made.

- Blades operate without frictional drag.
- A slip-stream, separates the flow passing through the rotor disc from outside the disc.
- The static pressure in and out of the slip-stream far ahead and far behind the rotor is equal to the undisturbed free-stream static pressure.
- Thrust loading is uniform over the rotor disc.
- No rotation is imparted to the flow by the disc.

Applying the momentum theorem to control volume in Figure 3.26 where the upstream and downstream control volume planes are infinitely far removed from the actuator disc plane, the thrust T is

$$T = \text{Momentum flux out} - \text{Momentum flux in}$$

or

$$T = \dot{m}(V_\infty - V_2) = \rho A U(V_\infty - V_2) \tag{3.9}$$

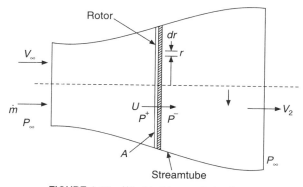

FIGURE 3.26 Wind turbine control volume.

Also from pressure conditions, the thrust can be expressed as

$$T = A(P^+ - P^-) \tag{3.10}$$

where
P^+ = pressure upstream
P^- = pressure downstream

Now applying Bernoulli's equation to flow upstream of the wind turbine,

$$\frac{1}{2}\rho V_\infty^2 + P_\infty = \frac{1}{2}\rho U^2 + P^+ \tag{3.11}$$

and downstream of the wind turbine

$$\frac{1}{2}\rho V_2^2 + P_\infty = \frac{1}{2}\rho U^2 + P^- \tag{3.12}$$

From Eqs. (3.11) and (3.12), we get

$$P^+ - P^- = \frac{1}{2}\rho(V_\infty^2 - V_2^2)$$

And by substituting for $(P^+ - P^-)$ into Eq. (3.10), we get

$$T = \frac{1}{2}\rho A(V_\infty^2 - V_2^2) \tag{3.13}$$

By equating Eqs. (3.9) and (3.13), we have

$$U = \frac{V_\infty + V_2}{2} \tag{3.14}$$

The result states that the velocity through the turbine is the average of the wind velocity ahead of the turbine and wake velocity after the turbine.

Now by defining axial interference factor a as, $a = \dfrac{V_\infty - U}{V_\infty}$

$$U = V_\infty(1 - a) \quad (3.15)$$

and using the definition of a in Eq. (3.14) gives,

$$V_\infty(1 - a) = \dfrac{V_\infty + V_2}{2}$$

The wake velocity, V_2, can then be expressed as

$$V_2 = V_\infty(1 - 2a) \quad (3.16)$$

Equation 3.16 implies that if the rotor absorbs all the energy, i.e. $V_2 = 0$, then a would have maximum value of 0.5. The limit $0 < a < 0.5$ is wind turbine state of propeller operation.

From Eqs. (3.13) and (3.16),

$$T = \dfrac{1}{2} \rho A V_\infty^2 \, 4a(1 - a) \quad (3.17)$$

A non-dimensional quantity, coefficient of thrust C_T is defined as the ratio of thrust Eq. (3.17) to the force exerted by wind on the actuator disc.

$$C_T = \dfrac{\dfrac{1}{2} \rho A V_\infty^2 \, 4a(1 - a)}{\dfrac{1}{2} \rho A V_\infty^2} = 4a(1 - a) \quad (3.18)$$

For C_T to be maximum,

$$\dfrac{dC_T}{da} = 0 = 4 - 8a$$

$$a = \dfrac{1}{2}$$

And $C_{Tmax} = 1$, variation of C_T vs axial interference factor 'a' is shown in Figure 3.27.

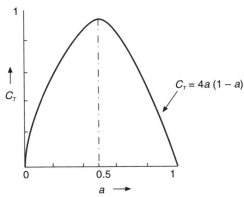

FIGURE 3.27 Variation of C_T vs axial induction factor a.

As the power is given by mass flow rate times the change in kinetic energy, the power P is

$$P = \dot{m}\Delta KE = \frac{1}{2}\rho AU(V_\infty^2 - V_2^2)$$

$$P = 2\rho AV_\infty^3 a(1-a)^2 = \frac{1}{2}\rho AV_\infty^3\, 4a(1-a)^2 \tag{3.19}$$

A non-dimensional quantity, coefficient of power C_P is defined as the ratio of power output of rotor Eq. (3.19) to the power in the wind,

$$C_P = \frac{\frac{1}{2}\rho AV_\infty^3 4a(1-a)^2}{\frac{1}{2}\rho AV_\infty^3} = 4a(1-a)^2 \tag{3.19}$$

Variation of C_P vs axial interference factor 'a' is shown in Figure 3.28.

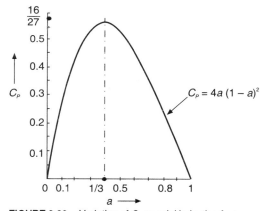

FIGURE 3.28 Variation of C_P vs axial induction factor α.

The maximum power would be obtained when

$$\frac{dC_P}{da} = 0$$

or

$$\frac{d}{da}4a(1-a)^2 = 1 - 4a + 3a^2 = 0$$

Solving quadratic equation gives, $a = 1$ or $1/3$. Since a cannot be greater than $1/2$ for wind turbine state of operation, therefore, $a = 1/3$. By substituting the value $a = 1/3$, in Eqn. (3.19),

and

$$P_{max} = \frac{16}{27}\left(\frac{1}{2}\rho AV_\infty^3\right) \tag{3.20}$$

Therefore, the maximum power that can be extracted is 16/27 times the power in wind. The fraction is known as *Betz's* coefficient, after German aerodynamist Albert Betz. The limit is caused due to flow of fluid. The stream tube to expand downstream of aetuator disc therefore free stream velocity of air must be smaller than upstream.

In practical operation, a good commercial wind energy conversion system may have a maximum power coefficient of about 0.4. This may be described as having an efficiency relative to the Betz criterion of 0.4/0.59 = 67.79 per cent.

The power coefficient is the efficiency of extracting power from the mass of air in the supposed stream tube passing through the activator disc, area A_1. This incident air passes through area A_0 upstream of the turbine. The power extracted per unit area of a cross section equal to A_0 upstream is greater than per unit area of A_1. Since $A_0 < A_1$, it can be shown that the maximum power extraction per unit of A_0 is 8/9 of the power in the wind.

By conservation of mass,

$$\rho A_0 V_\infty = \rho A_1 V_1$$

with $a = 1/3$ for maximum power extraction $2V_\infty = 3V_1$, so $A_1 = 3A_0/2$ and at maximum power extraction:

$$\frac{\text{Output power}}{\text{Input power}} = \frac{\frac{16}{27} A_1 V_0^3}{A_0 V_0^3} = \frac{8}{9}$$

And so the turbine has a maximum efficiency of 89 per cent considered in this way. Effects of this sort are important for array of wind turbines in a wind farm.

Betz Criterion Alternatively

For air-flow in a control volume of a tube of diameter D,

Power
$$P = \frac{dE}{dt} \tag{3.21}$$

Energy
$$E = \frac{1}{2} mV^2$$

Therefore,
$$P = \frac{d}{dt}\left(\frac{1}{2} mV^2\right) \tag{3.22}$$

Apply rule of chain differentiation, $P = \left(\frac{1}{2} m 2V \frac{dV}{dt} + V^2 \frac{dm}{dt}\right)$

For constant wind speed V, $\frac{dV}{dt} = 0$, Therefore,

$$P = \frac{1}{2} \frac{dm}{dt} V^2 = \frac{1}{2} \dot{m} V^2 \tag{3.23}$$

For a cross-sectional area, with air density ρ, and mass flow rate, $\dot{m} = \rho A V$

$$P = \frac{1}{2} \rho A V^3 = \frac{1}{2} \rho \frac{\pi}{4} D^2 V^3 \tag{3.24}$$

It shows $P \propto D^2$ and $P \propto V^3$. Betz model assumes infinite number of blades and when flow takes place it is too without any drag resistance. The variation of pressure and velocity in Betz model is shown in Figure 3.29 where $V_1 > V_2$ and $A_1 < A_2$.

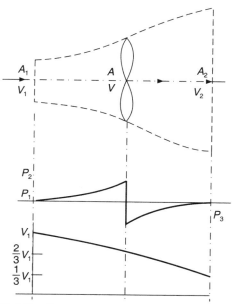

FIGURE 3.29 Pressure and velocity variation in Betz model.

As per continuity equation,
$$\dot{m} = \rho A V = \rho A_1 V_1 = \rho A_2 V_2$$

From Euler's equation, the thrust exerted by air on rotor is:

$$F = ma = m\frac{dV}{dt} = \dot{m}\Delta V = \dot{m}(V_1 - V_2) = \rho A V (V_1 - V_2) \qquad (3.25)$$

Incremental work done by air-stream,
$$dE = F dx$$

And power content,

$$P = \frac{dE}{dt} = F\frac{dx}{dt} = FV \qquad (3.26)$$

From Eq. (3.25), we get

$$P = \rho A V^2 (V_1 - V_2) \qquad (3.27)$$

The power is determined alternatively from the law of conservation of energy,

$$P = \frac{\Delta E}{\Delta t} = \frac{\frac{1}{2}mV_1^2 - \frac{1}{2}mV_2^2}{\Delta t} = \frac{1}{2}\dot{m}(V_1^2 - V_2^2) = \frac{1}{2}\rho A V (V_1^2 - V_2^2) \qquad (3.28)$$

From Eqs. (3.27) and (3.28),

$$V = \frac{1}{2}(V_1 + V_2) \tag{3.29}$$

From Eqs. (3.28) and (3.29),

$$P = \frac{\Delta E}{\Delta t} = \frac{\frac{1}{2}mV_1^2 - \frac{1}{2}mV_2^2}{\Delta t} = \frac{1}{2}\dot{m}(V_1^2 - V_2^2) = \frac{1}{2}\rho A V(V_1^2 - V_2^2) \tag{3.30}$$

Now, introducing interference parameter as $b = \dfrac{V_2}{V_1}$ (3.31)

Now Eqs. (3.30) becomes

$$P = \frac{1}{2}\rho A V_1^3 (1+b)(1-b^2) \tag{3.32}$$

Important observation of above abstraction is that the power extraction is a function of cube of undisturbed wind and interference factor.

Now, power flux (*term 'flux' is used for area [m²]*) is defined as:

$$P_f = \frac{P}{A} = \frac{\frac{1}{2}\rho A V^3}{A} = \frac{1}{2}\rho V^3 \; [\text{W/m}^2] \tag{3.33}$$

For determining coefficient of performance C_P, the kinetic power of air stream with $V = V_1$ and over a cross-sectional area A is to be considered,

$$C_P = \frac{\frac{1}{4}\rho A V_1^3 (1+b)(1-b^2)}{\frac{1}{2}\rho A V_1^3} = \frac{1}{2}(1+b)(1-b^2) \tag{3.34}$$

The variation of C_P with interference parameter is shown in Figure 3.30.

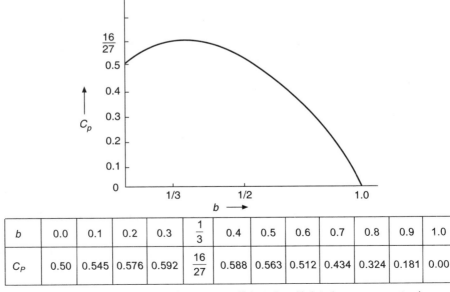

b	0.0	0.1	0.2	0.3	$\frac{1}{3}$	0.4	0.5	0.6	0.7	0.8	0.9	1.0
C_P	0.50	0.545	0.576	0.592	$\frac{16}{27}$	0.588	0.563	0.512	0.434	0.324	0.181	0.00

FIGURE 3.30 Variation of performance coefficient C_P with interference parameter b.

When $b = 0$, $V_2 = 0$, the rotor stops all the flow and $C_P = 0.5$, and when $b = 1$, $V_1 = V_2$, the wind flow is undisturbed and $C_P = 0.0$. The $C_p = \dfrac{16}{27}$ as maximum when $B = \dfrac{1}{3}$ with $V = \dfrac{2}{3}V_1$. For obtaining Betz limit, differentiate Eq. (3.34) with respect to b and equating it to zero.

$$\frac{dC_P}{db} = \frac{1}{2}\frac{d}{db}[(1+b)(1-b^2)] = 0$$

Now, using $\dfrac{d}{dx}(u,v) = u\dfrac{dv}{dx} + v\dfrac{du}{dx}$ rule,

$$\frac{dC_P}{db} = 0 = \frac{1}{2}[(1+b)(0-2b) + (1-b^2)] = (1+b)(1-3b)$$

Solution yields $b = -1$ is trivial result and $b = \dfrac{1}{3}$ yields $V_2 = \dfrac{1}{3}V_1$ or $V = \dfrac{1}{3}V_1$ and also from Eq. (3.34), $C_P = \dfrac{16}{27} \cong 0.59$. From continuity equation,

$$\dot{m} = \rho AV = \rho A_1 V_1 = \rho A_2 V_2 \text{ and } A_1 = \frac{2}{3}A \text{ or } A_1 = \frac{2}{3}A \text{ or } A = \frac{3}{2}A_1 \text{ or}$$

$$A_2 = A_1 \frac{V_1}{V_2} = A_1 \frac{V_1}{\frac{1}{3}V_1} = 3A_1$$

It means the area downstream the rotor expand to three times the upstream.

3.5 MOMENTUM THEORY FOR A ROTATING WAKE

The initial assumptions of the axial momentum theory considered that no rotation was imparted to the flow. But rotation represents loss of kinetic energy for the wind rotor. The loss is high for low speed rotors and low for rotors having high design tip speed ratio.

For analysis of momentum with wake rotation, an annular stream tube model is used as shown in Figure 3.31. With ring radius r and thickness dr, the cross-sectional area of the annular tube becomes $2\pi r\, dr$. From frame of reference of blades, by applying Bernoulli's equation, the pressure difference over the blades will be:

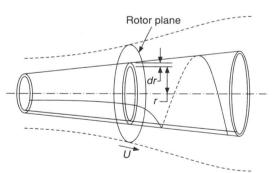

FIGURE 3.31 Stream tube model illustrating the rotation of wake.

$$P^+ - P^- = \frac{1}{2}\rho(\Omega + \omega)^2 r^2 - \frac{1}{2}\rho\Omega^2 r^2$$

$$P^+ - P^- = \rho\left(\Omega + \frac{1}{2}\omega\right)\omega r^2 \quad (3.35)$$

The relative angular velocity increases from Ω to $\Omega + \omega$, while the axial components of the velocity remain unchanged. The resulting thrust on the annular element of the rotor is

$$dT = \rho\left(\Omega + \frac{1}{2}\omega\right)\omega r^2 \, 2\pi r \, dr \qquad (3.36)$$

By defining the term angular speed interference factor a' as

$$a' = \frac{\omega}{2\Omega} \qquad (3.37)$$

the expression for thrust becomes

$$dT = 4a'(1 + a')\frac{1}{2}\rho\Omega^2 r^2 \, 2\pi r \, dr \qquad (3.38)$$

Similar expression for dT can be obtained from axial momentum theory, by substituting $A = 2\pi r \, dr$ and $V_2 = V_\infty(1 - 2a)$ in Eq. (3.13),

$$dT = 4a(1 - a)\frac{1}{2}\rho V_\infty^2 \, 2\pi r \, dr \qquad (3.39)$$

From Eqs. (3.38) and (3.39),

$$\frac{a(1-a)}{a'(1+a')} = \frac{\Omega^2 r^2}{V_\infty^2} = \lambda_r^2 \qquad (3.40)$$

where λ_r = local tip speed ratio.

From the conservation of angular momentum, the torque exerted must be equal to the angular momentum of the wake. Hence,

$$dQ = \rho U \, 2\pi r \, dr \, \omega r^2$$

By substituting for U and ω from Eqs. (3.15) and (3.37)

$$dQ = 4a'(1 - a)\frac{1}{2}\rho V_\infty \Omega r^2 \, 2\pi r \, dr \qquad (3.41)$$

The power output from the rotor is

$$P = \int \Omega \, dQ$$

Introducing local speed ratio λ_r as

$$\lambda_r = \frac{\Omega r}{V_\infty} = \frac{r\lambda}{R} \qquad (3.42)$$

The power output becomes

$$P = \frac{1}{2}\rho A V_\infty^3 \frac{8}{\lambda^2}\int a'(1-a)\lambda_r^3 \, d\lambda_r \qquad (3.43)$$

And the power coefficient becomes

$$C_P = \frac{8}{\lambda^2}\int a'(1-a)\lambda_r^3 \, d\lambda_r \qquad (3.44)$$

3.6 BLADE ELEMENT THEORY

The momentum theory cannot provide the necessary information on how to design the blades of a wind rotor. The blade element theory, combined with the momentum theory, provides the desired information. In blade element theory, by determining the forces acting on a differential element of the blade and then integrating over the length of the blade, torque and thrust loading of the rotor are determined analytically. The fundamental assumptions are that there is no interference between successive blade elements along the blade and that the forces acting on blade element are solely due to the lift and drag characteristics of the sectional profile of the blade element.

The modified blade element theory, sometimes termed *vortex theory*, is based on the assumption that vorticities originate from the trailing edge of the blade and conveyed down the slipstream with a velocity corresponding to the induced and free stream velocity in the form of rigid helical vortex sheet.

The flow around an element of a rotor blade is shown in Figure 3.32. The rotor has B number of blades of radius R. It is considered that a blade element of chord c and length dr at radius r is inclined at a blade setting angle β to the plane of rotation and rotates with an angular velocity Ω. The wind velocity is V_∞ at an infinite distance upstream of the rotor and the axial velocity at the rotor disc is $V_\infty(1 - a)$. The relative velocity of air in tangential direction to the element is $\Omega r(1 + a')$, where a and a' axial and angular speed interference factors, respectively. The resultant inflow relative to the blade element is W at an angle of attack α to the blade element and an angle ϕ to the plane of rotation of the rotor. Under these conditions, the aerodynamic forces acting on the blade element are as follows:

$$dL = \frac{1}{2} \rho W^2 c C_L \, dr \tag{3.45}$$

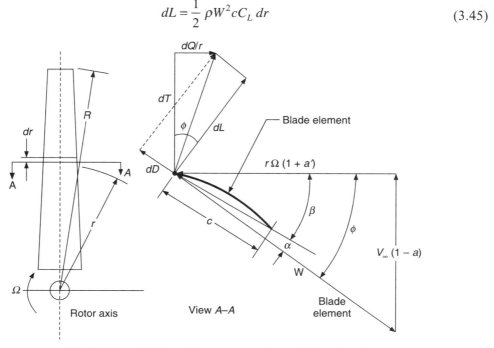

FIGURE 3.32 Velocity and force components on a blade element.

and
$$dD = \frac{1}{2}\rho W^2 c C_d \, dr \tag{3.46}$$

From Figure 3.32, the thrust and torque experienced by the blade element are:
$$dT = dL \cos \phi + dD \sin \phi \tag{3.47}$$
and
$$dQ = (dL \sin \phi - dD \cos \phi) \, r \tag{3.48}$$

From the flow geometry,
$$\tan \phi = \frac{(1-a)}{(1+a')\lambda_r} \tag{3.49}$$

where local speed ratio λ_r given by Eq. (3.42) is
$$\lambda_r = \frac{r\Omega}{V_\infty} = \frac{r\lambda}{R}$$

The local solidity ratio is defined as
$$\sigma = \frac{Bc}{2\pi r} \tag{3.50}$$

where B = number of blades.

From the above relationships, the results of the blade element theory transform into

$$dT = (1-a)^2 \frac{\sigma C_l \cos \phi}{\sin^2 \phi} \left(1 + \frac{C_d \tan \phi}{C_l}\right) \frac{1}{2} \rho V_\infty^2 \, 2\pi r \, dr \tag{3.51}$$

$$dQ = (1-a)^2 \frac{\sigma C_l}{\sin \phi} \left(1 - \frac{C_d}{C_l \tan \phi}\right) \frac{1}{2} \rho V_\infty^2 \, 2\pi r^2 \, dr \tag{3.52}$$

With the help of Eqs. (3.42) and (3.49),

$$dQ = (1+a')^2 \frac{\sigma C_l \sin \phi}{\cos^2 \phi} \left(1 - \frac{C_d}{C_l \tan \phi}\right) \frac{1}{2} \rho \Omega^2 r^3 \, 2\pi r^2 \, dr \tag{3.53}$$

3.7 STRIP THEORY

Utilizing both the axial momentum and blade element theories, a series of relationships can be developed to design a wind turbine and determine its performance. The strip theory is easy to programme, inexpensive to run, and readily adaptable to computers. The foremost assumption in strip theory is that individual stream tubes or strips can be analysed independently of the rest of the flow. Such an assumption works well for cases where the circulation distribution over the blade is relatively uniform so that most of the vorticity is shed at the blade root and at the blade tip. The second assumption associated with the development of strip theory is to neglect span wise flow. The strip theory does not predict any induced flow along the blades.

By combining Eqs. (3.39) and (3.51), one gets

$$\frac{4a}{1+a} = \sigma C_l \frac{\cos \phi}{\sin^2 \phi} \left(1 + \frac{C_d}{C_l} \tan \phi\right) \tag{3.54}$$

and by Eqs. (3.49), (3.52) and (3.53), one gets

$$\frac{4a'}{1+a} = \frac{\sigma C_l}{\cos \phi}\left(1 - \frac{C_d}{C_l \tan \phi}\right) \qquad (3.55)$$

3.8 TIP LOSSES

The strip theory, as developed in the preceding section, does not account for the interaction of shed vorticity with blade flow near the blade tip. The tip loss or circulation reduction near the tip can be explained by momentum theory. According to the momentum theory, with rotating wake, the wind imparts rotation to the rotor, thus dissipating some of its kinetic energy or velocity and creating a pressure difference between one side of the blade and the other. Because the pressure is greater on one side, air will flow around the blade tips, reducing the circulation at the tip. As a result of this, the torque is reduced. As the blade element forces at the tip contribute greatly to the torque and thus to the overall performance of a wind turbine, the tip flow is very important to the analysis.

Due to the tip losses, the velocities in the rotor plane with frame reference of blades and calculated with the aid of momentum theory are altered by the disturbed flow near the tips. The tip losses have been analysed by Prandtl. The Prandtl's model of tip loss correction is adopted due to the following reasons:

1. The difference in the results of the two models is small when the number of blades is more than three.
2. The Prandtl model predicts continuous change in the circulation, in qualitative agreement with the behaviour of the wind turbine.
3. The Prandtl model is easy to adopt on computers.
4. Overall calculations of power and thrust give good agreement with test results.
5. Comparisons of free wake vortex theory calculations with strip theory calculations using the Prandtl tip loss model show good agreement.

In the Prandtl model, a method has been developed to approximate the radial flow effect near the blade tip. The result of the approach is circulation reduction factor F such that

$$F = \frac{B\Gamma}{\Gamma_\infty}$$

where B is the number of blades, Γ is the circulation at a radial station, r, and Γ_∞ is the corresponding circulation for a rotor with an infinite number of blades. The factor F is a function of tip speed ratio, number of blades and radial position.

The basis for Prandtl's approximation was to replace the system of vortex sheets generated by the blade with a series of parallel planes at a uniform spacing s, where s is the normal distance between successive vortex sheets at the slip stream boundary, or

$$s = \frac{2\pi r}{B}\sin \phi_T$$

where B represents the number of blades and ϕ_T is the angle of the helical surface with the slip-stream boundary. Thus, the flow around the edges of the vortex sheets can then be approximated as the flow around the edges of a system of parallel planes.

The Prandtl tip loss factor is defined as

$$F = \frac{2}{\pi} \arccos e^{-f} \tag{3.56}$$

where

$$f = \frac{B(R-r)}{2R \sin \phi_T}$$

The expression for f can be suitably approximated by writing $r \sin \phi$ in place of $R \sin \phi_T$, because local relative wind angles ϕ are more convenient in calculation procedures. The Prandtl's approximation was developed for lightly loaded propeller (contraction negligible) assuming the vortex system to be rigid helix, an optimum condition for propellers. For non-contracting wake (in case of wind turbines, non-expanding wake), the terms a and a' can be neglected and hence for the rotor tip, from Figure 3.31,

$$\sin \phi = \frac{V_\infty}{\sqrt{(\Omega R)^2 + V_\infty^2}} = \frac{1}{\sqrt{1 + \lambda^2}} \tag{3.57}$$

Therefore, from Eqs. (3.56) and (3.57), for lightly loaded rotors, the Prandtl tip loss factor is

$$F = \frac{2}{\pi} \cos^{-1} e^{\left\{-\frac{B}{2}\left(1 - \frac{r}{R}\right)\sqrt{1 + \lambda^2}\right\}} \tag{3.58}$$

The variation of Prandtl tip loss factor with the number of blades for different tip speed ratios is shown in Figure 3.33. The values shown are obtained from Eq. (3.58), and are for 90 per cent

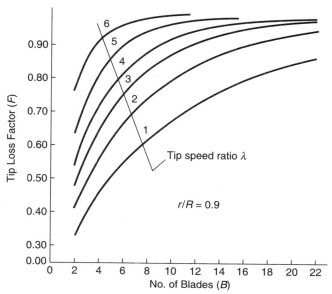

FIGURE 3.33 Variation of Prandtl tip loss factor with number of blades for different tip speed ratios.

radius of the rotor. It is seen from the figure that for the low speed rotors, i.e. $\lambda = 2.0$, the effect of the number of blades on tip loss factor is pronounced.

3.9 TIP LOSS CORRECTION

For tip loss correction, the flow at rotor is modified by the application of the tip loss factor in the momentum theory. The tip loss correction implies that the maximum decrease of axial velocity $2aV_\infty$ in the slip stream occurs at the vortex sheets, and the average decrease in axial velocity in the slip stream is only a percentage of velocity $2aV_\infty$. Therefore, the momentum Eqs. (3.39) and (3.41) assume the form

$$dT = 4Fa(1-a)\frac{1}{2}\rho V_\infty^2 \, 2\pi r \, dr \qquad (3.59)$$

and

$$dQ = 4Fa'(1-a)\frac{1}{2}\rho V_\infty \Omega r^2 \, 2\pi r \, dr \qquad (3.60)$$

The application of tip-loss factor in this manner has been termed *standard approach* or *first order approach*. They have also suggested a second order method of tip-loss correction through which the speed interference factors a and a' both have to be multiplied by F in Eqs. (3.39) and (3.41). Then, through the second order method of tip-loss correction, the equations for thrust and torque take the forms

$$dT = 4Fa(1-aF)\frac{1}{2}\rho V_\infty^2 \, r^2 \, 2\pi r \, dr \qquad (3.61)$$

and

$$dQ = 4Fa'(1-aF)\frac{1}{2}\rho V_\infty \Omega r^2 \, 2\pi r \, dr \qquad (3.62)$$

After tip-loss correction through the second order method, the strip theory Eqs. (3.54) and (3.55) become

$$\frac{4aF(1-F)}{(1-a)^2} = \frac{\sigma C_l \cos\phi}{\sin^2\phi}\left(1 + \frac{C_d}{C_l}\tan\phi\right) \qquad (3.63)$$

and

$$\frac{4a'F(1-aF)}{(1-a)^2} = \frac{\sigma C_l}{\sin\phi}\left(1 - \frac{C_d}{C_l \tan\phi}\right)\frac{1}{\lambda_r} \qquad (3.64)$$

From Eq. (3.64), the equation for power coefficient dC_p becomes

$$dC_p = \frac{8}{\lambda^2} a'F(1-aF)\lambda_r^3 \, d\lambda_r \qquad (3.65)$$

3.9.1 Condition for Maximum Power from Blade Element

From Eq. (3.65), it is seen that the term $a'(1-aF)$ should be maximum for maximum elemental power. To find out the value of $a'(1-aF)$,

$$F(a) = a'(1-aF) \qquad (3.66)$$

By differentiating Eq. (3.66) with respect to 'a',

$$\frac{d F(a)}{da} = \frac{da'}{da} - Fa' - Fa\frac{da'}{da} \tag{3.67}$$

From $\frac{dF(a)}{da} = 0$, one gets

$$a = \frac{1}{F}\left(1 - \frac{Fa'}{da'/da}\right) \tag{3.68}$$

To test for maximum and minimum, Eq. (3.67) is differentiated with respect to 'a',

$$\frac{d^2 F(a)}{da^2} = \frac{d^2 a'}{da^2}(1 - Fa) - 2F\frac{da'}{da} \tag{3.69}$$

From Eq. (3.68),

$$\frac{da'}{da} = \frac{Fa'}{(1 - Fa)} \tag{3.70}$$

By differentiating Eq. (3.70) with respect to 'a',

$$\frac{d^2 a'}{da^2} = \frac{(1 - Fa)F\frac{da'}{da} + F^2 a'}{(1 - Fa)^2} \tag{3.71}$$

By substituting the values of $\frac{da'}{da}$ and $\frac{d^2 a'}{da^2}$ from Eqs. (3.70) and (3.71) into Eq. (3.69), one gets

$$\frac{d^2 F(a)}{da^2} = 0$$

Hence, from second derivative, it is not possible to conclude whether the value of 'a' given by Eq. (3.68) is maximum or minimum. Therefore, the function has to be tested for maximum and minimum with its first derivative (Eq. 3.67) itself. For this, the values of 'a' smaller than and higher than that from Eq. (3.68) are put for the signs of the first derivative.

For putting value of 'a' smaller than that from Eq. (3.68),

$$a = \frac{1}{F}\left(1 - \frac{a'F}{da'/da}\right) - \Delta a \tag{3.72}$$

By substituting for 'a' from Eq. (3.72) into Eq. (3.67), one gets

$$\frac{dF(a)}{da} = \frac{da'}{da} - Fa' - \frac{Fda'}{da}\left\{\frac{1}{F}\left(1 - \frac{Fa'}{da'/da}\right) - \Delta a\right\}$$

$$= + F \Delta a \frac{da'}{da}$$

Since the values of F and da'/da cannot be negative, $\frac{dF(a)}{da} = +\text{ve}$. Similarly, by taking the value of 'a' higher than that from Eq. (3.68),

$$a = \frac{1}{F}\left(1 - \frac{a'F}{da'/da}\right) + \Delta a \quad (3.73)$$

By putting the values of 'a' from Eq. (3.73) into Eq. (3.67), one gets

$$\frac{dF(a)}{da} = -F\,\Delta a\,\frac{da'}{da}$$
$$= -ve$$

Since the value of 'a' smaller than that in Eq. (3.68) results in positive first derivative and value higher than that in Eq. (3.68) results in negative first derivative, the value of 'a' given by Eq. (3.68) is for maximum value of the function $a'(1 - aF)$. Therefore, from Eq. (3.68), the condition for maximum power is

$$\frac{da'}{da} = a'F/(1 - aF) \quad (3.74)$$

3.9.2 Optimum Design and Peak Performance

Optimum design of wind turbine rotor refers to the determination of blade chord length and twist at different radial locations so that the efficiency is maximum at design tip-speed ratio. By neglecting the effects of tip-losses and drag, a set of equations are obtained which give a closed form solution for blade design parameters. The equations are:

$$\lambda_r = \frac{r}{R}\lambda \quad (3.75)$$

$$\phi = \frac{2}{3}\tan^{-1}\left(\frac{1}{\lambda_r}\right) \quad (3.76)$$

$$c = \frac{8\pi r}{B.\,C_{ld}}(1 - \cos\phi) \quad (3.77)$$

and
$$\beta = \phi - \alpha \quad (3.78)$$

From the above equations, the values of blade chord and blade twist can be obtained at different radii of the rotor.

Shepherd* developed a simple method and performance prediction for wind turbine rotors.

The value of 'a' is used directly for the determination of elemental coefficient of performance dCp, which included the effects of drag and tip-losses.

$$dC_p = \frac{8}{\lambda^2}aF(1 - aF)\tan\phi\left(1 - \frac{C_d}{C_l\tan\phi}\right)\cdot \lambda_r^2 \cdot d\lambda_r \quad (3.79)$$

Wilson et al.** and Walker*** had given a method for optimum design and peak performance

---

* Shepherd, D.G. (1984), Note on a simplified approach to design point performance analysis of HAWT rotors, *Wind Engineering,* 8(2): 122–130.
** Wilson, R.E., P.B.S. Lissaman and S.N. Walker (1976), Aerodynamic performance of wind turbines, ERDA/NSF/04014-76/1, Oregon State Univ., p. 164.
*** Walker, S.N. (1976), Performance and optimum design analysis/computations for propeller type wind turbines, *Ph.D. Thesis,* Oregon State Univ., p. 216.

prediction of HAWT rotors. In this method, the design of the rotor is done by maximizing elemental power coefficient dC_p at a radial station, neglecting the effect of drag.

The speed interference factors corresponding to the maximum power coefficient are used to determine blade twist and chord at the radial station.

The integration of maximized elemental power over the blade gives peak performance of the rotor. The approach for maximizing the elemental power coefficient is illustrated in Figure 3.34. For given radia location r, local tip speed ratio λ_r, number of blades B and tip loss factor F at a given station, the steps are as follows:

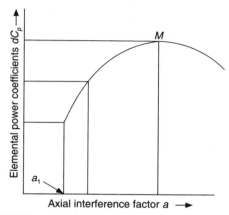

FIGURE 3.34 Optimum design approach for rotors.

(i) Value of 'a' is taken to be a_1 (= 0.30) and $a' = 0$ as initial values.
(ii) For $a = a_1$, corresponding values of a' and ϕ are found out by iterative process as these are interrelated. For this, Eqs. (3.39), (3.63) and (3.64) are used neglecting the drag term.
(iii) For $a = a_1$, dC_{p1} is computed using Eq. (3.65)
(iv) The a_1 is replaced by $a_1 + \Delta a$.
(v) Repeating the steps (ii) and (iii), dC_{p2} is computed and compared with dC_{p1}.
(vi) If $dC_{p\ new} > dC_{p\ previous}$, then the process is repeated by increasing the value of 'a' and repeating the steps (ii) to (v).
(vii) If $dC_{p\ new} \leq dC_{p\ previous}$, then the $dC_{p\ previous}$ corresponding to the optimum design and values of 'a' and a' are taken for design.

The solidity ratio σ for given design value of C_l is determined with the help of Eq. (3.65) after neglecting the drag term. The value of chord length is found out with Eq. (3.50). From 'a' and a', the values of ϕ and β are found out with the help of Eqs. (3.59) and (3.77).

3.9.3 Analytical Method for Optimum Design and Peak Performance Prediction*

Based on analytical approach, the effect of drag and tip-losses are taken into account for optimum

---

* Pandey, M.M., Studies on design and performance aspect of low speed horizontal axis wind turbine rotors, *Ph.D. Thesis*, 1988, IIT Kharagpur.

design and peak performance prediction. The elemental power coefficient as given by Eq. (3.65) is

$$dC_p = \frac{8}{\lambda^2} a'F(1-aF)\lambda_r^3 \cdot d\lambda_r$$

where

$$d\lambda_r = \frac{\Omega dr}{V_\infty}$$

At a given radial station and tip-speed ratio, the values of F and λ_r are constant. Therefore, at a given radial station, for dC_p to be maximum, the term $a'(1-aF)$ has to be maximized, and the condition for maximum value of $a'(1-aF)$ is given as in Eq. (3.74),

$$\frac{da'}{da} = a'F/(1-aF)$$

By dividing Eqs. (3.60) with Eq. (3.63), one gets

$$\frac{a'}{a} = \frac{\tan\phi - \varepsilon}{\lambda_r(1+\varepsilon\tan\phi)} \tag{3.80}$$

where

$$\varepsilon = \frac{C_d}{C_l}$$

By substituting for $\tan\phi$ from Eq. (3.49),

$$\frac{a'}{a} = \frac{(1-a) - \varepsilon(1+a')\cdot\lambda_r}{\lambda_r\{(1+a')\lambda_r + \varepsilon(1+a)\}}$$

or

$$a'^2\lambda_r^2 + a'\lambda_r(\lambda_r + \varepsilon) + a^2 + a(\varepsilon\lambda_r - 1) = 0 \tag{3.81}$$

Differentiation of Eq. (3.80) with respect to 'a' yields.

$$\lambda_r^2 \cdot 2a'\frac{da'}{da} + \lambda_r(\lambda_r + \varepsilon)\frac{da'}{da} + 2a + \varepsilon(\lambda_r - 1) = 0$$

or

$$\frac{da'}{da} = \frac{1 - 2a - \varepsilon\lambda_r}{\lambda_r(2\lambda_r \cdot a' + \lambda_r + \varepsilon)} \tag{3.82}$$

Comparing the values of $\frac{da'}{da}$ from Eq. (3.74) and Eqs. (3.81),

$$\frac{Fa'}{1-Fa} = \frac{1-2a-\varepsilon\lambda_r}{\lambda_r(2\lambda_r a' + \lambda_r + \varepsilon)}$$

or

$$2F\lambda_r^2 a'^2 + F\lambda_r(\lambda_r + \varepsilon)a' - (1-Fa)(1-2a-\varepsilon\lambda_r) = 0 \tag{3.83}$$

By substituting for $\lambda_r^2 \cdot a'^2$ from Eq. (3.80) into Eq. (3.82), one gets

$$2Fa(1-\varepsilon\lambda_r) - 2Fa^2 - 2F\lambda_r a'(\lambda_r + \varepsilon) + a'F\lambda_r(\lambda_r + \varepsilon) - (1-Fa)\cdot(1-2a-\varepsilon\lambda_r) = 0$$

$$a' = \frac{(1-\varepsilon\lambda_r - a)(3Fa - 1) + a(1-Fa)}{F\lambda_r(\lambda_r + \varepsilon)} \tag{3.84}$$

In Eq. (3.83), the relationship is shown among two speed interference factors, drag to lift ratio and tip-loss factor for maximum elemental power coefficient at a given radial station on blade.

(1) Speed interference factors

For the determination of 'a' and a', the following steps were followed:

(i) For the given radial station and number of blades, the tip loss factor F and local speed ratio λ_r are determined with the help of Eqs. (3.58) and (3.42)
(ii) The initial value of 'a' is taken to be 0.30.
(iii) The a' is determined with the help of Eq. (3.83).
(iv) For solution of Eq. (3.80) through Newton–Raphson method (Explained in Appendix C), the functions G_1 and G_2 are taken as follows:

$$G_1 = a'^2 \lambda_r^2 + a'\lambda_r(\lambda_r + \varepsilon) + a^2 + a(\varepsilon\lambda_r - 1) \tag{3.85}$$

and

$$G_2 = \frac{dG_1}{da}$$
$$= 2a - 1 + \varepsilon\lambda_r + \frac{da'}{da} \lambda_r(\lambda_r + 2a'\lambda_r + \varepsilon) \tag{3.86}$$

where $\dfrac{da'}{da}$ from Eq. (3.83) is,

$$\frac{da'}{da} = \frac{\frac{2}{F} + 3 - 3\varepsilon\lambda_r - 8a}{\lambda_r(\lambda_r + \varepsilon)} \tag{3.87}$$

G_1 and G_2 are determined from Eqs. (3.84), (3.85) and (3.86).

(v) New value of 'a' is determined as

$$a_n = a - \frac{G_1}{G_2}$$

(vi) Difference of old and new values of 'a' is computed as

$$da = |a - a_n|$$

(vii) Old value of 'a' is replaced with new value a_n.
(viii) If $da \geq 10^{-5}$, steps (iii) through (vii) are repeated.
(ix) For $da < 10^{-5}$, the values of a and a' were taken as the speed interference factors.

(2) Chord length and twist of blade

At the given radial station, the relative wind angle ϕ is determined with the help of Eq. (3.49). The twist of the blade β is $(\phi - \alpha)$, where α is the angle of attack for airfoil section of the blade at which the C_d/C_l ratio is minimum. The local solidity ratio σ is determined by Eq. (3.63) as

$$\sigma = \frac{4aF(1 - aF)\sin^2\phi}{(1-a)^2 C_{ld}(1 + \varepsilon\tan\phi)\cos\phi}$$

where C_{ld} is lift coefficient corresponding to minimum C_d/C_l ratio for the given airfoil section. After calculating σ, the chord length is determined with the help of Eq. (3.50) as

$$c = \frac{\sigma \, 2\pi r}{B}$$

(3) Power coefficient

The elemental power coefficient at the radial station is determined with the help of Eq. (3.65). It is modified to the following form:

$$dC_p = \frac{8}{\lambda} a' F (a - aF) \lambda_p^3 \cdot \frac{dr}{R}$$

Following Simpson's rule of numerical integration, the total power coefficient is determined by

$$C_p = \frac{1}{3} \sum_{i=1}^{i=I} dC_{pi} \cdot q_{ki} \qquad (3.88)$$

where $q_{ki} = 3 + (-1)^{i+1}$ and I are the number of radial stations. The above procedure is described in the form of flow chart for optimum design and peak performance prediction in Figure 3.35.

At the design value of tip-speed ratio, the resultant flow is such that the angle of attack is optimum, that is, the drag to lift ratio 'ε' is minimum. However, due to variation in wind speed and mismatching of rotor with load, wind turbine rotors quite often run at tip-speed ratios lower as well as higher than the design value. For off-design performance prediction, the airfoil data across a wide range of angle of attack is utilized. It is necessary to assess the performance of rotor models in wind tunnel at different wind speeds.

3.10 DRAG TRANSLATOR DEVICE

Let a drag translator device, shown in Figure 3.36, be subjected to wind with speed U. This is a device that utilizes the drag force of the wind acting on it to develop power. Let the device be moving at speed V under the wind forces acting on it in the direction same as that of the wind.

Then, the relative speed of the wind as seen by the device is $(U - V)$ and the drag force of the wind acting on a unit length of the drag translator device is given by,

$$D = C_d \, 1/2 \, \rho c \, (U - V)^2 \qquad (3.89)$$

where
  C_d = drag coefficient of the drag translator device
  c = width or chord of the drag translator device

And the power developed, P by the drag translator device is given by $(D \cdot V)$ or

$$P = C_d \, 1/2 \, \rho c \, (U - V)^2 V \qquad (3.90)$$

Coefficient of performance, C_p can then be written as,

$$C_p = C_d (U - V)^2 \, V/U^3 \qquad (3.91)$$

Maximizing this with respect to (V/U) gives,

$$(V/U) = 1/3$$

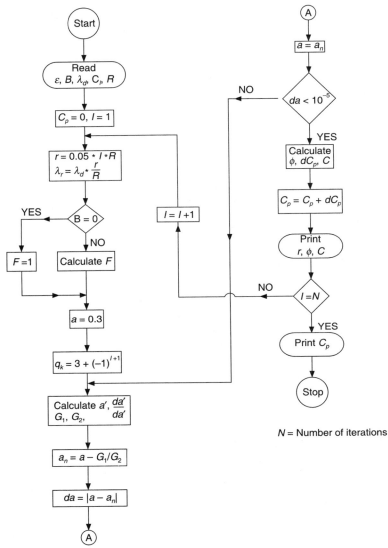

FIGURE 3.35 Flow chart for optimum design and peak performance prediction for H.A.W.T Rotors.

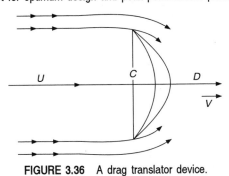

FIGURE 3.36 A drag translator device.

and
$$C_{P,\max} = (4/27)\, C_d \tag{3.92}$$
with
$$C_d = 1.0,\ C_{P,\max} = 0.148$$

3.11 WIND MACHINE CHARACTERISTICS

C_p–λ curves for typical wind machines are shown in Figure 3.37. Note the positions of the curves for Savonious machines and multi-bladed machines. Contrary to popular belief, the Savonious machines have been shown to have high C_p of the order of 0.3.

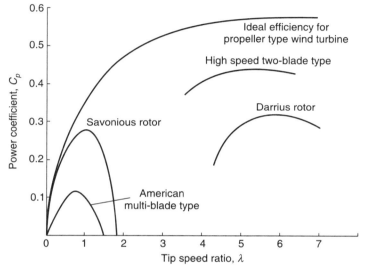

FIGURE 3.37 C_P–λ curves for practical wind machines.

For obtaining the minimum drag C_d to lift C_l ratio, a simple method is demonstrated in Figure 3.38 and the corresponding angle of attack. Draw the tangent from origin to C_l vs C_d curve, and from the common point, draw a line to find the corresponding angle of attack, corresponding point of intersection in C_l vs α characteristic is called the design angle for the purpose of optimum performance.

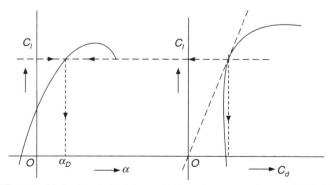

FIGURE 3.38 Method to find minimum C_d/C_l and corresponding angle of attack α.

Power developed by wind machine vs. its speed of rotation for different wind speed can be calculated for the given machine from its C_P–λ curve and is shown in Figure 3.39. Constant speed lines are marked on the graph.

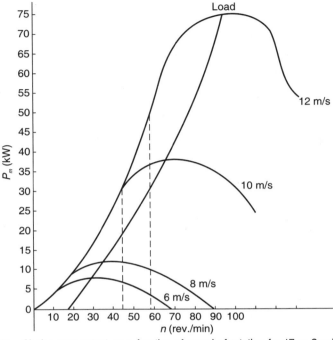

FIGURE 3.39 Shaft power output as a function of speed of rotation for 17 m Sandia Darrieus.

Figure 3.40 shows torque developed by a wind machine as a function of its rotational speed. The propeller type machine starts and stabilizes at high λ under no load. Machine stabilizes at appropriate λ matching with load conditions.

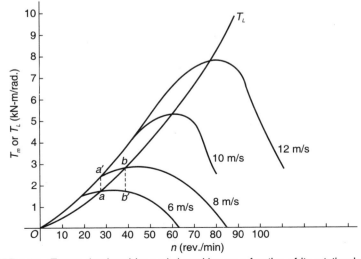

FIGURE 3.40 Torque developed by a wind machine as a function of its rotational speed.

3.11.1 Fixed Speed and Variable Speed Machines

Variable speed machines run at different RPM at different wind speeds. Load varies with machine RPM, e.g. reciprocating pump attached as load. Constant speed machines are designed to run at constant RPM. Load varies to match the power generated or blade angle is changed (pitched) to change the power developed. Two speed machines are designed to run at two different constant RPMs. They develop lower power at low wind speeds running at lower RPM. They develop higher power at high wind speeds running at higher RPM. The selection of operating speed is always an important design issue. Figure 3.41 shows power developed by a machine at different speeds of rotation. It can be seen that the machine develops more power at higher speed of rotation at higher wind speeds. However, at low wind speeds which are more frequent, a slower machine can develop higher power. Therefore, modem machines operate at two different rotational speeds with different ratings.

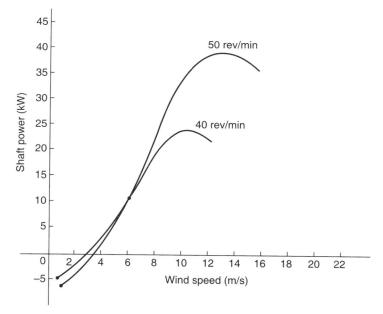

FIGURE 3.41 Showing power developed by a machine at different speeds of rotation.

3.11.2 Load Matching

An ideal load would be that for which the machine always develops maximum power or always operates at maximum C_p as shown in Figure 3.42. Any load that develops torque proportional to the square of its rotational speed can be matched with the machine requirement by appropriate gear selection.

In practical machines, generator has the rating and it cannot be operated beyond this limit. Hence, at higher wind speeds, wind machine blades are regulated to shed extra power. Thus, the machine operates at reduced C_p as shown in Figure 3.43.

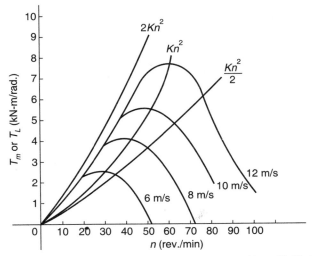

FIGURE 3.42 Torque vs. rotational speed for a wind machine with ideal load.

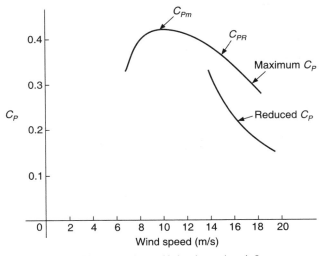

FIGURE 3.43 C_p vs. U showing reduced C_p.

Mechanical and electrical power output, as a function of wind velocity for constant speed machine are shown in Figure 3.44.

SOLVED EXAMPLES

1. A 60 m diameter rotor experiences an undisturbed wind speed of 10 m/s. If it is operating at maximum C_p, calculate:
   (a) Pressure difference across the disc using actuator disc analysis. Express this as percentage of the atmospheric pressure (1 kg/cm^2).
   (b) Thrust on the rotor for this condition.

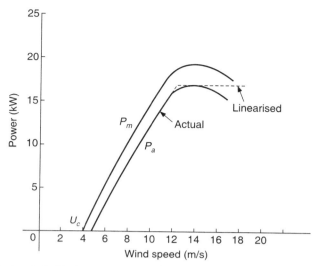

FIGURE 3.44 Mechanical and electrical power output as function of wind speed for 17 m Darriu turbine at 42 RPM.

(c) Maximum possible thrust on the rotor.
(d) Maximum power developed by the rotor.
(e) Torque developed at maximum power. Take air density = 1.165 kg/m³.

Solution: Given: $D = 60$ m, $V_\infty = 10$ m/s, $C_{P,\max} = 16/27$, $a = 1/3$, $p_{atm} = 1$ kg/m².

$$a = 1/3 = (V_\infty - V_D)/V_\infty = 1 - V_D/10$$
$$V_D = 20/3 \text{ m/s}$$
$$V_1 = V_\infty(1 - 2a) = 10(1 - 2/3) = 10/3 \text{ m/s}$$

(a) $\Delta p = \rho V_D (V_\infty - V_1) = 1.165 \times 20/3 \times (10 - 10/3) = 51.77$ N/m²
(b) Thrust T = Swept area $\times \Delta p = \pi/4 D^2 \Delta p = \pi/4 (60)^2\, 51.77 = 14.6376 \times 10^4$ N = 146 kN
(c) $T_{\max} = 0.5\, \rho A_D V_\infty^2 = 0.5 \times 1.165 \times \pi/4 (60)^2\, 10^2 = 164.69$ kN
(d) Maximum power $P_{\max} = C_{P,\max}\, 0.5\, \rho A_D V_\infty^3 = 16/27 \times 0.5 \times 1.165\, \pi/4 (60)^2\, 10^3$
     $= 975.98$ kW
(e) Torque developed at $P_{\max} = T\Omega$

[**Hint:** $C_Q = C_P/\lambda$, where λ is the tip speed ratio.]

$C_Q = 4a(1 - a) = 4/3(1 - 1/3) = 8/9$
$C_P = 16/27$
$\lambda = 2/3$
$\lambda = \Omega R/V_\infty = \Omega 30/10 = 2/3$
$\Omega = 2/9$
$P_{\max} = T\Omega$
$975.98 = T 2/9$
$T = 4391.91$ kNm

2. An electric power utility decides to add 50 MW of wind generation to its system. If the individual units are rated at 2 MW in a 13 m/s wind at standard conditions and have $C_P = 0.32$ and mechanical and electrical efficiencies of $\eta_m = 0.94$ and $\eta_e = 0.97$:
   (a) What is the required swept area of the rotor? What is the rotor diameter?
   (b) If the turbine is required to deliver rated power at 17.5 rpm and the generator rated speed is 1800 rpm, what is the average torque at the low speed (rotor side) and high speed (generator side) shaft?

Solution: Given: $V_\infty = 13$ m/s, $C_P = 0.32$, $\eta_m = 0.94$ and $\eta_e = 0.97$.

(a)
$$P = \tfrac{1}{2} C_P \rho A V_\infty^3$$
$2 \times 10^6 = \tfrac{1}{2}\, 0.32 \times 1.165 \times A(13)^3$
$A = 4.88 \times 10^3$ m^2
$D = 78.85$ m

(b) $N_R = 17.5$ rpm and $N_G = 1800$ rpm
$P = \eta T \omega$
$\omega_R = 2\pi N_R/60 = 2\pi\, 17.5/60 = 1.832$ rad/s
$\omega_G = 2\pi N_G/60 = 2\pi\, 1800/60 = 188.49$
$T_R = P/(\eta_m \omega_R) = 2 \times 10^6/(0.94 \times 1.832) = 1161.38$ kNm
$T_G = P/(\eta_m \eta_e \omega_G) = 2 \times 10^6/(0.94 \times 0.97 \times 188.49) = 11.63$ kNm

REVIEW QUESTIONS

1. Empirically, maximum power coefficient, $C_{p,\max}$ for a rotor may be given as function of the following expression:
$$C_{p,\max} = 0.593\,[R - Q]\,(C_d/C_l)$$
where
$R = \lambda \cdot B^{0.67}/[1.48 + (B^{0.67} - 0.04)\lambda \ldots + 0.0025\lambda^2]$
$Q = [(1.92\lambda^2 \cdot B)]/(1 + 2\lambda \ldots B)$
B = number of blades
C_d/C_l = ratio of drag coefficient and lift coefficient for the airfoil
   (a) Calculate values of $C_{p,\max}$ for $\lambda = 4$, 6 and 8 for $B = 3$ and $C_d/C_l = 0.1$.
   (b) Calculate power developed at the rotor, P for a rotor of diameter of 58.76 m, rotational speed of 13 rpm at wind speeds of 5, 6.67 and 10 m/s if it is operating at $C_{P,\max}$.

2. A 3-bladed propeller wind machine fixed with 220 kW generator has diameter = 30 m and tower height = 30 m. It is designed to have $C_p = 0.39$ at wind speed of 11 m/s when regulation starts. The cut-in and furling speeds for the machine are 3.5 m/s and 25 m/s, respectively. Mechanical and electrical efficiencies of the machine are 0.96 and 0.97, respectively.
   (a) What is the power developed by the rotor at the wind speed = 11 m/s when the regulation starts if it is calculated for mean air density at the surface = 1226 gm per cu.m?
   (b) If the tip speed ratio at this wind speed is 4.2, then what is the rpm of the rotor?

3. A constant speed pitch regulated wind machine with 60 m diameter, 14.32 rpm and 1.0 MW rated generator capacity is installed at a site with annual values of Weibull parameters, C and k given by 7.2 m/s and 1.8, respectively. It has cut-in speed of 4 m/s and cut-off speed of 27 m/s. It has 3 blades and the mechanical and electrical efficiencies for all wind speeds are given as 0.95 and 0.96, respectively. Its coefficient of performance, C_p as function of tip speed ratio, λ, is given by,
$$C_p = 0.089 + 0.0123\lambda^2 - 0.00091\lambda^3$$
Find the electrical power developed by the machine as function of speed from cut-in speed to cut-off speed in the steps of 1 m/s. Take air density = 1.165 kg/m³.

4. Calculate lift and drag forces acting on the following bodies when exposed to wind of speed 8.3 m/s with air density of 1200 gm/cu.m:
   (a) Cylindrical wire of length 10 m and dia 1 mm that has $C_L = 0$ and $C_D = 1.2$.
   (b) Flat plate of length 10 m, width 1 m and thickness 5 mm that has $C_L = 2\pi\alpha$ (α in radians) and $C_D = 0.4$ when inclined at angle of 5° w.r.t. on coming wind.
   (c) Asymmetric airfoil of length 10 m, chord 1 m and thickness 15 mm that has $C_L = 2\pi(\alpha + \alpha_o)$ (α and α_o in radians) with $\alpha_o = 2°$ and $C_D = 0.01$ when inclined at an angle of 10° w.r.t. on coming wind.

5. A long cylinder has coefficient of drag, $C_D = 1.0$ for flow with Reynolds number, R_e in the range $1000 < R_e < 5000$; whereas a symmetric airfoil with 15% thickness has $C_D = 0.01$ at zero angle of attack. If this airfoil has chord of 0.2 m, then:
   (a) Draw a symmetric airfoil and show chord c, thickness t, approach with velocity U and angle of attack α.
   (b) Calculate drag per unit span of the airfoil at zero angle of attack for $U = 0.4$ m/s.
   (c) Calculate the diameter of the cylinder having same drag per unit length.
   (d) Calculate the ratio of airfoil thickness to diameter of this cylinder.

6. What is Betz criterion? Derive the expression using momentum theory and show that it is 16/27 and plot the variation of coefficient of power with respect to speed interference factor.

7. Define Reynolds number. Explain its importance in fluid flow studies.

8. A wind turbine of 25 kW with rotor diameter of 10 m installed in an area of wind power density of 200 W/m². Calculate annual energy production (AEP). Estimated capacity factor is 0.25.
$$\text{AEP} = 0.25 \times 0.5 \times \text{Area} \times \text{WPD} \times 8760 \text{ kWh/year}$$
$$= 0.25 \times 0.5 \times \pi(5)^2 \times \frac{200}{1000} \times 8760$$
$$= 17200.219 \text{ kWh/year}$$

9. For a conventional HAWT, radius 50 m, estimate annual energy output for a good wind region (use class 3, 4, 5 and 6) from an Indian wind power map.

10. From the given power curve (use Figure 3.45) for annual energy, estimate the annual energy production for a region where the average wind speed is 9 m/s.

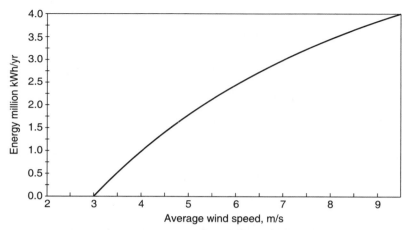

FIGURE 3.45 Cumulative energy production vs. Average wind speed.

11. Calculate power from Figure 3.46 at 20 m/s for the VAWT for the following conditions. Remember, rpm has to be converted to rad/s.
    (a) Wind turbine is operating at 160 rpm (line A).
    (b) Wind turbine is operating at maximum power coefficient (line B).
    (c) Wind turbine is operating at constant torque (line C) of 6,000 Nm.

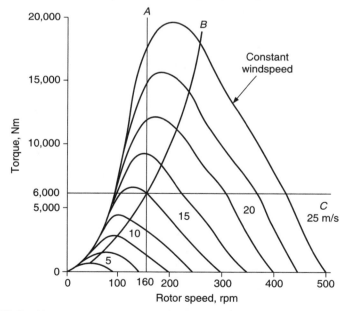

FIGURE 3.46 Wind turbine rotor output torque vs. rotational speed characteristics at different wind speeds.

12. From Figure 3.46, the design wind speed is 12.5 m/s (where line A, B and C cross). What is the torque? What is the rpm? What is the power?

13. Refer to Figure 3.47. What is the cut-in and rated wind speed for the 1,000 kW unit?
14. Refer to Figure 3.47. What is the cut-in and rated wind speed for the 400 kW unit?

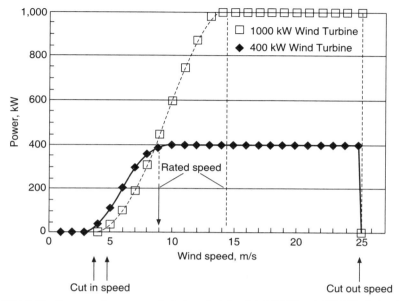

FIGURE 3.47 Power performance output characteristic of 1,000 kW and 400 kW wind turbine.

Chapter 4

Wind Turbine

4.1 INTRODUCTION

The extraction of power from wind with modern wind turbines and energy conversion system is an established industry. Machines are manufactured with a capacity from a few kilowatt to several megawatt. Most machines are built for electricity generation, either linked to a grid or in an autonomous mode. The wind turbine has evolved into a highly specialized device whose configuration, size, and technological sophistication are application-dependent. The horizontal axis orientation continues to be dominant in wind power production, as it has been for most of the modern era. Large-scale wind turbines, with rotor diameter larger than 45 m and/or power ratings of 1,000 kW or more, offer advantages that include: the ability to extract more wind energy per unit of land area when topography consists of one or more ridges and the wind blows predominantly from one direction, improved aerodynamic performance, because of higher Reynolds numbers associated with larger blade chord dimensions; and lower sensitivity of larger blades to dirt, rain and insects, because of larger blade thickness dimensions and potential economies of scale of some components, such as control system cost per unit of installed power.

Beginning with the birth of modern wind-driven electricity generators in the late 1970s, wind energy technology has improved dramatically upto the present. Capital costs have decreased, efficiency has increased, and reliability has improved. High-quality products are now routinely delivered by major suppliers of turbines around the world and complete wind generation plants are being engineered into the grid infrastructure to meet utility needs.

4.1.1 Historical Aspects

Some milestones in the history of wind machines are given in Table 4.1. Conceptual understanding of aerodynamic shape and position of centre of forces or zero moment reduced the structural problem of supporting the blade. This was in the second decade of twentieth century. This is an important milestone in the history of wind machines. Thereafter, longer blades of aerodynamic shape could be designed and used.

TABLE 4.1 Historical development of wind energy conversion system

Period	Machine	Application
640 AD	Persian wind mills	Grinding, etc.
Before 1200 AD	Chinese sail type wind mill	Grinding, water pumping, etc.
12th century AD	Dutch wind mills	Grinding, water pumping, etc.
1700 AD	Dutch wind/mill to America	Water pumping, etc.
1850–1930 AD	American multi-bladed	Water pumping, 35V DC power
1888 AD	Brush wind turbine; dia.17 m, tower 18.3 m	Electrical power
1925 AD	Jacob's 3-bladed propeller; Dia. 5 m, 125 to 225 RPM	Electrical power
1931 AD	Yaha propeller, Russia; 2-bladed, dia. 30 m	Electrical power
1941 AD	Smith–Putnam propeller 2-bladed, dia.58 m, 28 RPM	Electrical power
1925 AD	Savonius machine	Mechanical or electrical power
1931 AD	Darrius machine	Electrical power
1980s AD	2-bladed propeller 225 kN	Electrical power
2000 AD	HAWT, VAWT; 3 MW machine or more	Electrical power

4.1.2 Modern Wind Turbine

Modern wind turbines, which are currently being deployed around the world, have three-bladed rotors with diameter of 80 m or more mounted atop 80 m or more high towers as illustrated in Figure 4.1 and Figure 4.2. Turbine power output is controlled by rotating the blades about their long axis to change the angle of attack with respect to relative wind as the blades spin about the rotor hub. This is called *controlling the blade pitch*. The turbine is pointed towards the wind by rotating the nacelle about the tower. This is called *controlling the yaw*. Wind sensors on the nacelle tell the yaw controller where to point the turbine These wind sensors, along with sensors on the generator and drive train, also tell the blade pitch controller how to regulate the power output and rotor speed to prevent overloading the structural components. Generally, a turbine starts producing power in winds of about 3.0 m/s and reaches maximum power output at about 11.0 m/s. The turbine with pitch or feather the blades to stop power production and rotation at about 25.0 m/s. Figure 4.3 shows typical power output of a wind turbine versus wind speed with cut-in speed, rated speed and cut-out speed. The power output from cut-in wind speed to rated wind speed follows a cubic curve and from rated speed to cut-out speed, a constant power output by a variety of control mechanisms.

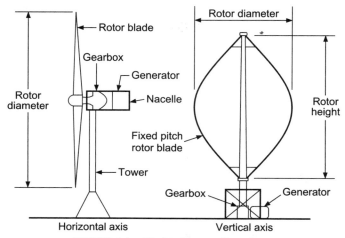

FIGURE 4.1 Wind turbine configurations.

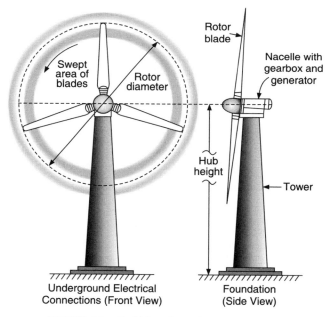

FIGURE 4.2 Typical modern wind turbine curve.

Most of the rotors on large-scale machines have an individual mechanism for pitch control; that is, the mechanism rotates the blade about its long axis to control the power in high winds. Blades can be rotated by pitch mechanism in high winds to feather them out of the wind. This reduces the maximum loads on the system when the machine is parked. Pitching the blades out of high winds and reduces operating loads, and the combination of pitchable blades with a variable-speed generator allows the turbine to maintain generation at a constant-speed power output.

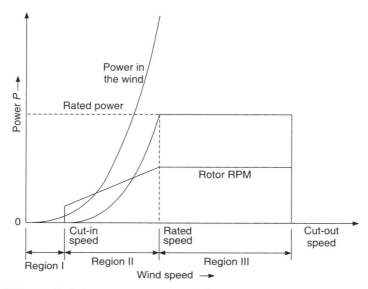

FIGURE 4.3 Typical power output versus wind speed characteristics of wind turbine.

Most utility-scale turbines are upwind machines, meaning that they operate with the blades upwind of the tower to avoid the blockage created by the tower.

The amount of energy in the wind available for extraction by the turbine increases with the cube of the wind speed; thus, a 10 per cent increase in wind speed creates a 33 per cent increase in the available energy. A turbine can capture only a portion of this cubic increase in energy, because power above the level for which the electrical system has been designed, referred to as the rated power, is allowed to pass through the rotor.

In general, the speed of the wind increases with height above the ground, which is why engineers have found ways to increase the height and the size of the wind turbines while minimizing the costs of materials. But land-based turbine size is not expected to grow as dramatically in the future as it has in the past. Larger sizes are physically possible; however, the logistical constraints of transporting the components via highways and of obtaining cranes large enough to lift the components present a major economic barrier that is difficult to overcome. The principle of wind turbine rotation is based on aerodynamic lift as shown in Figure 4.4.

The wind passes over both surfaces of the aerofoil shaped blade, but it passes more rapidly over the longer (upper) side of the airfoil, thus creating a lower-pressure area above the airfoil. The pressure differential between top and bottom surfaces results in aerodynamic lift. In an aircraft wing, this force causes the airfoil to rise, lifting the aircraft

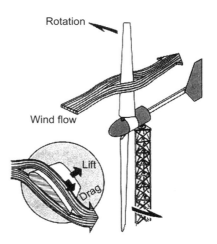

FIGURE 4.4 Principles of wind turbine aerodynamic lift.

off the ground. Since the blades of a wind turbine are constrained to move in a plane with hub as its centre, the lift force causes rotation, about the hub. In addition to the lift force, a drag force perpendicular to the lift force impedes rotor rotation. A prime objective in wind turbine design is for the blade to have a relatively high lift-to-drag ratio. This ratio can be varied along the length of the blade to optimize the turbine's energy output at various wind speeds.

4.1.3 Current Wind Turbine Size

In the recent past average wind turbine ratings have grown almost linearly, as illustrated by Figure 4.5. Each group of wind turbine designer has predicted that its latest machine is the largest a wind turbine can ever be. But with each new generation of wind turbines (roughly every five years), the size has grown along the linear curve and has achieved reductions in life-cycle cost of energy. This long-term drive to develop larger turbines is a direct result of the desire to improve energy capture by accessing the stronger winds at higher elevation. The increase in wind speed with elevation is referred to as *wind shear*. Although increase in turbine height is a major reason for increase in capacity factor over time, there are economic and logistic constraints to this continued growth to larger size.

The primary argument for limiting the size of the wind turbine is based on the square-cube law. This law roughly states that "as a wind turbine rotor grows in size, its energy output increases as the rotor swept area (the diameter squared), while the volume of material, and therefore its mass and cost, increases as the cube of the diameter. In other words, at some size, the cost for a larger turbine will grow faster than the resulting energy output revenue.

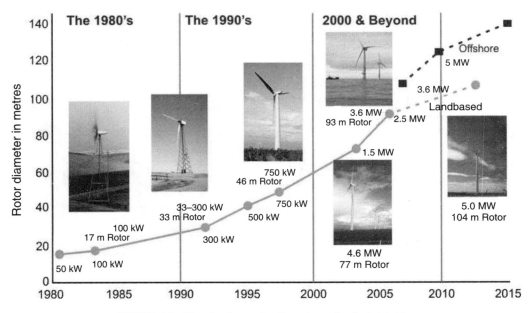

FIGURE 4.5 The development path and growth of wind turbines.

Turbine performance has clearly improved and the cost per unit of output has been reduced. The continued infusion of new technology will result in better design.

4.2 CLASSIFICATION OF WIND TURBINES

A classification of wind energy conversion systems (WECS) is given in Figure 4.6. This includes main types, but numerous other designs and adaptations occur.

4.2.1 Horizontal Axis Machines

The dominant driving force is lift. Blades on the rotor may be in front (upwind) or behind (downwind) the tower. Upwind turbines need a tail or some other mechanism to maintain orientation, such as side-facing fan tail rotors. Downwind turbines may be quite seriously affected by the tower, which produces wind shadow and turbulence in the blade path. Perturbations of this kind cause cyclic stresses on the structure, noise and output fluctuations. Wind may be expected to veer frequently in horizontal plane, and the rotor must turn (yaw) to follow the wind without oscillations.

Three-bladed rotors are common for electricity generation. The three-bladed rotors operate smoothly and may be cross-linked for greater rigidity. Gearing and generators are usually at the top of the tower in a nacelle as shown in Figure 4.7. It is possible to run a shaft down the tower for power generation at the ground level, but the complications are usually thought to outweigh the advantages. Multi blade rotors, having high starting torque in light wind, are used for water pumping and other low frequency mechanical power.

4.2.2 Vertical Axis Machines

By turbing with vertical axis, a machine can accept with from any direction without adjustment. The other main benefit is that gearing and generators can be directly coupled to the axis at the ground level. The principal disadvantages are: many vertical axis machines have suffered from fatigue failuers arising from the many natural resonances in the structure, and the rotational torque from the wind varies periodically within each cycle, and thus unwanted power periodicities appear at the output. As a result, a greater majority of working machines are horizontal axis, not vertical. In Figure 4.6, different vertical axis machines are shown.

- **Cup anemometer:** This device rotates by drag force. The shape of this cup produces a nearly linear relationship between rotational frequency and wind speed, see Figure 4.6.
- **Savonious rotor:** There is a complicated motion of the wind through and around the two-curved sheet airfoils. The driving force is principally drag. The construction is simple and inexpensive. The high solidity produces high starting torque, so savonious rotors are used for water pumping, see Figure 4.6.
- **Darrieus rotor:** This rotor has two or three thin curved blades with an airfoil section. The driving forces are lift, with maximum torque occurring when a blade is moving across the wind at a speed much greater than the wind speed. Its uses are for electricity generation. The rotor is not usually self-starting. Therefore, movement may be initiated with the electrical generator used as a motor. A modern vertical axis machine is shown in Figure 4.8.

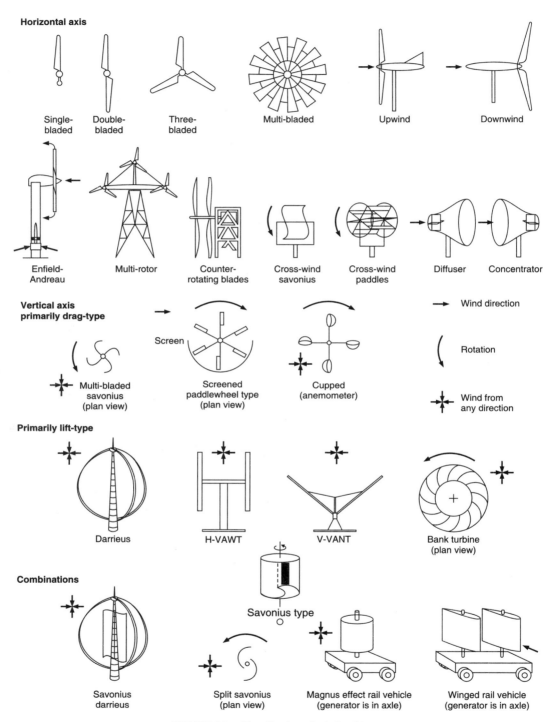

FIGURE 4.6 Classification of wind turbines.

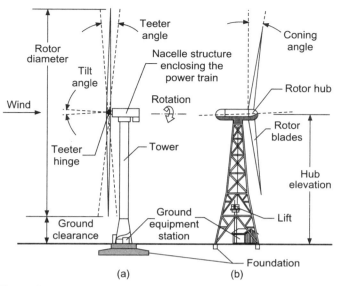

FIGURE 4.7 Principal subsystems of a HAWT. (a) Upwind rotor, (b) Downwind rotor.

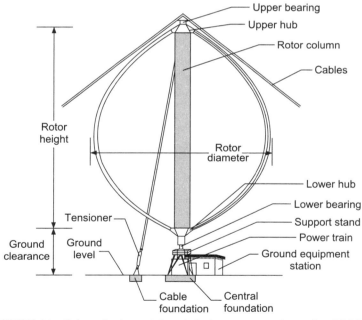

FIGURE 4.8 Schematic views of the principal components of a modern VAWT.

- **Concentrators:** Turbines draw power from the intercepted wind, and it may be advantageous to funnel or concentrate wind into the turbine from outside the rotor section. Various systems have been developed or suggested for horizontal axis propeller turbines. Various blade designs and adaptations are able to draw air into the rotor section, and hence harness power from a cross-section greater than the rotor area. Funnel shapes and deflectors fixed statically around the turbine draw the wind into the rotor. A typical concentrator is shown in Figure 4.9.

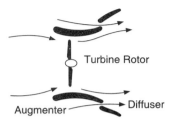

FIGURE 4.9 A typical concentrator.

The wind turbines are classified, as shown in Table 4.2, according to their diameter and/or their rated power.

TABLE 4.2 Scale classification of wind turbine

Scale	*Rotor diameter*	*Power rating*
Small	Less than 12 m	Less than 40 kW
Medium	12 m–45 m	40 kW–999 kW
Large	46 m and larger	1.0 MW and larger

Figure 4.8 presents the general configuration of a modern VAWT of the Darrieus type. Its principal subsystems are: (i) the rotor, (ii) the power train, (iii) the support structure, (iv) the foundations, and (v) the ground equipment station. Symmetry about its vertical axis allows a VAWT to receive winds from any direction, so no yaw mechanism is needed. This is one of its primary advantages. The rotor consists of curved blades with ends fastened to rigid upper and lower hubs separated by the rotor column. To minimise internal bending stresses during rotation, blades are shaped to approximate a troposkien (turning rope), a shape with zero bending stress. VAWT rotor contains two or three fixed-pitch blades, usually symmetrical in cross-section and without twist or taper. As with a HAWT, the swept area of a VAWT is defined by the projection on a vertical plane of the surface generated by the moving blades. Rotor diameter is the width of the swept area at its equator. Rotor height is the distance between upper and lower hubs and is usually 15 per cent–30 per cent larger than the diameter.

Darrieus rotors are stall-controlled, because pitch-change mechanism has not been found to be cost-effective. Motoring of the generator is the usual method for starting Darrieus rotors, since the blades develop lift and torque only through a superposition of the rotational (forward) speed and the wind speed and, therefore, are not normally self-starting. VAWT rotors are usually stopped by applying a rotor brake in the power train, although trailing edge flaps have also been used for this purpose.

4.3 TURBINE COMPONENTS

Rotor: The number one targets for advancement is the means by which the energy is initially captured—the rotor. There are considerable incentives to use better materials and innovative controls to build enlarged rotors that sweep a greater area for the same or lower loads.

Blades: Larger rotors with longer blades sweep a greater area, increasing energy capture. Simply lengthening a blade without changing the fundamental design, however, would make the blade much heavier. In addition, the blade would incur greater structural loads because of its weight and longer moment arm. Blade weight and resultant gravity–induced loads can be controlled by using advanced materials with higher strength-to-weight ratios. Because high-performance materials such as carbon fibres are more expensive, they would be included in design only when the pay-off is maximized. These innovative airfoil shapes hold the promise of maintaining excellent power performance, but have yet to be demonstrated in full-scale operation.

- Rotor blade fabrication process requires signifcant hand labour. Due to this, it gives lot of opportunity of deviation to fibre placement.
- Due to pressure of composite and sandwiched materials in blade making, it is prone to structural instabilities of buckilng and wrinkling.
- Buckling of blade leads to delamination which reduces the fatigue life and ultimately collapse the structure.

Rotor blades account for about quarter of total cost of a wind turbine. Currently blades are built by *vacuum infusion process*. In this process, two blade halves are reinforced with fibre glass or matting of carbon fibre in a vacuum environment. It is entirely manual process. Resin is injected to bond the material by hardening. Blades are varnished afterwards. Increased electricity production is ensured by making blades larger and tower of wind turbine taller to capture more wind at low speeds. For profetibilty bigger swept area offers more energy capture.

One of the concepts is to build passive means of reducing loads directly into the blade structure. By tailoring the structural properties of the blade using the unique attributes of composite materials, the internal structure of the blade can be built in a way that allows the outer portion of the blade to twist as it bends. 'Flab-pitch' or 'bend-twist' coupling, illustrated in Figure 4.10, is accomplished

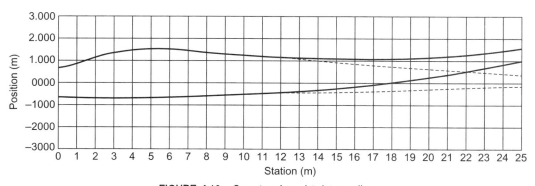

FIGURE 4.10 Curvature-based twist coupling.

by orienting the fibreglass and carbon plies within the composite layer of the blade. If properly designed, the resulting twisting changes the angle of attack over much of the blade, reducing the lift as wind gusts begin to load the blade and thus passively reducing the fatigue loads. Yet another approach to achieve flap-pitch coupling is to build the blade in a curved shape, as shown in Figure 4.11, so that aerodynamic loads apply a twisting action to the blade, which varies the angle of attack as the aerodynamic loads fluctuate. To reduce transportation costs, concepts such as on-site manufacturing and segmented blades are also being explored. The length of blade increase in recent years is shown in Figure 4.12.

Towers: Turbines could sit on even taller towers than those in current use if it can be figured out how to make them with less steel. Using other materials (e.g. carbon fibre) in place of steel could be a better option if there are no significant increase in costs. Active controls that damp out tower vibrations might be another enabling technology. Tower diameter greater than 4 m would incur sever overland transportation costs. Tower diameter and material requirements conflict directly with tower design goals—a larger diameter is beneficial because it spreads out the load and actually requires less material because its walls are thinner. The main design impact of taller tower is not on the tower itself, but on the dynamics of a system with the bulk of its mass atop a longer, more slender structure. Reducing tower-top weight improves the dynamics of such a flexible system.

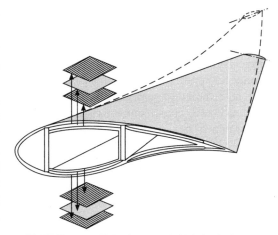

FIGURE 4.11 Twist-flap coupled blade design (material based twist coupling).

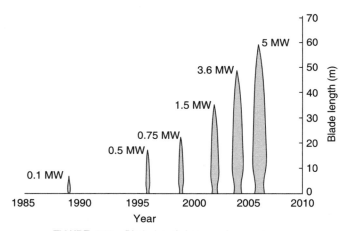

FIGURE 4.12 Blade length increase in recent years.

4.3.1 Power Train Subsystem

The power train of a wind turbine consists of the series of mechanical and electrical components required to convert the mechanical power received from the rotor hub to electrical power. In a HAWT, this equipment is atop the tower, so low maintenance is an important design requirement. Examples of small power train are given in Figure 4.13 and examples of power train for medium and large HAWT are given in Figure 4.14 and Figure 4.15. A typical HAWT power train consists of a turbine shaft assembly (also called a low-speed or primary shaft), a speed increasing gear box, a generator drive shaft (also called a high-speed or secondary shaft), a rotor brake, and an electrical generator, plus auxiliary equipment for control, lubrication and cooling functions. In some small-scale and with permanent magnet type generator with medium- and large-scale HAWT have a direct-drive from the turbine to the generator, with no gear box, turbine shaft, or generator-drive shaft.

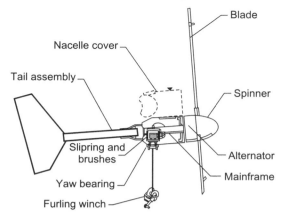

FIGURE 4.13 Typical small HAWT power train.

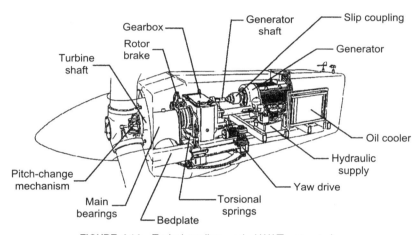

FIGURE 4.14 Typical medium scale HAWT power train.

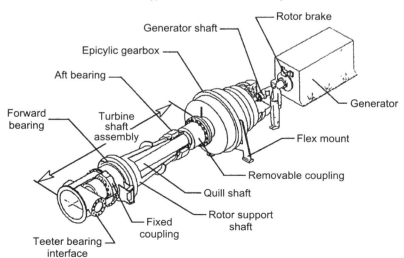

FIGURE 4.15 Typical large scale HAWT power train.

122 Wind Energy: Theory and Practice

The turbine shaft assembly is one of the most critical components in a HAWT because of its dual structural/mechanical function. Rotor weight, thrust, torque, and lateral forces all cause fatigue loading on this component whose design lifetime usually equals or exceeds that of the total system.

A HAWT speed-increasing gear box has a step-up ratio (equal to the generator shaft speed divided by turbine shaft speed) that may vary from 1.0 (in a direct-drive power train) to 100 (in a large scale HAWT). Blade tip speed, rotor diameter and generator design determine the step-up ratio. Parallel axis, epicyclic or planetary, and hybrid designs are used. The generator drive shaft is a conventional machine element with bolted flange on both ends. If there is a pitch control mechanism for stopping the rotor, the rotor brake is usually used only for parking and maintenance.

The function of the low-speed shaft is the transmission of the drive torque from the rotor hub to the gear box, and the transfer of all other rotor loadings to the nacelle structure. The mounting of the low-speed shaft on fore and aft bearing has allowed these two functions to be catered for separately; the gearbox is hung on the rear end of the shaft projecting beyond the rear bearing and the drive torque is resisted by a torque arm. The front bearing is positioned as close as possible to the shaft/hub flange connection, in order to minimize the gravity moment due to cantilevered rotor mass, which usually governs shaft fatigue design. As illustrated in Figure 4.16, the spacing between the two bearings will normally be greater than that between front bearing and rotor hub in order to moderate the bearing loads due to shaft moment.

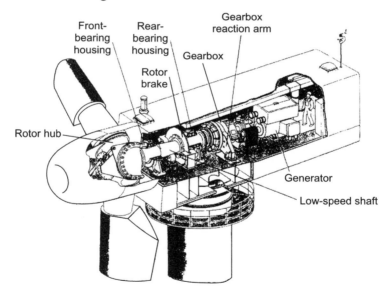

FIGURE 4.16 View of nacelle showing drive shaft arrangement.

In other configuration, shown in Figure 4.17, only the rear-low speed shaft bearing is absorbed into the gearbox. The gearbox is usually set well back from the front bearing in order to reduce the rear bearing loads, and is rigidly fixed to supporting pedestals positioned on either side of the nacelle.

Direct drive systems include innovative power electronics architectures and large scale use of permanent–magnet generators to eliminate gear box. The use of rare-earth permanent magnets in

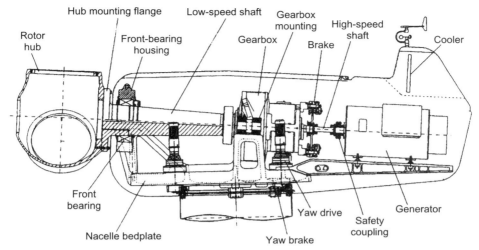

FIGURE 4.17 Nacelle arrangement for a large modern turbine.

generator rotors eliminates much of the weight associated with copper windings, eliminates problems associated with insulation degradation and shorting, and reduces electrical losses. Rare-earth magnets cannot be subjected to elevated temperatures, however, without permanently degrading magnetic field strength, which imposes corresponding demands on generator cooling reliability. Power electronics has already achieved elevated performance and reliability levels, but opportunities for significant improvement remain. New Silicon Carbide (SiC) devices allow operation at higher temperature and higher frequency, while improving reliability, lowering cost, or both. New circuit topologies could furnish better control of power quality, enable higher voltages to be used and increase overall converter efficiency.

In the case of wind turbines with direct-drive generators, the low-speed shaft arrangement is different. It is illustrated in Figure 4.18. The low-speed shaft, which now connects the rotor hub to the rotor of the generator, is hollow, so that it can be mounted on a concentric fixed shaft cantilevered out from the nacelle bedplate.

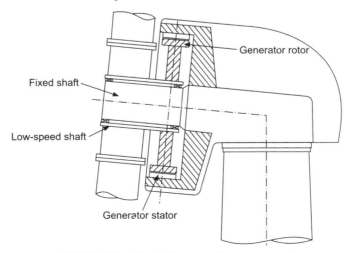

FIGURE 4.18 Direct-drive generator arrangement.

REVIEW QUESTIONS

1. Write a short essay covering materials used for construction of wind turbine blades of various wind turbines.
2. Write a short essay on historical development of wind mills and turbines. Draw sketches of various designs of wind energy conversion systems.
3. Write the chronological development of wind technology development starting from historical wind mill to modern wind turbine.
4. Explain with a line diagram the principal subsystems of a HAWT. Draw the diagram of a nacelle of a modern wind turbine and show the arrangement of different components.
5. Explain with a line diagram the principal subsystems of a VAWT (Darrius type and Savonious type wind turbines).
6. Explain the chronology of development and application of rotor blade materials for wind mills and turbines.
7. Write the classification of wind energy conversion systems with the help of sketches and classification tree.
8. Go to three leading manufacturers' websites and find out what is their largest commercial wind turbine state diameter rated power and other major specifications and generator type.
9. What are the applications for small wind turbines?
10. Explain the following mechanisms with sketches:
    (i) Yaw mechanism,
    (ii) Aerodynamics braking,
    (iii) Teetering mechanism, and
    (iv) Pitch mechanism.

CHAPTER 5

Wind Turbine Design

5.1 INTRODUCTION

In wind turbine design, many areas require attention. Table 5.1 shows some of the major requirements. One of the first tasks is to spell out the basic requirements for the machine to be built. The design process can then proceed creatively to meet these needs in the least expensive and most efficient way. It is not possible to satisfy all of them on the first try. As with any design, the process is iterative.

TABLE 5.1 Design tasks

Rotor:	Tip-speed ratio, solidity, number of blades, aerodynamic optimization, static and dynamic operating loads, parked rotor loads, material selection, manufacturing process, structural dynamics, fatigue, starting torque vs. friction torque, primary over speed control, secondary over speed control, blade tower clearance, brake system, yaw control, etc.
Tower:	Height, type: pole, truss or tilt up, structural loads, strength, structural dynamics, tower shadow, erosion protection, etc.
Generator:	Type: ac (synchronous, 3ϕ, 1ϕ), or dc (alternator, generator), size, weight, efficiency curves, speed-torque characteristics, power conditioning, excitation, etc.
Gearbox:	Ratio, maximum speed, torque capacity, strength and load deflections, noise, structural dynamics, lubrication, etc.

(Contd.)

TABLE 5.1 Design tasks (*Contd.*)

Control system:	Mechanical and/or electrical system, control algorithm, power supply, consequence of failure, start-up and shut-down transients, wind speed and direction sensors, reliability, failure analysis, lightning protection, etc.
General:	Can it be simplified?, system dynamics, shipping and erection, installation method, maintenance, aesthetics, safety, sensitivity to vandalism, corrosion protection, specifications and quality control, etc.
Cost:	Design life, development cost, cost per kWh of power produced, cost per kW installed, tax benefits, rate of return, and pay back period.

The design of a wind turbine is an exciting creative process. Extracting energy from wind is basically a simple conversion of energy, but the mastery of essential disciplines required to build a truly efficient, reliable, and competitively priced wind turbine is an immense undertaking. It is most convenient to have the support of technical experts in all the allied fields, such as aerodynamics, electrical machinery, structural dynamics, manufacturing processes, and the like.

5.2 ROTOR TORQUE AND POWER

The aerodynamic lift (and drag) forces on the spanwise elements of radius r and length dr of the several blades of a wind turbine rotor are responsible for the rate of change of axial and angular momentum of all of the air which passes through the annulus swept by the blade elements. In addition, the force on the blade elements caused by the drop in pressure associated with the rotational velocity in the wake must also be provided by the aerodynamic lift and drag. Figure 5.1 shows an element of a blade of a three bladed horizontal axis wind turbine.

It is assumed that the forces on a blade element can be calculated by means of two-dimensional aerofoil characteristics using an angle of attack determined from the incident resultant velocity in the cross-sectional plane of the element; the velocity component in the span-wise direction is ignored. Having information about how the aerofoil characteristic coefficients C_d and C_l vary with the angle of attack, the forces on the blades for given values of a and a' can be determined.

Consider a turbine with B blades of tip radius R, each with chord c and set pitch angle β measured between the aerofoil zero lift line and the plane of the disc. Both the chord length and pitch angle vary along the blade span. Let the blades be rotating at angular velocity Ω and let the wind speed be U_∞. The tangential velocity of the wake $a'\Omega r$ means that the net tangential flow velocity experienced by the blade element is $(1 + a')\Omega r$. Figure 5.2 shows all the velocities and forces relative to the blade chord line at radius r.

From Figure 5.2, the resultant relative velocity at the blade is

$$W = \sqrt{U_\infty^2 (1-a)^2 + \Omega^2 r^2 (1+a')^2} \tag{5.1}$$

which acts at an angle ϕ to the plane of rotation, such that

$$\sin \phi = \frac{U_\infty (1-a)}{W} \quad \text{and} \quad \cos \phi = \frac{\Omega r (1+a')}{W} \tag{5.2}$$

The angle of attack α is given by

$$\alpha = \phi - \beta \tag{5.3}$$

Wind Turbine Design

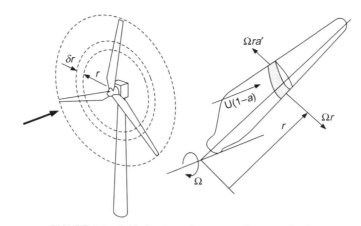

FIGURE 5.1 A blade element sweeps out an annular ring.

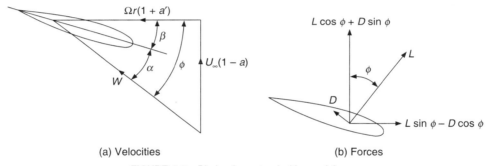

(a) Velocities (b) Forces

FIGURE 5.2 Blade element velocities and forces.

The lift force on a span-wise length δr of each blade, normal to the direction of W, is therefore

$$\delta L = \frac{1}{2}\rho W^2 c C_l\, \delta r \tag{5.4}$$

and the drag force parallel to W is

$$\delta D = \frac{1}{2}\rho W^2 c C_d\, r\, \delta r \tag{5.5}$$

The component of aerodynamic force on B blade elements resolved in the axial direction is

$$\delta L \cos\phi + \delta D \sin\phi = \frac{1}{2}\rho W^2 Bc\,(C_l \cos\phi + C_d \sin\phi)\,\delta r \tag{5.6}$$

The rate of change of axial momentum of the air passing through the swept annulus is

$$\rho U_\infty (1-a)2\pi r\, \delta r\, 2aU_\infty = 4\pi \rho U_\infty^2\, a(1-a)\, r\, \delta r \tag{5.7}$$

The drop in wake pressure caused by wake rotation is equal to the increase in dynamic head, which is

$$\frac{1}{2}\rho(2a'\Omega r)^2$$

Therefore, the additional axial force on the annulus is

$$\frac{1}{2}\rho(2a'\Omega r)^2 \, 2\pi r \, \delta r$$

Thus,

$$\frac{1}{2}\rho W^2 Bc(C_l \cos\phi + C_d \sin\phi)\,\delta r = 4\pi P[U_\infty^2 a(1-a) + (a'\Omega r)^2]r\,\delta r \qquad (5.8)$$

By simplifying, we get

$$\frac{W^2}{U_\infty^2} B \frac{c}{R}(C_l \cos\phi + C_d \sin\phi) = 8\pi\{a(1-a) + (a'\lambda\mu)^2\}\mu \qquad (5.9)$$

where parameter $\mu = \dfrac{r}{R}$.

The element of axial rotor torque caused by aerodynamic forces on the blade element is

$$(\delta L \sin\phi - \delta D \cos\phi)r = \frac{1}{2}\rho W^2 Bc(C_l \sin\phi - C_d \cos\phi)\,r\,dr \qquad (5.10)$$

The rate of change of angular momentum of the air passing through the annulus is

$$\rho U_\infty (1-a)\Omega r\, 2a'r\, 2\pi r\,\delta r = 4\pi \rho U_\infty (\Omega r) a'(1-a) r^2\, \delta r$$

Equating the two moments, we have

$$\frac{1}{2}\rho W^2 Bc(C_l \sin\phi - C_d \cos\phi)r\,\delta r = 4\pi \rho U_\infty (\Omega r) a'(1-a)r^2\,\delta r \qquad (5.11)$$

On simplification, we have

$$\frac{W^2}{U_\infty^2} B \frac{c}{R}(C_l \sin\phi - C_d \cos\phi) = 8\pi \lambda \mu^2 a'(1-a) \qquad (5.12)$$

It is convenient to put

$$C_l \cos\phi + C_d \sin\phi = C_x$$

and
$$C_l \sin\phi - C_d \cos\phi = C_y$$

Solving Eqs. (5.9) and (5.12) to obtain values for the flow induction factor a and a' using two-dimensional aerofoil characteristics requires an iterative process. It gives,

$$\frac{a}{1-a} = \frac{\sigma_r}{4\sin\phi^2}\left[(C_x) - \frac{\sigma_r}{4\sin^2\phi}C_Y^2\right] \qquad (5.13)$$

$$\frac{a}{1+a'} = \frac{\sigma_r C_Y}{4\sin\phi \cos\phi} \qquad (5.14)$$

Blade solidity σ is defined as the total blade area divided by the rotor disc area and is a primary parameter in determining rotor performance chord solidity σ_r is defined as the total blade chord length at a given radius divided by the circumferential length at that radius, i.e.

$$\sigma_r = \frac{B}{2\pi}\frac{c}{r} = \frac{N}{2\pi\mu}\frac{c}{R} \qquad (5.15)$$

The calculation of torque and power developed by a rotor requires a knowledge of the flow induction factors, which are obtained by solving Eqs. (5.13) and (5.14). The iterative procedure is to assume a and a' to be zero initially, determining ϕ, C_l and C_d on that basis, and then to calculate new values of the flow factors using Eqs. (5.13) and (5.14). The iteration is repeated until convergence is achieved.

From Eq. (5.13), the torque developed by the blade element of span-wise length δr is

$$\delta Q = 4\pi \rho U_\infty (\Omega r) a'(1-a) r^2 \, \delta r$$

If drag, or part of the drag, has been excluded from the determination of the flow induction factors, then its effects must be introduced when the torque caused by drag is calculated from blade element forces (see Eq. (5.10)), i.e.

$$\delta Q = 4\pi \rho U_\infty (\Omega r) a'(1-a) r^2 \, \delta r - \frac{1}{2}\rho W^2 B c\, C_d \cos\phi \, r \, \delta r$$

The complete rotor, therefore, develops a total torque Q:

$$Q = \frac{1}{2}\rho U_\infty^2 \pi R^3 \lambda \left[\int_0^R \mu^2 \left\{ 8a'(1-a)\mu - \frac{W}{U_\infty}\frac{B(c/R)}{\pi} C_d (1+a') \right\} d\mu \right] \qquad (5.16)$$

The power developed by the rotor is

$$P = Q\Omega$$

The power coefficient is

$$C_P = \frac{P}{\frac{1}{2}\rho U_\infty^3 \pi R^2}$$

Solving the blade element momentum, Eqs. (5.13) and (5.14) for a given, suitable blade geometry and aerodynamic design yields a series of values for the power and torque coefficients, which are function of the tip speed ratio. A typical performance curve for a modern, high-speed wind turbine is shown in Figure 5.3.

The maximum power coefficient occurs at a tip speed ratio for which the axial flow induction factor a, which in general varies with radius, approximates most closely to the Betz limit value of 1/3. At lower tip speed ratios, the axis flow induction factor can be much less than 1/3 and aerofoil angles of attack are high leading to stalled conditions. At high tip speed ratios, a is high, angles of attack are low, and drag begins to predominate. At both high and low tip speed ratios, therefore, drag is high and the general level of a is non-optimum so the power coefficient is low. Clearly, it would be best if a turbine can be operated at all wind speeds at a tip speed ratio close to that which gives the maximum power coefficient.

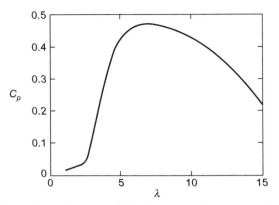

FIGURE 5.3 Power coefficient-tip speed performance curve.

5.2.1 Glauert Momentum Vortex Theory

Glauert momentum vortex theory assumes that small radial sections of the blade can be analysed independently. Consider an annular section of the rotor and examine a small section of length Δr of one blade. The net effect on air flowing this annular section of the rotor results from the forces and moments on all the blades. Let B represent the number of blades.

The geometry of a blade element showing the velocities, forces and moments is depicted in Figure 5.4. The relative wind at the rotor W varies with blade radius r and consists of an axial component U and a rotational component $(r\Omega a' + r\Omega)$. The term $r\Omega$ represents the velocity caused by blade rotation, whereas $(r\Omega a')$ portrays the swirl velocity of the air. The angle of the blade chord with respect to the plane of rotation is denoted by θ. The angle of attack of the airfoil with respect to the local relative wind, W, is denoted by α, and the angle of the relative wind with respect to the rotor plane, by ϕ. The drag force, D, is aligned with the relative wind, W, while the lift force, L is perpendicular to W.

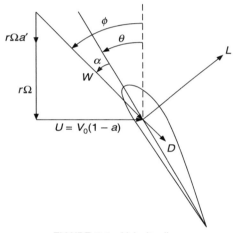

FIGURE 5.4 Velocity diagram.

The force component,

$$F_Q = L \sin \phi - D \cos \phi$$

in the direction of blade rotation generates useful torque, whereas the component,

$$F_T = L \cos \phi + D \sin \phi$$

in the downwind direction exerts a thrust load on the rotor. In terms of dimensionless coefficients C_l and C_d the net force, power and torque caused by B blades, each of local chord c, are as follows: For torque:

$$\Delta Q = \frac{1}{2}\rho W^2 r [C_l \sin \phi - C_d \cos \phi] Bc \, \Delta r \qquad (5.17)$$

For power:
$$\Delta P = \Omega \Delta Q = \frac{1}{2}\rho W^2 r\Omega[C_l \sin\phi - C_d \cos\phi] Bc\,\Delta r \tag{5.18}$$

For thrust:
$$\Delta T = \frac{1}{2}\rho W^2 [C_l \cos\phi + C_d \sin\phi] Bc\,\Delta r \tag{5.19}$$

where
$$W = \frac{U}{\sin\phi} = \frac{r\Omega + r\Omega a'}{\cos\phi} \tag{5.20}$$

In the elementary actuator disc theory, we found that the induced velocity was twice as large in the far wake as in the rotor plane, and this same behaviour is assumed for each streamtube. It is assumed that the air has acquired half of its final rotational (swirl) velocity when it reaches the rotor.

With these assumptions, the linear momentum loss gives the incremental thrust force ΔT on the rotor as
$$\Delta T = \rho u (2\pi r\, dr)\, 2(V_0 - U) \tag{5.21}$$

and the incremental torque ΔQ in terms of the rate of change of rotation momentum as
$$\Delta Q = \rho u (2\pi r\, dr)\, 2Wr \tag{5.22}$$

Equating the expressions for thrust and torque from momentum theory to those from aerodynamic forces on a blade element leads to
$$\frac{V_0 - U}{U} = \left(\frac{Bc}{8\pi r}\right)\left(\frac{C_l \cos\phi + C_d \sin\phi}{\sin^2 \phi}\right) \tag{5.23}$$

and
$$\frac{W}{r\Omega a' + \Omega r} = \left(\frac{Bc}{8\pi r}\right)\left(\frac{C_l \sin\phi - C_d \cos\phi}{\sin\phi \cos\phi}\right) \tag{5.24}$$

Using dimensionless axial and radial induction factors, $a = \frac{V_0 - U}{V_0}$ and $a' = \frac{W}{\Omega r}$ and solidity $\sigma = \frac{Bc}{\pi R}$, Eqs. (5.23) and (5.24) become

$$\frac{a}{(1-a)} = \left(\frac{\sigma R}{8r}\right)\left(\frac{C_l \cos\phi + C_d \sin\phi}{\sin^2\phi}\right) \tag{5.25}$$

and
$$\frac{a'}{(1+a')} = \left(\frac{\sigma R}{8r}\right)\left(\frac{C_l \sin\phi - C_d \cos\phi}{\sin\phi \cos\phi}\right) \tag{5.26}$$

Also,
$$\tan\phi = \frac{U}{\Omega r + r\Omega a'} = \frac{V_0(1-a)}{\Omega r (1+a')} = \frac{1-a}{x(1+a')} \tag{5.27}$$

where $x = \Omega r/V_0$ is the local speed ratio. At the end of the blade, r becomes R, and the tip speed ratio (TSR),

$$X = \frac{R\Omega}{V_0} \tag{5.28}$$

Using X,

$$\tan\phi = \left(\frac{R}{rX}\right)\left[\frac{(1-a)}{(1+a')}\right] \tag{5.29}$$

The two-dimensional lift and drag coefficients and a set of relations may be solved iteratively to find a and a' for any pitch angle θ. Instead of average solidity, define a symbol called the blade loading coefficient, $\lambda = BcC_l/8\pi r$. The parameter λ is one-fourth of the average retarding pressure the blade exerts on the air flowing through the annular streamtube, normalized by the relative dynamic pressure at the blade element.

From Eq. (5.19),

$$\Delta T = \frac{1}{2}\rho W^2 [C_l \cos\phi + C_d \sin\phi] Bc\Delta r$$

$$= \frac{1}{2}\rho W^2 C_l \left[\cos\phi + \left(\frac{C_d}{C_l}\right)\sin\phi\right] Bc\Delta r$$

If

$$\varepsilon = \frac{C_d}{C_l}$$

$$\Delta T = \frac{1}{2}\rho W^2 (BcC_l)[\cos\phi + \varepsilon \sin\phi]\Delta r$$

For small ϕ and small ε, $\cos\phi \cong 1$ and $\varepsilon \sin\phi \cong 0$. Thus,

$$\Delta T \cong \frac{1}{2}\rho W^2 (BcC_l)\Delta r$$

as the retarding force exerted by the blade.

In this, ΔT is divided by the annular area, $2\pi r\Delta r$, to get the average pressure exerted by the blades on the annulus, and normalized by $\frac{1}{2}\rho W^2$, so

$$\frac{\Delta T}{(2\pi r \Delta r)\left(\frac{1}{2}\rho W^2\right)} = \frac{BcC_l}{2\pi r} = 4\lambda$$

Here λ is blade loading coefficient $\lambda = \dfrac{BcC_l}{8\pi r}$

The corresponding normalized average local pressure on the blade element would be

$$\frac{4\lambda(2\pi r \Delta r)}{Bc\Delta r} = \frac{4\lambda}{\sigma_l} = C_l$$

where σ_L is local solidity defined as $\sigma_L = \dfrac{Bc}{\pi r}$

Thus, increasing λ increases the amount of slowing of the air and the pressure on the blade, whereas when λ goes to zero, the blade does not retard the wind. Using λ and ε and dividing the right-hand side of Eqs. (5.25) and (5.26) by sin ϕ and cos ϕ, respectively, we obtain

$$\frac{a}{(1-a)} = \lambda \frac{(\cot \phi + \varepsilon)}{\sin \phi} \quad (5.30)$$

and

$$\frac{a'}{1+a'} = \lambda (\tan \phi - \varepsilon)/\sin \phi \quad (5.31)$$

These equations are simplest possible and are quite insensitive to the value of ε for high L/D airfoils. For $\varepsilon = 0$ gives values of a with less than one per cent error for $0 < \phi < 40$ degree when compared with using the exact equation with $\varepsilon = 0.01$. These relationships are shown graphically in Figure 5.5 plotted for $\varepsilon = 0.01$. It is very useful for quick estimates of changes in rotor behaviour, such as sudden loss of C_l, start-up when stalled, etc.

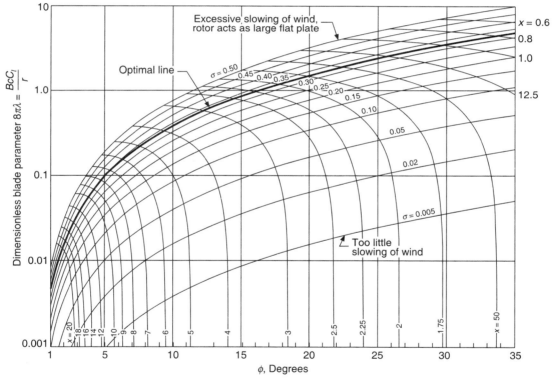

FIGURE 5.5 Blade element parameters.

The iterative solution procedures, given r/R, c, $C_l(\alpha)$, $C_d(\alpha)$, θ and V_o, are as follows:
- Guess values of a and a'.
- Calculate ϕ from Eq. (5.29).
- Calculate $\alpha = \phi - \theta$ and hence C_l and C_d.
- Update a and a' from Eqs. (5.25) and (5.26).

The procedure allows for the evaluation of local forces and moments on the rotor, which can then be integrated to find the total torque and loads. This provides a method of estimating performance once a candidate blade layout is known. One can iteratively change the blade layout and optimize the blade.

The thrust and torque coefficients are defined as

$$C_T = \frac{T}{\frac{1}{2}(\rho V_0^2)(\pi R^2)} \tag{5.32}$$

and

$$C_Q = \frac{Q}{\rho(\pi R^2)(\Omega R)^2 R} \tag{5.33}$$

Following Miller et al., these equations can be reduced to the following form for integrating along the blade length:

$$dC_T = 8a(1-a)\left(\frac{r}{R}\right)d\left(\frac{r}{R}\right) \tag{5.34}$$

and

$$dC_Q = \left(\frac{4}{X}\right)a'(1-a)\left(\frac{r}{R}\right)^3 d\left(\frac{r}{R}\right) \tag{5.35}$$

The Prandtl correction function is conveniently expressed as follows:

$$F = \left(\frac{2}{\pi}\right)\cos^{-1}\exp(-f) \tag{5.36}$$

where

$$f(r) = -\left(\frac{1}{2}R\right)(R-r)B\sqrt{(1+X^2)} \tag{5.37}$$

The function, $F(r)$, approaches unity for $1-(r/R) < 2/XB$ but goes rapidly to zero at tips where $r = R$.

The Prandtl correction for finite number of blades requires that the velocity at the rotor disc be

$$U = V_0(1 - aF)$$

and the axial and rotational interference factors a and a' must now be determined from the following equations:

$$a = \frac{p_1}{1 + p_1} \tag{5.38}$$

$$a' = \frac{p_2}{1 - p_2} \tag{5.39}$$

where

$$p_1 = \frac{\sigma(C_l \cos\phi + C_d \sin\phi)}{8F \sin^2\phi} \tag{5.40}$$

$$p_2 = \frac{\sigma(C_l \sin\phi - C_d \cos\phi)}{8F \sin\phi \cos\phi} \tag{5.41}$$

5.2.2 Optimal Rotors

If C_d is assumed to be zero, dividing Eq. (5.26) by (5.25) gives

$$\frac{a'(1-a)}{a(1+a)} = \tan^2\phi \tag{5.42}$$

and comparing this with Eq. (5.27) gives

$$\frac{a'(1+a')}{a(1-a)} = \frac{V_0^2}{\Omega^2 r^2} = \frac{1}{x^2} \tag{5.43}$$

At each radius, the right side of Eq. (5.43) is constant, and therefore the left side must remain constant also, while the power is maximized if the quantity $a'(1-a)$ from Eq. (5.42) is maximized. Performing these operations using Lagrange multipliers yields:

$$a' = \frac{(1-3a)}{(4a-1)} \tag{5.44}$$

In the wind turbine state, a' must be positive for positive output torque. Thus, for small x, a approaches 1/4 and a' becomes large, whereas for large x, a approaches 1/3 and a' approaches zero.

Substituting a' from Eq. (5.44) into Eq. (5.43) yields the required relationship between a and local speed ratio x as follows:

$$x = (4a-1)\sqrt{\frac{1-a}{1-3a}} \tag{5.45}$$

Miller et al. give a power series in $1/x^2$ as an approximation for the inverse relationship:

$$a \cong \left(\frac{1}{3}\right) - \left(\frac{2}{81x^2}\right) + \left(\frac{10}{729x^2}\right) - \left(\frac{418}{59049x^2}\right) + \ldots \tag{5.46}$$

The corresponding relative wind angle ϕ, can be found from Eq. (5.42) as given by,

$$\tan \phi = \left(\frac{1}{a}\right)\sqrt{(1-a)(1-3a)} \tag{5.47}$$

The optimum blade layout in terms of the product of chord c and lift coefficient C_l can be found from Eq. (5.24) with $C_D = 0$

$$\frac{\sigma C_l X}{8} = \frac{\sigma_L C_l x}{4} = \left(\frac{B\Omega}{8\pi V_0}\right)(cC_l) = \left[\frac{(4a-1)}{(1-2a)}\right]\sqrt{(1-a)(1-3a)} \tag{5.48}$$

As mentioned previously, there is still some design freedom in that both c and C_l may be varied while their product satisfies Eq. (5.48). Thus, if chord c is held constant, C_l (and hence α and θ) will follow from Eq. (5.48). Likewise, if C_l (and hence α) is held constant, chord c must vary according to Eq. (5.48), and twist angle θ has been specified as a function of r.

Miller et al. also integrate the C_P relation along the blade to give closed form and series solution for C_P. Thus, from Eqs. (5.31), (5.44) and (5.45),

$$dC_P = \left(\frac{12}{X^2}\right)\left[\frac{(4a-1)(1-a)(1-2a)}{(1-3a)}\right]^2 da \tag{5.49}$$

which must be integrated from $a = 1/4$ at $r = 0$ to the value of a at the tip, a_T, given by

$$X = (4a_T - 1)\sqrt{\frac{(1-a_T)}{(1-3a_T)}} \tag{5.50}$$

The integration can be done in closed form, resulting in

$$C_P = \left(\frac{4}{729X^2}\right)\left[\left(\frac{4}{z}\right) - 12\ln\left(\frac{1}{4z}\right) - \left(\frac{4363}{160}\right) + 63z - 38z^2 - 124z^3 - 74z^4 - \left(\frac{65}{5}\right)z^5\right] \quad (5.51)$$

where $z = 1 - 3a_T$. A series expansion valid for tip speed ratio X is

$$C_P = \left(\frac{8}{27}\right)\left[1 - \left(\frac{2}{9X^2}\right)\ln\left(\frac{(27X^2 + 15)}{8}\right) + \frac{0.0529}{X^2} + \ldots\right] \quad (5.52)$$

For any tip speed ratio, X, the local speed ratio is $x = Xr/R$. The parameter a is determined from Eq. (5.46), and therefore ϕ from Eq. (5.47), the product cC_l from Eq. (5.48), and C_P from one of the equations (5.49) through (5.52).

To obtain a single-point optimum including the effects of drag, Stewart begins by deriving a local power coefficient C_P' as

$$C_P' = \frac{\Delta P}{\left(\frac{1}{2}\rho V_0^3 \Delta A\right)} = \frac{\frac{1}{2}\rho W^2 r \Omega Bc(C_l \sin\phi - C_d \cos\phi)\Delta r}{\left[\frac{1}{2}\rho V_0^3 (2\pi\Delta r)\right]} \quad (5.53)$$

which, by using $W^2 = V_0^2(1-a)^2 + r^2\Omega^2(1+a')^2$ and Eq. (5.27) to eliminate x^2, reduces to

$$C_P' = (1-a)^2(1+\cot^2\phi)\, 4x\lambda(\sin\phi - \varepsilon\cos\phi)$$

Then, eliminating λ using Eq. (5.30) and expanding $1/(\cot\phi + \varepsilon)$ in Taylor's series of two terms, there results.

$$C_P' = 4xa(1-a)(\tan\phi - \varepsilon)(1 + \varepsilon\tan\phi) \quad (5.54)$$

The last term, $(1 + \varepsilon\tan\phi)$ is quite small, improving the accuracy less than 1 per cent for $\varepsilon = 0.01$ and $\phi = 40$ degree. Thus, in most cases it may be neglected, leaving

$$C_P' = 4xa(1-a)(\tan\phi - \varepsilon) \quad (5.55)$$

By defining a local Froude efficiency, $\eta_F = \left(\frac{27}{16}\right)C_P$, we can relate the performance of each blade to the ideal value of unity.

By dividing Eq. (5.31) into Eq. (5.30) and using Eq. (5.27) to eliminate a', there results

$$x\tan\phi = 1 - a\left[1 + \left(\frac{\tan\phi - \varepsilon}{\cot\phi + \varepsilon}\right)\right] \quad (5.56)$$

For small ε, this is closely given by

$$x\tan\phi = 1 - a\sec^2\phi \quad (5.57)$$

If the function, $F = a(1-a)(\tan\phi - \varepsilon)$, is then maximised using $dF/d\phi = 0$, with relation between a and ϕ given implicity by Eq. (5.56), there results a quadratic equation as follows:

$$\left[\frac{-a\tan^2\phi}{(1-a)}\right] + \left[\frac{(1-a)}{a}\right] = 2 + \left[\frac{\varepsilon\sec^2\phi}{(\tan\phi - \varepsilon)}\right] \quad (5.58)$$

Solving this quadratic equation, the optimal value of a inducing the effect of drag is given by

$$\left(\frac{1}{a}\right) = 2 + \sec\phi\left(G + \sqrt{G + H^2}\right) \qquad (5.59)$$

where $G = \dfrac{\varepsilon \sec \phi}{2(\tan\phi - \varepsilon)}$ and $H = \dfrac{\tan\phi}{(\tan\phi - \varepsilon)}$.

Thus, a complete single-point optimization of a rotor-blade element, given ϕ and ε, proceeds as follows:

1. Find a from Eq. (5.59).

2. Calculate λ from $\lambda = \dfrac{a \sin\phi}{(1-a)(\cot\phi + \varepsilon)}$, Eq. (5.30)

3. Find a' from $a' = \dfrac{a(\tan\phi - \varepsilon)}{(1-a)(\cot\phi + \varepsilon) - a(\tan\phi - \varepsilon)}$, modified Eq. (5.31)

4. Then, $x = \dfrac{(1-a)}{\tan\phi]}\left[1 + \left(\dfrac{\tan\phi - \varepsilon}{\cot\phi + \varepsilon}\right)\right]$, Eq. (5.56)

5. $C_P' = 4xa(1-a)(\tan\phi - \varepsilon)(1 + \varepsilon \tan\phi)$

6. $\eta_F = \left(\dfrac{27}{16}\right) C_P'$

Since η_F is the local Froude efficiency of the blade element, the optimization of the power output from this one blade element is then complete. A plot of η_F versus x from various drag/lift ratio, ε, is given in Figure 5.6.

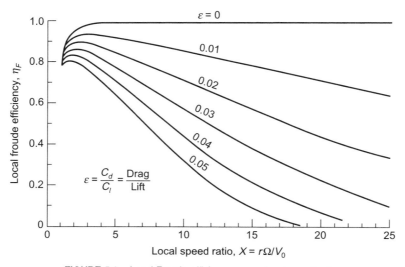

FIGURE 5.6 Local Froude efficiency versus local speed ratio.

Miller et al. give an analysis for optimal rotor including the drag effect but neglecting wake rotation. In their analysis, a drag function, $C_d = C_l + C_2\alpha^2$, is given, and the following two parameters are defined:

$$\zeta = \frac{16C_2}{\sigma X A^2} \quad \text{and} \quad \delta = \frac{\sigma C_L X^3}{8} \tag{5.60}$$

The results of the optimization give

$$a_{opt} = \frac{2}{3 + \zeta + \sqrt{(3+\zeta)^2 - 8\zeta}} \tag{5.61}$$

$$\theta_{opt} = \left(\frac{R}{Xr}\right)(1 - a_{opt})\left[1 - a_{opt}\left(\frac{8}{\sigma XA}\right)\right] \tag{5.62}$$

$$\phi_{opt} = \left(\frac{R}{Xr}\right)(1 - a_{opt}) \tag{5.63}$$

$$C_{P,\max} = 4a_{opt}(1 - a_{opt})\left(1 - \frac{1}{2}\zeta a_{opt}\right) - 2\delta \tag{5.64}$$

5.2.3 Dual Optimum

To find a blade chord c such that a blade element will satisfy the optimal conditions at two speed ratios, x_1 and x_2, where $x_1 > x_2$. Then, if $\Delta\lambda = \lambda_2 - \lambda_1$ and $\Delta\phi = \phi_2 - \phi_1$.

$$\Delta\lambda = \left(\frac{Bc}{8\pi r}\right)(C_{L2} - C_{L1}) = \left(\frac{Bc}{8\pi r}\right)a_o(\Delta\phi + \tau) \tag{5.65}$$

where a_o is the lift-curve slope of the airfoil in appropriate units, τ is a possible change in the pitch angle of the entire blade between the two conditions, and a value of $\tau > 0$ implies that $\lambda_2 > \lambda_1$.

Since we can calculate the ideal λ and ϕ for each speed ratio x, we can calculate $\Delta\lambda$ and $\Delta\phi$. This permits determination of an optimal value of $Bc/8\pi r$, and hence the dual optimum c and the desired lift coefficients at x_1 and x_2 may be found.

EXAMPLE 5.1 Suppose at the tip we wish to have a blade chord optimal at both $X_1 = 6$ and $X_2 = 4$. The calculations are best done in the tabular form shown in Table 5.2(a). Here all angles are in degrees so that the lift-curve slope is taken to be approximate $a_o \cong 0.1$. For convenience, the values of r/R along the blade are in the ratio $4/5 = 0.8$. The angle ϕ for a particular value of x can be found from Eq. (5.57). Since the inverse of this relation is needed, table of $x = x(\phi)$ would be helpful, or a short programme using Newton's method might be used. The last two rows of Table 5.2(a) give the differences, $\Delta\lambda$ and $\Delta\phi$, for the two conditions.

Once these values have been found, blade chords satisfying the dual optimum conditions are easily calculated, as in Table 5.2(b) for $\tau = 0$ and Table 5.2(c) for $\tau = 2$ degrees. The $Bc/8\pi R$ row shows the variation of chord with blade radius. By allowing a pitch change, a blade with a smaller chord and a desirable taper can be found, as in Table 5.2(c). The design lift coefficients of this blade are higher in proportion to the reduction of chord from those of Table 5.2(b).

TABLE 5.2(a) Optimum parameters for $X_1 = 6$, $X_2 = 4$, $\varepsilon = 0$

r/R	1	0.8	0.64	0.512	0.4096	0.32768
x_1	6	4.8	3.84	3.0720	2.4576	1.9661
λ_1	0.006055	0.009360	0.014388	0.021928	0.033003	0.048796
ϕ_1	6.3082	7.8455	9.7311	12.0207	14.7610	17.9727
x_2	4	3.2	2.56	2.048	1.6384	1.3107
λ_2	0.013307	0.020317	0.030660	0.045500	0.065994	0.92914
ϕ_2	9.3574	11.5693	14.2245	17.3502	20.9319	24.8943
$\Delta\lambda$	0.007252	0.010957	0.016272	0.023572	0.032991	0.044118
$\Delta\phi$	3.0492	3.7238	4.4934	5.3295	6.1709	6.9216

TABLE 5.2(b) Optimum blading for $X_1 = 6$, $X_2 = 4$, $\varepsilon = 0$, $\tau = 0$

r/R	1	0.8	0.64	0.512	0.4096	0.32768
$Bc/8\pi r$	0.02378	0.02942	0.03621	0.04423	0.05346	0.06374
$Bc/8\pi R$	0.02378	0.02354	0.02317	0.02265	0.02190	0.02089
C_{l1}	0.255	0.318	0.397	0.496	0.617	0.766
C_{l2}	0.560	0.788	0.847	1.029	1.234	1.458

TABLE 5.2(c) Optimum blading for $X_1 = 6$, $X_2 = 4$, $\varepsilon = 0$, $\tau = 2$ deg

r/R	1	0.8	0.64	0.512	0.4096	0.32768
$Bc/8\pi r$	0.01436	0.01914	0.02506	0.03216	0.04038	0.04945
$Bc/8\pi R$	0.01436	0.01531	0.01604	0.01647	0.01654	0.01620
C_{l1}	0.422	0.489	0.574	0.682	0.817	0.987
C_{l2}	0.927	1.061	1.223	1.415	1.634	1.879

5.3 OPTIMUM DESIGN FOR VARIABLE OPERATION

A turbine operating at variable speed can maintain the constant tip speed ratio required for the maximum power coefficient to be developed regardless of wind speed. To develop the maximum possible coefficient requires a suitable blade geometry, the conditions for which will be derived.

For a chosen tip speed ratio λ, the torque developed at each blade station is maximum if

$$\frac{d}{da'} 8\pi\lambda\mu^2 a'(1-a) = 0$$

Equation (5.12), giving

$$\frac{d}{da'} a = \frac{1-a}{a'} \tag{5.66}$$

From Eqs. (5.9) and (5.12), a relationship between the flow induction factors can be obtained. Dividing Eqs. (5.9) and (5.12), we get

$$\frac{\dfrac{C_l}{C_d}\tan\phi - 1}{\dfrac{C_l}{C_d} + \tan\phi} = \frac{\lambda\mu a'(1-a)}{a(1-a) + (a'\lambda\mu)^2} \tag{5.67}$$

The flow angle ϕ is given by

$$\tan\phi = \frac{1-a}{\lambda\mu(1+a')} \qquad (5.68)$$

Substituting Eq. (5.68) in Eq. (5.67) gives

$$\frac{\dfrac{C_l}{C_d}\dfrac{1-a}{\lambda\mu(1+a')}-1}{\dfrac{C_l}{C_d}+\dfrac{1-a}{\lambda\mu(1+a')}} = \frac{\lambda\mu a'(1-a)}{a(1-a)+(a'\lambda\mu)^2}$$

By simplifying, we get

$$\frac{\dfrac{C_l}{C_d}(1-a)-\lambda\mu(1+a')}{\lambda\mu(1+a')\dfrac{C_l}{C_d}+(1-a)} = \frac{\lambda\mu a'(1-a)}{a(1-a)+(a'\lambda\mu)^2}$$

$$\left\{\frac{C_l}{C_d}(1-a)-\lambda\mu(1+a')\right\}\left\{a(1-a)+(a'\lambda\mu)^2\right\}$$

$$=\left\{\lambda\mu(1+a')\frac{C_l}{C_d}+(1-a)\right\}\lambda\mu a'(1-a) \qquad (5.69)$$

At this stage, the process is made easier to follow if drag is ignored. Equation (5.69) then reduces to

$$a(1-a)-\lambda^2\mu^2 a' = 0 \qquad (5.70)$$

Differentiating Eq. (5.70) with respect to a' gives

$$(1-2a)\frac{d}{da'}a - \lambda^2\mu^2 = 0 \qquad (5.71)$$

and substituting Eq. (5.66) into Eq. (5.71) gives

$$(1-2a)(1-a) - \lambda^2\mu^2 a' = 0 \qquad (5.72)$$

Equations (5.70) and (5.72), together, give the flow induction factors for optimized operation,

$$a = \frac{1}{3} \quad \text{and} \quad a' = \frac{a(1-a)}{\lambda^2\mu^2} \qquad (5.73)$$

which agree exactly with the momentum theory prediction because no losses, such as aerodynamic drag, have been induced and the number of blades is assumed to be large; every fluid particle which passes through the rotor disc interacts with a blade resulting in a uniform axial velocity over the area of the disc.

To achieve the optimum conditions, the blade design has to be specific and can be determined from either of the fundamental Eqs. (5.9) and (5.11). Choosing Eq. (5.11), because it is simpler, and ignoring the drag, the torque developed in optimized operation is

$$\delta Q = 4\pi \rho U_\infty (\Omega r) a'(1-a) r^2 \, \delta r = 4\pi \rho \frac{U_\infty^3}{\Omega} a(1-a)^2 \, r \, \delta r$$

The component of lift per unit span in the tangential direction is therefore

$$L \sin \phi = 4\pi \rho \frac{U_\infty^3}{\Omega} a(1-a)^2$$

By the Kutta–Joukowski theorem, the lift per unit span is

$$L = \rho W \Gamma$$

where Γ is the sum of the individual blade circulations.

Consequently,

$$\rho W \Gamma \sin \phi = \rho \Gamma U_\infty (1-a) = 4\pi \rho \frac{U_\infty^3}{\Omega} a(1-a)^2 \tag{5.74}$$

So,

$$\Gamma = 4\pi \frac{U_\infty^2}{\Omega} a(1-a) \tag{5.75}$$

The circulation is therefore uniform along the blade span and this is a condition for optimized operation.

To determine the blade geometry, that is, how should the chord size vary along the blade and what pitch angle β distribution is necessary, (see Eq. (5.12)).

$$\frac{W^2}{U_\infty^2} B \frac{c}{R} C_l \sin \phi = 8\pi \mu^2 \lambda a'(1-a)$$

Substituting for $\sin \phi$ gives

$$\frac{W}{U_\infty} B \frac{c}{R} C_l (1-a) = 8\pi \lambda \mu^2 a'(1-a) \tag{5.76}$$

From which is derived

$$\frac{B}{2\pi} \frac{c}{R} = \frac{4 \lambda \mu^2 a'}{\dfrac{W}{U_\infty} C_l}$$

The only unknown on the right-hand side of the above equation is the value of the lift coefficient C_l and so it is common to include it on the left-side of the equation with the chord solidity as a blade geometry parameter. The lift coefficient can be chosen as that value which corresponds to the maximum lift/drag ratio C_l/C_d as this will minimize drag losses; even though drag has been ignored in the determination of the optimum flow induction factors and blade geometry, it cannot be ignored in the calculation of torque and power. Blade geometry also depends upon the tip speed ratio λ, so it is also included in the blade geometry parameter. Hence

$$\sigma_r \lambda C_l = \frac{N}{2\pi} \frac{c}{R} \lambda C_l = \frac{4 \lambda^2 \mu^2 a'}{\sqrt{(1-a)^2 + \{\lambda \mu (1+a')\}^2}} \tag{5.77}$$

Introducing the optimum conditions of Eq. (5.73),

$$\sigma_r \lambda C_l = \frac{8/9}{\sqrt{\left(1-\frac{1}{3}\right)^2 + \lambda^2 \mu^2 \left\{1 + \frac{2}{9(\lambda^2 \mu^2)}\right\}^2}} \qquad (5.77a)$$

The parameter $\lambda\mu$ is called the local speed ratio and is equal to the tip speed ratio when $\mu = 1$. Over the outboard, half of the blade, which produces the bulk of the power, the local speed ratio $\lambda\mu$, will normally be large enough to enable the denominator to be approximated as $\lambda\mu$, giving

$$\sigma_r \lambda C_l = \frac{Bc(\mu)}{2\pi R} \lambda C_l = \frac{8}{9\lambda\mu} \qquad (5.77b)$$

where B is the number of blades. After rearrangement, this gives

$$c(\mu)\left(\frac{\Omega R}{U_\infty}\right)^2 = \frac{16\pi R}{9 C_l B \mu} \qquad (5.77c)$$

Hence it can be seen that, for a family of designs optimized for different rotational speeds at the same wind speed, the blade chord at a particular radius is inversely proportional to the square of the rotational speed, assuming that N and R are fixed and the lift coefficient is maintained at a constant value by altering the local blade pitch to maintain a constant angle of attack.

If for a given design C_l is held constant, then Figure 5.7 shows the blade plan form for increasing tip speed ratio. A high design tip speed ratio would require a long, slender blade (high aspect ratio), while a low design tip speed ratio would need a short, fat blade. The design tip speed ratio is that at which optimum performance is achieved. Operating a rotor at other than the design tip speed ratio gives a less than optimum performance even in ideal drag free conditions.

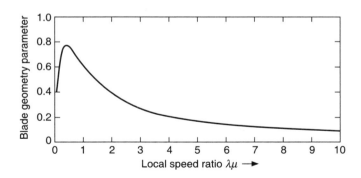

FIGURE 5.7 Variation of blade geometry parameter with local speed ratio.

In off-optimum operation, the axial inflow factor is not uniformly equal to 1/3. In fact, it is not uniform at all.

The local inflow angle ϕ at each blade station also varies along the blade span as shown in Eq. (5.78) and Figure 5.8.

$$\tan \phi = \frac{1-a}{\lambda\mu(1+a')} \qquad (5.78)$$

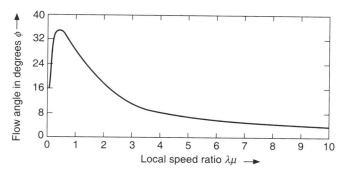

FIGURE 5.8 Variation of inflow angle with local speed ratio.

which, for optimum operation, is

$$\tan \phi = \frac{1 - \frac{1}{3}}{\lambda \mu \left(1 + \frac{2}{3\lambda^2 \mu^2}\right)} \tag{5.79}$$

EXAMPLE 5.2 Take NACA 4412 blade aerofoil, popular for hand-built wind turbines because the bottom (high pressure) side of the profile is almost flat, which facilitates manufacture. At a Reynolds number of about 5×10^5, the maximum lift/drag ratio occurs at a lift coefficient of about 0.7 and an angle of attack of about 3°. Assuming both C_l and α are to be held constant along each blade and there are to be three blades operating at a tip speed ratio of 6, then the blade design in plan-form and pitch (twist) variation are as shown in Figure 5.9(a) and (b), respectively.

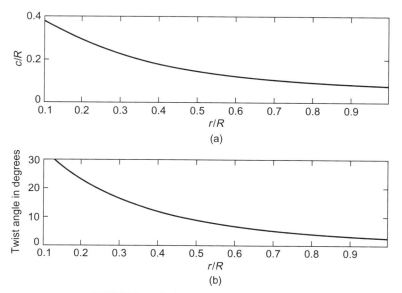

FIGURE 5.9 Optimum blade design with $\lambda = 6$.

The blade design of Figure 5.9 is efficient but complex to build. A straight line drawn through the 70 per cent and 90 per cent span points as shown in Figure 5.10 not only simplifies the planform but removes a lot of materials close to the root.

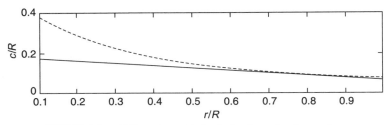

FIGURE 5.10 Uniform taper blade design for optimal operation.

$$\frac{cU_\infty}{R} = \frac{8}{9\lambda 0.8}\left(2 - \frac{\lambda\mu}{\lambda 0.8}\right)\frac{2\pi}{C_l \lambda B} \quad (5.80)$$

The 0.8 in Eq. (5.80) refers to the 80 per cent point, midway between the target points.

Equations (5.77) and (5.80) can be combined to give the required span-wise variation of C_l for optimal operation.

$$C_l = \frac{8}{9} \frac{1}{\frac{BcU_\infty \lambda}{2\pi}\sqrt{\left(1-\frac{1}{3}\right)^2 + \lambda^2\mu^2\left\{1 - \frac{2}{9(\lambda^2\mu^2)}\right\}^2}}$$

Close to the blade root, the lift coefficient approaches the stalled condition and drag is high, but the penalty is small because the adverse torque is small in that region.

Assuming that stall does not occur and that for the aerofoil in question, which has 4 per cent camber the lift coefficient is given approximately by,

$$C_l = 0.1\,(\alpha + 4\text{ deg})$$

where α is in degrees, so

$$\alpha = \frac{C_l}{0.1} - 4\text{ deg}$$

The blade twist distribution can now be determined from Eqs. (5.79) and (5.80). The twist angle close to the root is still high but lower than for the constant C_L blade. The span-wise distribution of the lift coefficient required for the linear taper blade is shown in Figure 5.11 and distribution of the twist of blade is shown in Figure 5.12.

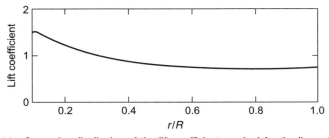

FIGURE 5.11 Span-wise distribution of the lift coefficient required for the linear taper blade.

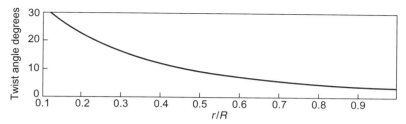

FIGURE 5.12 Span-wise distribution of the twist required for the linear taper blade.

For grid connected wind turbines, the rotational speed is maintained at constant values. Due to variation in wind speed, the tip speed ratio continuously changes. In view of this, a blade optimized for fixed tip speed ratio would not be appropriate. No simple solution is available for optimum design of a blade or rotor running at constant rotational speed. With known wind speed distribution at a particular site, a non-linear programming method could be applied for maximizing energy capture for specified duration of time, say, in a year. Alternatively, a design tip speed ratio could be chosen which corresponds to the representative wind speed at a site or average wind speed at a site. More practical approach would be adjustment of pitch angle of blade for different ranges of wind speeds to maximize energy capture. The variation of maximum coefficient of power would turbine rotor with design tip speed ratio for a range of lift to drag ratio is plotted in Figure 5.13.

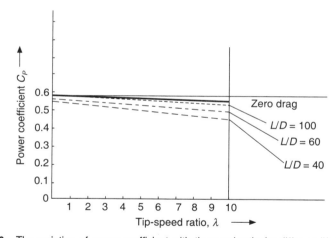

FIGURE 5.13 The variation of power coefficient with tip-speed ratio for different lift to drag ratios.

EXAMPLE 5.3 The turbine blade is designed for optimum performance at a tip speed ratio of 7, with an angle of attack as 8° as an illustrated example. Assuming the maximum lift to drag ratio for these conditions for NACA 632xx aerofoil. Further, the turbine is operating at constant rotational speed with fixed pitch angle. The geometrical characteristics of aerofoil are given in Table 5.3 and aerodynamic characteristics are shown in Figure 5.14. The power performance characteristics of the variation of coefficient C_p with blade tip ratio λ is shown in Figure 5.3.

TABLE 5.3 Blade Design of a 8.5 m radius rotor

Radius, r (mm)	r/R	Chord, c (mm)	Pitch, β (degree)	Thickness/chord (%)
1700	0.20	1085	15.0	24.6
2150	0.30	1005	9.5	20.7
3400	0.40	925	6.1	18.7
4250	0.50	845	3.9	17.6
5100	0.60	765	2.4	16.6
5950	0.70	685	1.5	15.6
6800	0.80	605	0.9	14.6
7225	0.90	525	0.4	13.6
8500	1.00	445	0.0	12.6

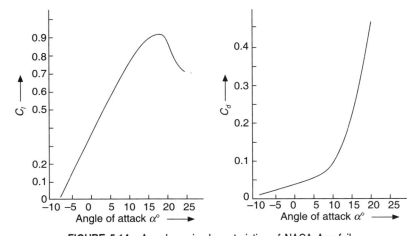

FIGURE 5.14 Aerodynamic characteristics of NACA Aerofoil.

At lower tip speed ratio, the entire blade is stalled as shown in Figure 5.15. For a rotational speed of 60 r.p.m., the corresponding wind speed is 25 m/s, which is cut-out speed of turbine. For the highest tip speed ratio, the Corresponding wind speed is 4.2 m/s, which is cut-in speed of turbine. Maximum power is developed at a tip speed ratio of 5 at a wind speed of 14 m/s. The variation of flow factors is shown in Figure 5.16. The axial flow factor increases with tip speed ratio, while the tangential flow factor decreases with tip speed ratio.

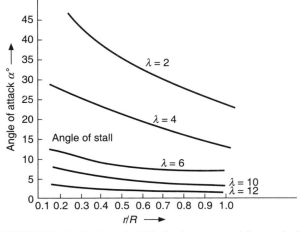

FIGURE 5.15 Angle of attack distribution for a range of tip speed ratio.

Wind Turbine Design   147

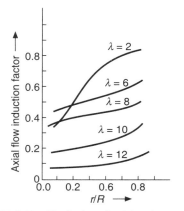

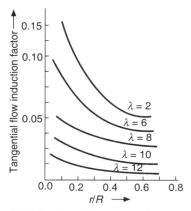

FIGURE 5.16 Distribution of axial and tangential flow induction factor for a range of tip speed ratio.

At high tip speed ratio of $\lambda = 12$, the effect of drag reduces torque as the square of the local speed ratio and also causes significant loss of power. The dramatic effect of stall is illustrated for the differences in torque distribution for the tip speed ratio of 2 and 4 in Figure 5.17

It must be noted that the actual thrust force acting on the blade will increase with wind speed as illustrated by Figure 5.18(a) and axial force coefficient with increasing tip speed ratio in Figure 5.18(b).

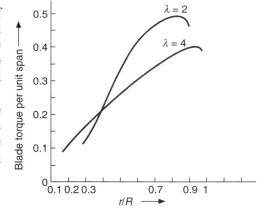

FIGURE 5.17 Difference in torque distribution for tip ratio of 2 and 4.

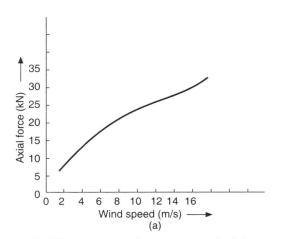

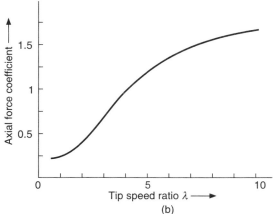

FIGURE 5.18 Axial force variation with wind speed and axial force coefficient variation with tip speed ratio.

5.4 INFLUENCE OF REYNOLDS NUMBER

The lift coefficient increases linearly with increasing angle of attack below the stall condition. With the increase of Reynolds number the stall angle will increase and at the same time the maximum value of the lift coefficient will also rises. The variation of Reynolds number determines the flow around an aerofoil and significantly affects the values of lift and drag coefficients. The value of drag coefficient will rise with decreasing value of Reynolds number. Below a critical Reynolds number of about 200000, the boundary layer remains laminar and it causes rapid increase in the drag coefficient. The influence of Reynolds number is illustrated in Figure 5.19 and Figure 5.20.

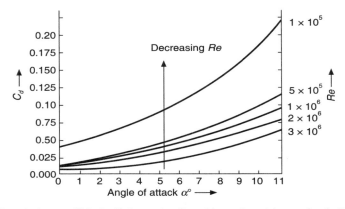

FIGURE 5.19 Variation of drag coefficient with increasing Reynolds number at low angle of attack for NACA aerofoil.

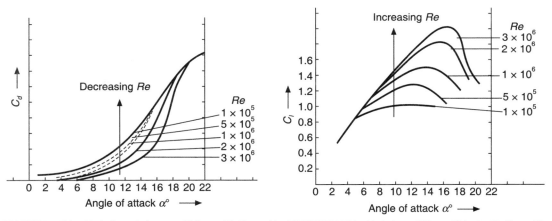

FIGURE 5.20(a) Variation of drag coefficient with Reynold number in stall region for NACA aerofoil.

FIGURE 5.20(b) Variation of lift coefficient with Reynolds number in stall region for NACA aerofoil.

5.5 CAMBERED AEROFOILS

Cambered aerofoils are asymmetrical about central chord line, which is a curved line to allow for producing lift even at zero angle of attack. Cambered aerofoils have higher lift to drag ratio as compared to symmetrical aerofoils for positive angle of attack, and therefore are used in wind turbine blades. A typical cambered aerofoil NACA 4412 is shown in Figure 5.21. The four-digit designation used for NACA aerofoil from left to right is as follows: the first digit represents the amount of camber as a percentage of chord length, the second digit the percentage chord position in units of 10 per cent at which the maximum camber occurs and the last two digits are the maximum thickness to chord ratio, as a percentage of the chord length. The curved camber line is made of two parabolic arcs which join smoothly at maximum camber point. A straight line joining the ends of the camber line is called chord line and angle of attack a is measured from it. Figure 5.22 shows the scheme of designation for 4-digit NACA aerofoil.

FIGURE 5.21 Profile of NACA 4412 aerofoil.

For cambered aerofoils, zero lift occurs at a small negative angle of attack of about 4°. The characteristic behaviour of NACA aerofoil is shown in Figure 5.23.

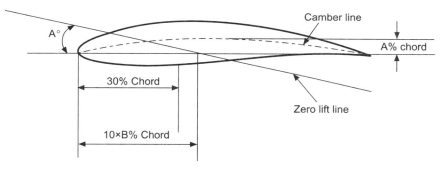

FIGURE 5.22 Designation of NACA xxxx aerofoil.

For asymmetrical aerofoil, the centre of pressure lies at 1/4 of chord from the leading edge or nose. For cambered aerofoils, the centre of pressure lies aft of 1/4 chord position and with increasing angle of attack, further moves towards the trailing edge. The resultant force acting through the centre of pressure point gives rise to pitching moment. When the pitching moment has the tendency of moving the nose up, it is assumed to be positive by convention. A pitching moment coefficient is defined as

$$C_m = \frac{\text{Pitching/Unit span}}{\frac{1}{2}\rho U_c^2} \tag{5.81}$$

There will be a position, called the aerodynamic centre, for which

$$\frac{dC_m}{dC_l} = 0$$

For all practical purposes, the aerodynamic centre lies at the 1/4 chord position. The value of C_m depends upon the degree of camber that is asymmetry and its value is 0.1 for NACA 4412. Above the stall condition, the pitching moment becomes independent of α and, therefore, pre-stall position continues to be used to determine the pitching moment coefficient.

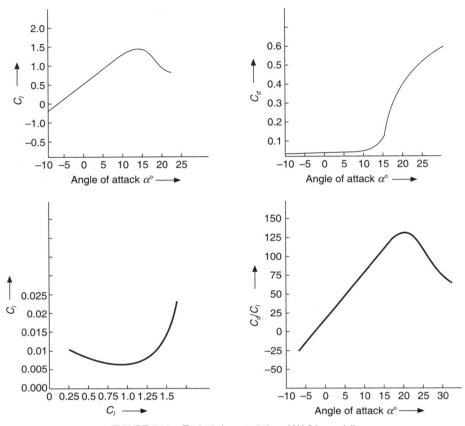

FIGURE 5.23 Typical characteristics of NACA aerofoil.

5.5.1 Rotor Sizing

In a high-wind speed area, a relatively small rotor can develop a large amount of power. The power loading per unit area of the rotor in watt/square metre is a useful index in comparing wind turbines. Preliminary rotor sizing can be done with the elementary actuator disc momentum theory. The power output of an ideal rotor is given by

$$P = \frac{1}{2}\rho A C_P V_0^3 \tag{5.82}$$

where
 ρ = air density
 V_0 = wind speed
 A = rotor swept area
 C_P = power coefficient

Air density ρ varies with altitude and temperature. In SI units, this is as follows:

$$\rho = \rho_0 \exp\left(\frac{-0.297 h}{3048}\right) \tag{5.83}$$

where ρ = 1.22496 kg/m³ and h in metres.

The temperature defined for the standard atmosphere is a linearly decreasing function of altitude given by

$$T = 15 - 1.983\left(\frac{h}{304.8}\right) °C \qquad (5.84)$$

where h is in metres.

The maximum rotor power coefficient is $C_P = 16/27$, whereas real rotors achieve considerably lower power. With very precise, smooth airfoil blades and tip coefficient speed ratios above 10, rotor power coefficients as high as 0.45 have been reported. For most machines, a power coefficients of 0.3 to 0.35 would be possible with good design. With a drive train efficiency, η_d, and a generator efficiency, η_g, the actual power output would be

$$P_{out} = \frac{1}{2}\rho A C_P \eta_d \eta_g V_0^3 \qquad (5.85)$$

when this expression is solved for the rotor-swept area, A, we have

$$A = \frac{2 P_{out}}{\rho C_P \eta_d \eta_g V_0^3} \qquad (5.86)$$

EXAMPLE 5.4 Find the area required for a rotor to generate 4.5 kW in a 5 m/s. Assume a drive train efficiency of 0.85, a generator efficiency of 0.75, and an altitude of 92 m.

Solution: The standard atmospheric air density at 92 m would be

$$\rho = 1.225 \exp\left[\frac{(-0.297)(92)}{3048}\right]$$
$$= 1.214 \text{ kg/m}^3$$

and the standard atmosphere temperature would be

$$T = 15 - 1.983\left(\frac{92}{304.8}\right) = 14.4°C$$

The required area would be

$$A = \frac{2(4500)}{(1.214)(0.35)(0.85)(0.75)(5^3)}$$
$$= 265 \text{ m}^2 = \pi R^2$$
$$D = 2R = 18.4 \text{ m}.$$

5.6 LOAD CALCULATION

The simplified method for load calculation is from the Danish Standard DS 472. The dimensioning loads of the wind turbine in the operational mode, production operation, may be calculated according to a simplified method as examined in Danish standard, provided that the following requirements are met. The calculated load range distributions are used to assess the fatigue limit states of the wind turbine. In addition, an ultimate load limit state must be examined, whereby the average value of the parameters studied is combined with the corresponding half maximum load range found in the load range distribution of the parameter.

The following assumptions define the type of wind turbine, which may be dealt with by means of simplified calculations:

1. The wind turbine shall have 3 blades, generate electricity and be stall-regulated/pitch regulated with fixed blade angle, with no hub hinge.
2. The wind turbine shall produce electricity by means of an induction generator, connected to the power network at a fixed frequency.
3. The wind turbine rotor shall be placed on the windward side (upwind rotor) with a maximum 10° tilt, radius $R < 15, 5$ m, solidity between 5 and 15 per cent and a minimum distance from the blade to the tower during production operation of half of the local tower diameter for the outer half of the blade.
4. Defined maximum power of the wind turbine P_{max} may exceed the maximum long-term average power (nominal power) P_{nom} (maximum point on the power curve between V_{min} and V_{max}) by 15 per cent at the most (Figure 5.24).

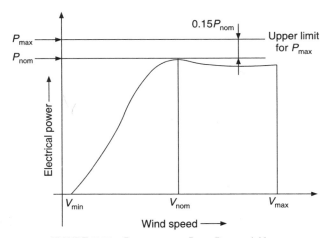

FIGURE 5.24 Power curve, P_{max}, P_{nom} and V_{nom}.

5. The nominal generator power of the electric generator of the wind turbine $P_{nom, g}$ must be at least P_{nom}.
6. The maximum operating frequency of rotation of the wind turbine $n_{r, max}$ shall be limited, so that $n_{r, max} < 1.05\ n_{r, syn}$, where $n_{r, syn}$ is the operating frequency of rotation corresponding to idling (synchronous frequency of rotation) on the primary generator.
7. The yawing speed of the wind turbine shall not exceed $\omega_k = 1°$/second.
8. At an arbitrary given wind speed, V the 10 minute yaw-error in the wind turbine must be distributed with a maximum 10° average yaw-error and a maximum 10° standard deviation.
9. The stop wind velocity of the wind turbine V_{max} shall not exceed 25 m/s.
10. The lowest blade and tower natural frequencies n_o during operation shall differ from the operating frequency of rotation n_r by at least 10 per cent.
11. The lowest natural frequency of the tower for bending shall be less than $2.5\ n_r$.
12. The Weibull k parameter must be between 1.85 and 2.00.
13. For wind turbines in wind farms or groups, the distance between wind turbines must be $> 5 \times D$.

The *characteristic aerodynamic load* value, P_o shall be calculated as a basis for the load calculations. The characteristic load is defined as the y' component of the aerodynamic load per unit of length of the blade acting at the 2/3-point of the blade (i.e. where $r = 2/3R$ and at right angles to the rotor plane, see definition of the rotor coordinates system in Table 5.6 and blade geometry in Figure 5.25(a) and 5.25(b)).

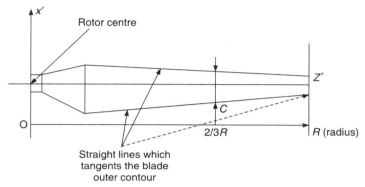

FIGURE 5.25(a) Definition of rotor radius at the 2/3-point and characteristic chord line length.

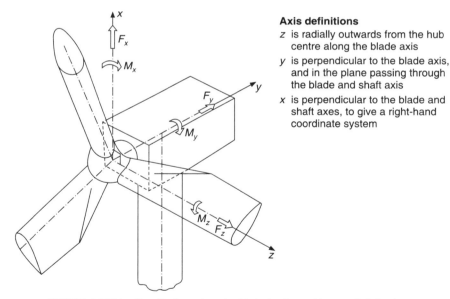

Axis definitions
- z is radially outwards from the hub centre along the blade axis
- y is perpendicular to the blade axis, and in the plane passing through the blade and shaft axis
- x is perpendicular to the blade and shaft axes, to give a right-hand coordinate system

FIGURE 5.25(b) Coordinate system for blade loads, positions and deflections.

P_o is calculated from

$$P_o = \frac{1}{2} \rho W^2 c C_l \qquad (5.87)$$

where
  C_l is the lift coefficient at the 2/3-point

W is the resulting wind velocity

ρ is air density

c is the characteristic chord line length at $2/3\ R$ (see Figure 5.25(b))

As lift coefficient, the maximum for the blade profile concerned shall be used, but it must be at least $C_l = 1.5$. The resulting wind velocity W shall be determined for blade loads by means of,

$$W^2 = \left(\frac{2}{3} R * 2\pi n_r\right)^2 + V_0^2 \qquad (5.88)$$

where n_r is the rotor frequency of rotation in Hz with the primary generator connected, and R is the rotor radius.

V_0 is the nominal stall wind velocity defined as the lower of the two following wind velocities: nominal wind velocity V_{nom}, or 10-min average wind velocity, where just the whole blade is stalled with airflow parallel with the rotor shaft. The wind velocity is defined at hub height. The wind velocity corresponding to the 10-minute average, at which the stall just extends to the whole blade, can be determined by means of

$$V_{0,stall} = 2\pi n_r, Rtg\ (\alpha_{stall,tip} + \theta_{b,tip}) \qquad (5.89)$$

where $\alpha_{stall,tip}$ is angle of attack for the tip profile at max C_l and $\theta_{b,tip}$ is blade angle for tip.

As a rule, a linear line load distribution outwards along blade of $p_0 r/R$ (triangular load) can be assumed for the y' direction. It may be advantageous to replace this with an alternative distribution:

$$P(r) = \text{const} * \{(2\pi n_r r)^2 + V_0^2\} c(r) C_l(r) \qquad (5.90)$$

where the local relative wind velocity and the exact chord line distribution along the blade is included. The constant in the expression for $p(r)$ is adjusted so that the moment in the rotor centre has the same value as for the distribution $p_0 r/R$, namely $P_0 R^2/3$. In the case of fatigue calculations, loads are generally described by means of a constant average value overlaid with an accumulated load range distribution $F\Delta(N_v)$ (see Figure 5.27).

An accumulated load range distribution is described in terms of a distribution function $F\Delta(N_v)$. This function is defined in such a way that $F\Delta(N_v)$ is the load range which is exceeded N_v times in the service life of the wind turbine. The model used for this presupposes that the blade line loads are divided into two types. The first type (called *deterministic*) describes the effect of the gravitational force on the rotating blade. The corresponding load range distribution includes only load ranges of one constant size to a number, N_f, which is equal to the total number of rotor revolutions during production operation in the design service life of the wind turbine, t_f (see Figure 5.27, curve b).

The second type (called *stochastic*) includes all other loads, e.g. variable aerodynamic loads. This type of distribution is described with the help of a standardized load range distribution, with a load distribution $F\Delta(N_v)$, of this type being expressed by means of a dimensioning constant multiplied by the standardized distribution $^\wedge F\Delta(N_v)$ (see Figure 5.27, curve a).

The standardized load range distribution is defined by means of,

$$^\wedge F\Delta(N_v) = \beta \log (N_f) - \log (N_v)) + 0.18 \quad \text{until } F\Delta(N_v) = 2.0 k_\beta \qquad (5.91)$$

where

$$\beta = 0.11 k_\beta\ (I + 0.1)\ (A + 4.4) \qquad (5.92)$$

I = the turbulence intensity at hub height in the direction of the wind at the actual site
k_β = a correction factor for the slope of the distribution
N_f = number of widths during the lifetime of the turbine corresponding to the characteristic load frequency, given as

$$N_f = n_c t_f \left\{ \exp - \left(\frac{V_{min}}{A}\right)^k \right\} - \exp\left\{-\left(\frac{V_{max}}{A}\right)^k\right\} \quad (5.93)$$

where

A, k = scale parameter and shape parameter, respectively, for the Weibull distribution
n_c = the characteristic load frequency
t_f = the design service life of the wind turbine.

The correction factor k_β for the bias in the distribution is usually 1. Only for calculating the rotor pressure F_y (see Table 5.7) shall a standard range distribution with k_β = 2.5 be used. In connection with wind farms or groups, it is necessary to consider how far the park configuration affects the average wind velocity (and therefore the A parameter) and the turbulence intensity I. The formula for N_f, takes into account how often wind velocity is between V_{min} and V_{max}.

The standardized load range distribution is shown in Figure 5.26. The distributions, which are defined for different loads by means of this standardized load range distribution, are described as a constant times the standard distribution. There are three parameters which affect a given standard distribution: N_f is the total number of ranges in the service life of the wind turbine. A change in N_f will involve parallel displacement of the distribution along the frequency axis, β changes the slope of the load distribution, but the greatest loads remain unchanged, and k_β also changes the bias. In this case, it also increases the greatest loads. The dimensioning constant, which is multiplied by the standard distribution, is independent of N_v, but does depend on wind turbine geometry, climate etc. This means that if, for example, it is necessary to find range distributions as the integral along the blade of load range distributions, the integration variable is contained in the dimensioning constant, while the standard distribution $^\wedge F\Delta(N_v)$ acts as a constant as far as the integration is concerned.

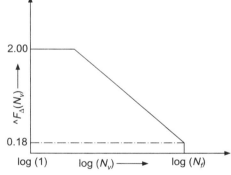

FIGURE 5.26 Standardized stochastic load range distribution.

When load range distributions are calculated for sums of loads of the two types, stochastic and deterministic, the following shall apply. Deterministic loads are compounded first, with the following dependency of time t being assumed (the blade vertically upwards, when $t = 0$):

$$P^d(t, P^d \Delta c, P^d \Delta s, n_c) = \bar{P} + 1/2 P^d \Delta c * \cos(2\pi n_c t) + 1/2 P^d \Delta s * \sin(2\pi n_c t) \quad (5.94)$$

When stochastic loads are compounded with deterministic or other stochastic loads, it is necessary to be conservative in the calculations, as the phase relationships for stochastic loads are not known. Load ranges from each of two distributions are added directly one at a time. The largest from each distribution is taken first, and then the smaller ones are taken in succession according to size. Each load range is included only once. Therefore, it is assumed that each load range in the first distribution occurs the same time as the largest still unused range from the second distribution.

Figure 5.27 shows the accumulated load distribution for p_x. Curve (a) is the stochastic range distribution, the number of ranges $N_a = 3N_r$. Curve (b) is the deterministic (gravity induced) distribution with constant range 2 mg and number of ranges $N_b = N_r$. The solid-line curve (c) is the sum distribution, as it can be shown that the conservative summation method specified above can be achieved by direct addition of the two accumulated distribution functions (a) and (b) for frequencies $< N_r$. (This is only correct for accumulated distributions.) For frequencies $> N_r$, the sum is equal to the stochastic one, because the deterministic ranges are 'used up'.

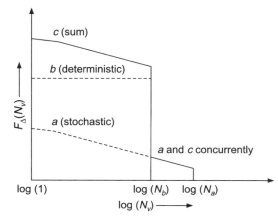

FIGURE 5.27 Compounding of stochastic and deterministic loads.

5.6.1 Blade Loads

Table 5.5 sums up the blade line loads used according to blade coordinates (Table 5.4). However, the deterministic contribution is divided into a cosine and a sine element, which must be used when compounding several deterministic loads. The last series in the table indicates the characteristic load frequency for the load type in question.

TABLE 5.4 Description of blade coordinates

Direction	Description
x'	Rotation direction and perpendicular to the main shaft
y'	Parallel to main shaft in wind direction
z'	Following the blade length perpendicular to the main shaft

$$m = m(r) = \text{blade mass per unit of length at distance } r$$
$$g = \text{gravitational acceleration}$$
$$M_{nom} = P_{nom}/2\pi n_r \eta_0 = \text{nominal driving torque corresponding to nominal power}$$
$$\eta_0 = \text{nominal efficiency } (\leq 0.9)$$
$$F_\Delta(N_v) = \text{standardized distribution, see above}$$
$$\Delta c, \Delta s = \text{indices for load ranges for cosine and sine elements, respectively}$$

TABLE 5.5 Blade line load

Direction	Average value	Load range distributions		
		Deterministic		Stochastic
	$\bar{P}$	$P^d \Delta c$	$P^d \Delta s$	$P^s \Delta$
x'	$2M_{\text{nom}}/(3R^2)$	0	$+2\,mg$	$0.3F_\Delta(N_v)P_o$
y'	$1.5P_o r/R$*	0	0	$F_\Delta(N_v)P_o r/R$
z'	$(2\pi n_r)^2 mr$	$-2\,mg$	0	0
Frequency n_c		n_r	n_r	$3n_r$

* $\bar{P} = P_o$ for $r \geq 2/3\,R$

Hub loads

When the internal loads on the rotor hub are calculated, allowance should be made for blade loads with reciprocal phase lag of 120°. For the purposes of the calculation, the hub is assumed to be rigidly fixed at the shaft attachment, where the rotor loads are crucial. It should be pointed out that the rotor loads are defined in the fixed x, y, z system (Table 5.6) and therefore must be converted to the hub system, with allowance being made also for dead weight.

TABLE 5.6 Description of blade coordinates

Direction	Description
x	Horizontal direction and perpendicular to the main shaft
y	Parallel to main shaft in wind direction
z	'Vertical' and perpendicular to the main shaft

5.6.2 Rotor Loads

The rotor loads are defined on the basis of the characteristic aerodynamic line load P_o, the standardized load range distribution form $F_\Delta(N_v)$ as well as the nominal driving torque M_{nom} defined in Table 5.5.

The average value, amplitude and frequency of rotor loads are given in Table 5.7, with the individual load components being separately compounded as

$$F = \bar{F} + \frac{1}{2} F_\Delta \cos(2\pi n_c t) \tag{5.95}$$

where t is time in seconds.

In Table 5.7 and Figure 5.28,

M = rotor mass

k_R = a correction factor which for M_x and M_z takes into account the position of relevant resonance frequencies in relation to the frequency content of the external influences. The k_R factor is read from Figure 5.28 as a function of the resonance frequency

n_o = lowest resonance frequency for the oscillation form in question

n_R = lowest resonance frequency for the collective asymmetric rotor oscillation at standstill, where one blade oscillates out of phase to the two others. The oscillation will link with nacelle/tower, for which reason its frequency in yawing and tilting, respectively, may be different

n_T = lowest resonance frequency for tower in bending ($< 2.5 n_r$).

TABLE 5.7 Rotor loads

Load F	Average F	Range F_Δ	Frequency n_c	Resonance frequency (oscillation from), n_o
F_x	0	0	0	
F_y	$1.5 P_o R$	$0.5 P_o R F_\Delta(N_v)$	Stator: n_T Rotor: $3 n_R$	n_T (tower, bending)
F_z	Mg	0		
M_x	0	$1/3 k_R P_o R^2 F_\Delta(N_v)$	$3 n_r$	n_R (rotor, tilt, asymmetry)
M_y**		$0.45 P_o R^2 F_\Delta(N_v)$	$3 n_r$	
M_z		$1/3 k_R P_o R^2 F_\Delta(N_v)$	$3 n_r$	n_R (rotor, yaw, asymmetry)

** It may be assumed that M_y is always between 0 and $1.3 M_{nom}$.

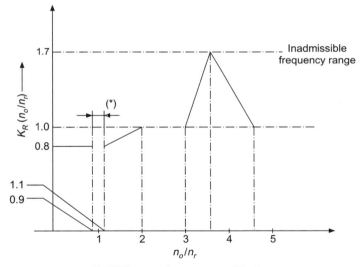

FIGURE 5.28 Response amplification.

It should be pointed out that, as far as gearbox calculations are concerned, it is necessary to use a duration curve for shaft moment. For example, it may be conservatively assumed here that the moment is constant at the upper limit indicated. Note that for F_y, it is necessary to use a modified standard distribution with the bias β multiplied by $k_\beta = 2.5$ as mentioned in the definition of the standard distribution above. The two frequencies which are indicated for F_y are used for F_y distributions in, respectively, the stationary ($n_c = n_T$) and the rotating part of the wind turbine ($n_c = 3 n_r$), thus taking into account that the oscillations are damped when it is transmitted from the rotating part to the stationary part. When the load range distributions are determined, it is necessary to be conservative in calculations when adding the individual components and distributions, as discussed in connection with Figure 5.28. In the calculations shown, the wind load on the tower can be assumed to have been included.

5.6.3 Codes and Standards

The IEC 61400-1 wind turbine generator systems–Part 1 safety requirements identifies four different classes of wind turbine to suit differing site wind conditions, with increasing class designation number corresponding to reducing wind speed. The wind speed parameters for each class are given in Table 5.8.

TABLE 5.8 Wind speed parameters for wind turbine classes

Parameters	Class I	Class II	Class III	Class IV
Reference wind speed, U_{ref} (m/s)	50	42.5	37.5	30
Annual average wind speed, U_{ave} (m/s)	10	8.5	7.5	6
50-year return gust speed, 1.4 U_{ref} (m/s)	70	59.5	52.5	42
1 year return gust speed, 1.05 U_{ref} (m/s)	52.5	44.6	39.4	31.5

The reference wind speed is defined as the 10 min mean wind speed at hub-height with a 50-year return period. To allow for other sites where conditions do not conform to any of these classes, a fifth class is provided for in which the basic wind parameters are to be specified by the manufacturer. The normal value of air density is specified as 1.225 kg/m³. A crucial parameter for wind turbine design is the turbulence intensity, which is defined as the ratio of the standard deviation of wind speed fluctuations to the mean. The standard specifies two levels of turbulence intensity, designated category A (higher) and category B (lower), which are independent of the wind speed classes above. In each case, the turbulence varies with hub height mean wind speed, $\bar{U}$, according to the formula,

$$I_U = I_{15} \frac{\left(a + \dfrac{15}{\bar{U}}\right)}{(a+1)} \tag{5.96}$$

where I_{15} is the turbulence intensity at a mean wind speed of 15 m/s, defined as 18 per cent for category A and 16 per cent for category B. The constant a takes the values 2 and 3 for categories A and B, respectively.

Germanischer Lloyd's Regulation for the classification of Wind Energy Conversion systems, commonly referred to as *GL rules*, specifies a single value of hub-height turbulence intensity of 20 per cent.

Danish Standard DS 472 bases the derivation of design-extreme wind speeds on four terrain classes, ranging from very smooth (expanse of water) to the very rough (e.g. built-up areas). The base wind velocity is taken to be the same all over Denmark, so the result is four alternative profiles of wind speed variation with height.

5.6.4 Design Loads

The sources of loading to be taken into account may be categorized as follows:

- Aerodynamic loads

- Gravitational loads
- Inertia loads (including centrifugal and gyroscopic effects)
- Operational loads arising from actions of the control system (e.g. braking, yawing, blade-pitch control, generator disconnection).

The load cases selected for ultimate load design must cover realistic combinations of a wide range of external wind conditions and machine states. It is a common practice to distinguish between normal and extreme wind conditions on the one hand, and between normal machine states and fault states, on the other. The load cases for design are then chosen from:

- Normal wind conditions in combination with normal machine states
- Normal wind conditions in combination with machine fault states
- Extreme wind conditions in combination with normal machine states.

Extreme and normal wind conditions are generally defined in terms of the worst condition occurring with a 50-year and 1 year return period, respectively. It is assumed that machine fault states arise only rarely and are uncorrelated with extreme wind conditions, so that the occurrence of a machine fault in combination with an extreme wind condition is an event with such a high return period that it need not be considered as a load case. However, IEC 61400-1 wisely stipulates that if there is some correlation between an extreme external condition and a fault state, then the combination should be considered as a design case.

A typical wind turbine is subjected to a severe fatigue loading regime. The rotor of a typical wind turbine will rotate some 2×10^8 times during a 20-year life, with each revolution causing a complete gravity stress reversal in the low speed shaft and in each blade, together with a cycle of blade out-of-plane loading due to the combined effects of wind shear, yaw error, shaft tilt, tower shadow and turbulence. It is therefore hardly surprising that the design of many wind turbine components is often governed by fatigue rather than by ultimate load.

The partial safety factors for ultimate loads stipulated by IEC 61400-1 the GL rules and DS 472 are given in Table 5.9.

TABLE 5.9 Partial safety factors for loads

Sources of loading	Unfavourable loads Types of loading						Favourable loads		
	Normal and extreme			Abnormal					
	IEC	GL	DS	IEC	GL and DS		IEC	GL	DS
		Normal	Extreme						
Aerodynamic	1.35	1.2	1.5	1.3	1.1	1.0	0.9	–	–
Operational	1.35	1.35	1.2	1.3	1.1	1.0	0.9	–	–
Gravity	1.1*	1.1*	1.1*	1.0	1.1	1.0	0.9	1.0	–
Inertia	1.25	1.1*	1.1*	1.0	1.1	1.0	0.9	1.0	–

*Factor increased to 1.35 if masses are not determined by weighing.

Normally, the critical operating parameters are:
- Turbine operational speed
- Power output

- Vibration level
- Twist of pendant cables running up into nacelle.

The fluctuation of the wind speed about the short-term mean, or turbulence, naturally has a major impact on the design loadings, as it is the source of both the extreme gust loading and a large part of the blade fatigue loading. The latter is exacerbated by the gust slicing effect in which a blade slices through a localized gust repeatedly in the course of several revolutions. Within a wind farm, turbines operating in the wake of other turbines experience increased turbulence and reduced mean velocities. In general, a downwind turbine will lie off-centre with respect to the wake of the turbine immediately upwind, leading to horizontal wind shear. The consensus is not emerged for the models which are describing velocity deficit and the increase in turbulence intensity due to turbine wake are yet to be used in design calculations.

A non-operational machine state is defined as one in which the machine is neither generating power, nor starting up, nor shutting down. It may be stationary, i.e. parked or idling. The design wind speed for this load case is commonly taken as the gust speed with a return period of 50 years. The magnitude of the 50-year return gust depends on the gust duration chosen, which, in turn, should be based on the size of the loaded area. IEC 61400-1 and the GL rules specify the use of gust durations of 3s and 5s, respectively, regardless of the turbine size.

5.7 COST MODELLING

The sensitivity of the cost of energy to changes in the values of parameters governing turbine design can be examined with the aid of a model of the way component costs vary in response. It is usually sufficiently accurate to represent the relationship between component cost and mass as a linear one with a fixed component:

$$C(x) = C_B \left(\mu \frac{m(x)}{m_B} + (1 - \mu) \right) \tag{5.97}$$

where $C(x)$ and $m(x)$ are the cost and mass of the component, respectively, when the design parameter takes the value x, and C_B and m_B are the baseline values; μ is the proportion of the cost that varies with mass, which will obviously differ for different baseline machine sizes.

5.7.1 Simplified Cost Model

The baseline machine design is taken as a 60 m diameter, 1.5 MW turbine, with the costs of the various components given as a percentage of the total. Machine designs for the other diameters are obtained by scalling all dimensions of all components in the same proportion, except in the case of gear box, generator, grid connection and controller. Rotational speed is kept inversely proportional to rotor diameter to maintain constant tip speed, and hence constant tip speed ratio at a given wind speed. As a result, all machine designs reach rated power at the same wind speed, so that rated power is proportional to diameter squared. Consequently, the low-speed shaft torque increases as diameter cubed, which is the basis for assuming that the gear box mass increases as the cube of rotor diameter, even though the gear box ratio changes. Hence, if a blanket value of μ of 0.9 is adopted for simplicity, the cost of all components, apart from generator, controller and grid connection, for a machine of diameter D, is given by:

$$C_1(D) = 0.8C_T(60)\left\{0.9\left(\frac{D}{60}\right)^3 + 0.1\right\} \quad (5.98)$$

where $C_T(60)$ is the total cost of the base line machine. The rating of the generator and the grid connection is proportional. Only to the diameter squared. It is assumed that Eq. (5.97) applies to the cost of these components, but with mass replaced by rating. Thus if μ is taken as 0.9 once more, the costs of generator and grid connection are given by:

$$C_2(D) = 0.158C_T(60)\left\{0.9\left(\frac{D}{60}\right)^2 + 0.1\right\} \quad (5.99)$$

The controller cost is assumed to be fixed. Hence, the resulting turbine cost as a function of diameter is:

$$C_T(D) = C_T(60)\left[0.8\left\{0.9\left(\frac{D}{60}\right)^3 + 0.1\right\} + 0.158\left\{0.9\left(\frac{D}{60}\right)^2 + 0.1\right\} + 0.042\right]$$

$$= C_T(60)\left\{0.72\left(\frac{D}{60}\right)^3 + 0.1422\left(\frac{D}{60}\right)^2 + 0.1378\right\} \quad (5.100)$$

As the tower height, along with all other dimensions is assumed to increase in proportion to rotor diameter, the annual mean wind speed at hub height will increase with rotor diameter because of wind shear. The energy yield should thus be calculated taking this effect into account. The cost of energy (excluding operation and maintenance costs) can then be calculated in dollar/kWh/annum by dividing the turbine cost by the annual energy yield. The variation of energy cost with diameter, calculated according to the assumptions described above, is plotted in Figure 5.29 for two levels of

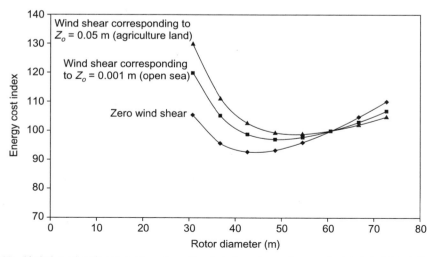

FIGURE 5.29 Variation of optimum turbine size with wind shear (assuming constant hub height of diameter ratio).

wind shear, corresponding to roughness lengths, Z_o of 0.001 m and 0.05 m, the hub-height mean wind speed being scaled according to the relation

$$\bar{U}(Z) \alpha \ln\left(\frac{Z}{Z_o}\right)$$

i.e. the wind shear is then given by a logarithmic wind profile, where Z is the height above ground and Z_o is the surface roughness length (tabulated in Table 5.10).

TABLE 5.10 Typical surface roughness lengths

Type of terrain	Roughness length $Z_o(m)$
Cities, forests	0.7
Suburbs, wooded countryside	0.3
Villages, countryside with trees and hedges	0.1
Open farmland, few trees and buildings	0.03
Flat grassy plains	0.01
Flat desert, rough sea	0.001

In Figure 5.29, it also included a plot for the case of zero wind shear. It is apparent that the level of wind shear has a noticeable effect on the optimum machine diameter, which varies from 44 m for zero wind shear to 52 m for the wind shear corresponding to a surface roughness length of 0.05 m, which is applicable to the farmland with boundary hedges and occasional buildings. Strictly, the impact of the increased annual mean wind speed with hub height on the fatigue design of the rotor and other components should also be taken into account, which would reduce the optimum machine size slightly.

It should be emphasized that the optimum sizes derived above depend critically on the value of μ adopted. For example, if μ were taken as 0.8 instead of 0.9, the optimum diameter in the absence of wind shear would increase to 54 m, although the minimum cost of energy would alter by only 0.3 per cent. A more sophisticated approach would allocate different values of μ to different components.

The cost model outlined above provides a straight forward means of investigating scale effects on machine economics for a chosen machine design. In practice, the use of different materials or different machine configurations may prove more economical at different machine sizes and will yield a series of alternative cost versus diameter curves.

5.8 POWER CONTROL

5.8.1 Passive Stall Control

The simplest form of power control is passive stall control, which makes use of the post-stall reduction in lift coefficient and associated increase in drag coefficient to place a ceiling on output power as wind speed increases, without the need for any changes in blade geometry. The fixed-blade pitch is chosen so that the turbine reaches its maximum or rated power at the desired wind speed. Stall-regulated machines suffer from the disadvantage of uncertainties of aerodynamic behaviour post-stall, which can result in inaccurate prediction of power levels and blade loadings at rated wind speed and above.

5.8.2 Active Pitch Control

Active pitch control achieves power limitations above rated wind speed by rotating all parts of each blade about its axis in the direction which reduces the angle of attack and hence the lift coefficient—a process known as *blade feathering* shown in Figure 5.30. The main benefits of active pitch control are increased energy capture, the aerodynamic braking facility it provides, and the reduced extreme loads on the turbine when shut-down.

FIGURE 5.30 Blade pitching system using a separate electric motor for each blade.

Although full-span pitch control is the option favoured by the overwhelming majority of manufacturers, power control can still be fully effective even if only the outer 15 per cent of the blade is pitched. The principal benefits are that the duty of the pitch actuator is significantly reduced, and that the inboard portion of the blade remains in stall, significantly reducing the blade load fluctuations. On the other hand, partial–span pitch control has several disadvantages as follows:

- The introduction of extra weight near the tip
- The difficulty of physically accommodating the actuator within the blade profile
- The high bending moment to be carried by the tip-blade shaft
- The need to design the equipment for the high centrifugal loadings found at large radii
- The difficulty of access for maintenance.

5.8.3 Passive Pitch Control

An attractive alternative to active control of the blade pitch to limit power is to design the blade and/or its hub mountings to twist under the action of loads. On the blades, in order to achieve the desired pitch changes at higher wind speeds. Unfortunately, the principle is easy to state but difficult to achieve in practice because of the required variation in blade load. In the case of stand-alone wind turbines, the optimization of energy yield is not the key objective, so passive pitch control is sometimes adopted, but the concept has not been utilized as yet for many grid connected machines.

5.8.4 Active Stall Control

Active stall control achieves power limitation above the rated wind speed by pitching the blades initially into stall, i.e. in the opposite direction to that employed for active pitch control, and is thus sometimes known as *negative pitch control*. At higher wind speeds, however, it is usually necessary to pitch the blades back towards feather in order to maintain power output at rated wind speed.

A significant advantage of active stall control is that the blade remains essentially stalled above the rated wind speed, so that gust slicing results in much smaller cyclic fluctuations in blade loads and power output. It is found that only small changes of pitch angle are required to maintain the power output at rated, so pitch rates do not need to be as large as for positive pitch control.

5.8.5 Yaw Control

As most horizontal-axis wind turbines employ a yaw drive mechanism to keep the turbine headed into the wind, the use of the same mechanism to yaw the turbine out of wind to limit power output is obviously an attractive alternative. However, there are two factors which militate against the rapid response of such a system to limit power : first, the large amount of inertia of the nacelle and rotor about the yaw axis and second, the cosine relationship between the component of wind speed perpendicular to the rotor disc and the yaw angle. The latter factor means that, at small initial yaw angles, yaw changes of, say, 10°C only bring about reductions in power of a few per cent, whereas blade pitch changes of this magnitude can easily halve the power output. Thus, active yaw control is only practicable for variable speed machines where the extra energy of a wind gust can be stored a rotor kinetic energy until the yaw drive has made the necessary yaw correction.

5.8.6 Teetering

Two-bladed rotors are often mounted on a teeter hinge—with hinge axis perpendicular to the shaft axis, but not necessarily perpendicular to the longitudinal axis of the blades—in order to prevent differential blade root out-of-plane bending moments arising during operation. Instead, differential aerodynamic loads on the two-blades result in rotor angular acceleration about the teeter axis, with large teeter excursions being prevented by the restoring moment generated by centrifugal forces.

5.9 BRAKING SYSTEMS

The normal practice is to provide both aerodynamic and mechanical braking. However, if independent aerodynamic braking systems are provided on each blade, and each has the capacity to decelerate the rotor after the worst-case grid loss, then the mechanical brake will not necessarily be designed to do this. The function of the mechanical brake in this case is solely to bring the rotor to rest, i.e. to park it, as aerodynamic braking is unable to do this.

5.9.1 Active Pitch Control Braking

Blade pitching to feather (i.e. to align the blade chord with the wind direction) provides a highly effective means of aerodynamic braking. Blade pitch rates of 10° per second are generally found adequate, and this is of the same order as the pitch rate required for power control. The utilization of the blade pitch system for start up and power control means that it is regularly exercised with the result that the existence of a dormant fault is highly unlikely. In machines relying solely on blade pitching for emergency braking, independent actuation of each blade is required, together with fail-safe operation should power or hydraulic supplies passing through a hallow low-speed shaft from the nacelle be interrupted. In the case of hydraulic actuators, oil at pressure is commonly stored in accumulators in the hub for this purpose.

5.9.2 Braking by Pitching Blade Tips

Blade tips which pitch to feather have become the standard form of aerodynamic braking for stall-regulated turbines. Typically, the blade tip is mounted on a tip shaft, as illustrated in Figure 5.31, and held in against centrifugal force during normal operation by a hydraulic cylinder.

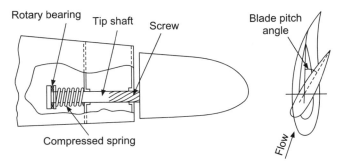

FIGURE 5.31 Passive control for tip blade, using screw on tip shaft and spring.

On release of the hydraulic pressure (which is triggered by a control system, or directly by an overspeed sensor, the tip blade flies outward under the action of centrifugal force, pitching to feather simultaneously on the shaft screw. The length of the tip blade is commonly some 15 per cent of the tip radius.

The penalty is that the low-speed shaft needs to be hollow to accommodate the feed to the hydraulic cylinder.

5.10 TURBINE BLADE DESIGN

A successful blade design must satisfy a wide range of objectives, some of which are in conflict. These objectives are:

1. Maximise annual energy yield for the specified wind speed distribution.
2. Limit maximum power output (in the case of stall regulated machines.
3. Resist extreme and fatigue loads.
4. Restrict tip deflections to avoid blade/tower collisions (in the case of upwind machines).
5. Avoid resonances.
6. Minimise weight and cost.

The design process can be divided into two stages: the aerodynamic design, in which objectives (1) and (2) are satisfied, and the structural design. The aerodynamic design addresses the selection of optimum geometry of the blade external surface—normally simply referred to as the blade geometry—which is defined by the aerofoil family and the chord, twist and thickness distributions. The structural design consists of blade material selection and the determination of a structural cross section or spar within the external envelope that meets objectives (3) to (6). Inevitably, there is interaction between the two stages, as the blade thickness needs to be large enough to accommodate a spar which is structurally efficient.

The aerodynamic design encompasses the selection of aerofoil family and optimization of the chord and twist distribution. The variation of thickness to chord ratio along the blade also has to be considered, but this ratio is usually set at the minimum value permitted by structural design considerations, as this minimizes drag losses.

The process for optimizing the blade design of machines operating at a fixed tip speed ratio is described with analytical expressions for the blade geometry parameter,

$$\sigma_r \lambda C_l = \frac{Bc(\mu)}{2\pi R} \lambda C_l$$

and the local inflow angle, ϕ, are derived as a function of the local tip speed ratio, $\lambda\mu = \lambda r/R$. If $\lambda\mu \gg 1$, the expressions can be approximated by

$$\sigma_r \lambda C_l = \frac{Bc(\mu)}{2\pi R} \lambda C_l = \frac{8}{9\lambda\mu}$$

and
$$\phi = \frac{2}{2\lambda\mu}$$

If it is decided to maintain the angle of attack, α, and hence the lift coefficient, C_L, constant along the blade, then these relations translate to

$$c(\mu) = \frac{16\pi R}{9 C_l B \lambda^2 \mu}$$

and
$$\beta = \frac{2}{3\lambda\mu} - \alpha$$

so that both the chord and twist are inversely proportional to radius.

The blade root area is normally circular in cross section in order to match up with the pitch bearing in the case of pitchable blades, or to allow pitch angle adjustment at the bolted flange (to compensate for non-standard air density) in the case of stall regulated blades.

5.10.1 Form of Blade Structure

A hollow shell corresponding to the defined blade envelope clearly provides a simple, efficient structure to resist flexural and torsional loads and some blade manufacturers adopt this form of construction (see Figures 5.32, 5.33 and 5.34). However, in the case of small and medium size machines, where the out-of-plane loads dominate, there is greater benefit in concentrating skin material in the forward half of the blade, where the blade thickness is maximum, so that it acts more efficiently in resisting out-of-plane bending moments. The weekened areas of the shell

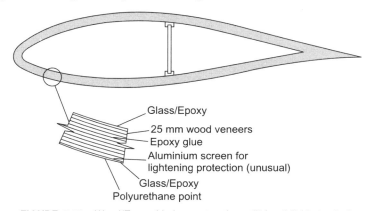

FIGURE 5.32 Wood/Epoxy blade construction utilizing full blade shell.

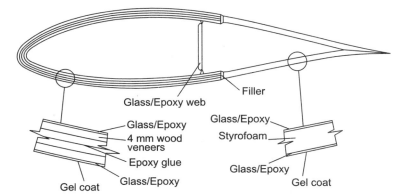

FIGURE 5.33 Wood/Epoxy blade construction utilizing forward half of blade shell.

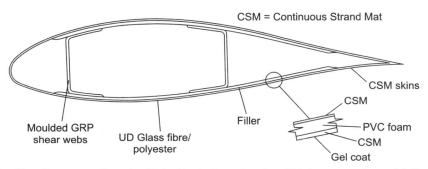

FIGURE 5.34 Glass-fibre construction using blade skins in forward portion of blade cross section and linking shear webs.

towards the trailing edge are then typically stiffened by means of sandwich construction utilizing a PVC foam filling.

5.10.2 Blade Materials

The ideal material for blade construction will combine the necessary properties—namely high strength to weight ratio, fatigue life and stiffness—with low cost and the ability to be formed into desired aerofoil shape.

Table 5.11 lists the structrual properties of the materials in general use for blade manufacture and those of some other candidate materials. For comparative purposes values are also presented of:

- Compressive strength-to-weight ratio
- Fatigue strength as a percentage of compressive strength
- Stiffness-to-weight ratio
- A panel stability parameter, $E/(UCS)^2$ (UCS is ultimate compressive strength).

It is evident that glass and carbon-fibre composites (GFRP and CFRP) have a substantially higher compressive strength-to-weight ratio compared with the other materials. Figure 5.35 shows a typical GFRP blade. However, this apparent advantage is not as decisive as it appears, for two reasons. First of all, the fibres of some of the plies making up the laminated blade shell have to be

TABLE 5.11 Structural properties of materials for wind-turbine blades

Material (NB: UD denotes unidirectional fibres – i.e., all fibres running longitudinally)	Ultimate tensile strength (UTS) (MPa)	Ultimate compressive strength (UCS) (MPa)	Specific gravity (s.g.)	Compressive strength to weight ratio UCS/s.g.	Mean fatigue strength at 10^7 cycles (amplitude) (MPa)	Mean fatigue strength as percentage of UCS	Young's modulus, E (GPa)	Stiffness to weight ratio E/s.g. (GPa)	Panel stability parameter $E/(UCS)^2$ $(MPa)^{-1}$
	(Mean for composites, minimum for metals)								
1. Glass/polyester ply with 50 per cent fibre volume fraction and UD lay-up	860–900 [1] [2]	~720 [1]	1.85	390	140 [3]	19%	38 [12]	20.5	0.07
2. Glass/epoxy ply with 50 per cent fibre volume fraction and UD lay-up	Properties are generally very close to those for GRP given above								
3. Glass/polyester laminate with 50 per cent fibre volume fraction and 80 per cent of fibres running longitudinally	690–720	~580	1.85	310	120	21%	33.5	18	0.1
4. Carbon fibre/epoxy ply with 60 per cent fibre volume fraction and UD lay-up	1830 [4]	1100 [4]	1.58	700	350 [5]	32%	142 [4]	90	0.12
5. Khaya ivorensis/ epoxy laminate	82 [6]	50 [6]	0.55	90	15 [7]	30%	10 [8]	18	4
6. Birch/epoxy laminate	117 [9]	81 [10]	0.67	121	16.5 [7]	20%	15 [10]	22.5	2.3
7. High yield steel (Grade Fe 510)	510	510	7.85	65	50 [11]	10%	210	27	0.81
8. Weldable aluminium alloy AA6082 (formerly H30)	295 [12]	295 [12]	2.71	109	17 [13]	6%	69 [12]	25.5	0.79

Wind Turbine Design   169

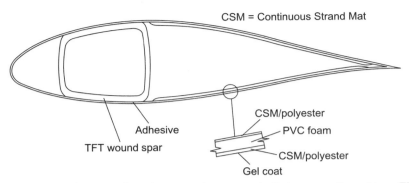

FIGURE 5.35 Glass-fibre blade construction using compact spar wound with transverse filament type (TFT) on mandrel.

aligned off-axis (typically at ± 45°) to resist shear loads, giving reduced strengths in the axial direction. Secondly, the relatively low Young's modulus of these composites means that resistance to buckling of the thin skins governs the design rather than simple compression yielding. The likelihood given in the last column of the blade, so that materials with high values, such as wood composites will be least sensitive to buckling. As a result, wood composite blades are generally lighter than equivalent glass-fibre composite blades. Table 5.12 lists different wind turbine blade materials with design strength-to-stiffness ratios.

TABLE 5.12 Design strength-to-stiffness ratios for different wind turbine blade materials

Material	Ultimate compression strength, σ_{cu} (MPa)	Partial safety factor for strength, γ_{mu}	Compression design strength, σ_{cd} (MPa)	Young's modulus, E (GPa)	Strength-to-stiffness ratio, $(\sigma_{cd}/E) \times 10^3$
Glass/polyester *laminate* with 50 per cent fibre volume fraction and 80 per cent UD	580	2.94	197 (ignoring) (buckling)	32.5	6.1
Carbon fibre/epoxy *ply* with 60 per cent fibre volume fraction and UD lay-up	1100	2.94	374	142	2.6
Khaya/epoxy laminate	50	1.5	33	10	3.3
Birch/epoxy laminate	81	1.5	54	15	3.6
	Yield strength, σ_y	γ_{my}			
High-yield steel (Grade Fe 510)	355	1.1	323	210	1.54
Weldable aluminium alloy AA6082	240	1.1	218	69	3.2

The properties of glass/polyester and glass/epoxy plies with the same fibre volume fraction and lay-up are generally very similar, i.e. the influence of the matrix is slight. They will therefore be treated as the same material except in relation to fatigue, where some differences have been noted. The glass used in blade construction is E-glass, which has good structure properties in relation to its cost.

The plate elements forming the spar of a GFRP blade are normally laminates consisting of several plies, with fibres in different orientations to resist the design loads. Within a ply (typically 0.5–1.0 mm in thickness), the fibres may all be arranged in the same direction, i.e. UD (unidirectional) or they may run in two directions at right angles in a wide variety of woven or non-woven fabrics. Although the strength and stiffness properties of the fibres and matrix are well-defined, only some of the properties of a ply can be derived from them using simple rules. Thus, for a ply reinforced by UD fibres, the longitudinal stiffness modulus E_1, can be derived accurately from the rule of mixtures formula

$$E_1 = E_f V_f + E_m (1 - V_f) \quad (5.101)$$

where E_f is the fibre modulus (72.3 GPa for E-Glass), E_m is the matrix modulus (in the range 2.7–3.4 GPa), and V_f is the fibre volume fraction. On the other hand, the inverse form of this formula, e.g.,

$$\frac{1}{E_2} = \frac{(1 - V_f)}{E_m} + \frac{V_f}{E_f} \quad (5.102)$$

significantly underestimates the transverse modulus, E_2 and the in-plane shear modulus, G_{12}. The longitudinal tensile strength of a ply reinforced by UD fibres, σ_{lt}, can be estimated from:

$$\sigma_{lt} = \sigma_{fu} \left[V_f + \frac{E_m}{E_f} (1 - V_f) \right] \quad (5.103)$$

where σ_{fu} is the ultimate tensile strength of the fibres. However, the tensile strengths of E-glass single fibres (3.45 GPa) cannot be realized in a composite, where fibre strength reductions of upto 50 per cent have been measured. Accordingly, a value of σ_{fu} of 1750 MPa should be used in Eq. (5.103).

5.10.3 SERI Blade Sections

The design requirements for airfoil sections of a wind machine are different from those of airplanes.

In wind machines, fouling is a major problem because once the machine goes on line, it has to run round the day and round the year preferably without any break. Also, the blades of the wind machine are inaccessible once installed and cannot be easily cleaned.

Further, dirt deposited on the blades affects the performance of the blade and, in turn, that of the machine significantly as shown in Figure 5.36.

Another consideration is the thickness of the blade at root. Typical NACA sections, common for airplanes, used for wind machine are shown in Figure 5.37. They have high C_L and high (C_L/C_D) ratio. But they are comparatively thinner. Structurally, a thick blade at root will always be easier to design to take loads.

Based on these considerations, blades have been designed by National Renewable Energy Laboratory (NREL) in USA. A set of these blades used for tip, middle span and root is shown in Figure 5.38.

The comparative performance of wind machine with NACA and SERI blades when they are clean and fouled (dirty) is shown in Figure 5.39 and Figure 5.40.

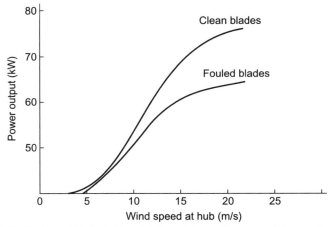

FIGURE 5.36 Effect of dirt on blade and the performance of the machine.

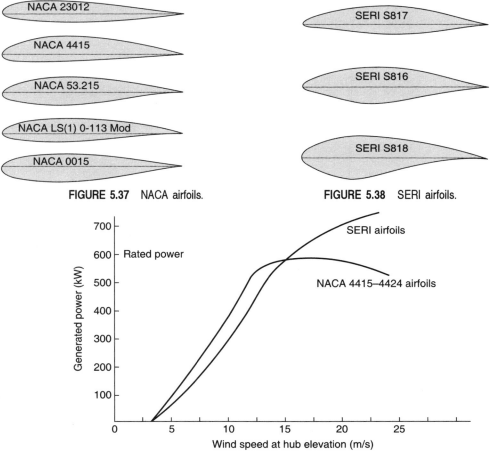

FIGURE 5.37 NACA airfoils.

FIGURE 5.38 SERI airfoils.

FIGURE 5.39 Comparative performance of wind machine with clean NACA and SERI blades.

If the aerofoil is sensitive to roughness, good performance is lost if the wings are contaminated by dust, rain particles or insects. On a wind turbine, this could alter the performance with time if, for example, turbine is sited in an area with many insects. If a wind turbine is situated near the coast, salt might built-up on the blades, if the winds comes from the sea and, if the aerofoil used are sensitive to roughness, the power output from the turbine will become dependent on the direction of the wind.

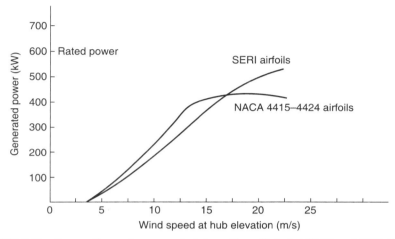

FIGURE 5.40 Comparative performance of wind machine with fouled NACA and SERI blades.

5.11 ROTOR HUB

The relatively complex three-dimensional geometry of rotor hubs favours the use of casting in their manufacture, with spheroidal graphite iron being the material generally chosen. Two distinct shapes of hub for three-bladed machines can be identified: tri-cylindrical or spherical The former consists of three cylindrical shells concentric with the blade axes, which flare into each other where they meet, while the latter consists simply of a spherical shell with cut-outs at the three-blade mounting positions. Diagrams of both types are shown in Figure 5.41, while an actual spherical hub is illustrated in Figure 5.42.

The structural actions of the hub in resisting three loadings are as follows:

1. *Symmetrical rotor thrust loading:*   The blade root bending moments due to symmetric rotor thrust loading put the front of the hub in bi-axial tension near the rotor axis and the rear in bi-axial compression, while the thrust itself generates out-of-plane bending stresses in the hub shell adjacent to low speed shaft flange connection.

2. *Thrust loading in single blade:*   This generates out-of-plane bending stresses in the hub shell at the rear, and in-plane tensile stresses around a curved load path between the upwind side of the blade bearing and the portion of the low-speed shaft flange connection remote from the blade. The resultant lateral loads will result in out-of-plane bending.

3. *Blade gravity moments:*   On the tri-cylindrical hub, equal and opposite blade gravity moments are communicated via the cylindrical shells to areas near the rotor axis at front

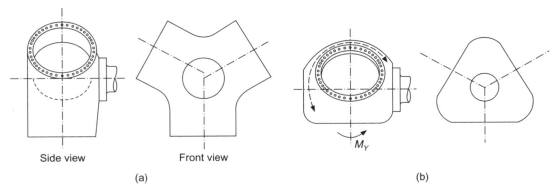

FIGURE 5.41 (a) Tri-cylindrical hub, (b) Spherical hub.

FIGURE 5.42 Rotor hub, view of spherical-shaped rotor hub.

and rear where they cancel each other out. It is less straight forward to visualize the corresponding load paths on the spherical hub, as out-of-plane bending is likely to be mobilized.

The complexity of the stress states arising from the latter two types of loading renders finite-element analysis of rotor hubs more or less mandatory. At the most, six load cases need to be analyzed, corresponding to the separate application of moments about the three axes and forces along the three axes at a single hub/blade interface. Then the distribution of hub stresses due to combination of loadings on different blades can be obtained by superposition. Similarly, the fluctuation of hub stresses over time can be derived by inputting the time histories of the blade loads obtained from a wind simulation.

The critical stresses for hub design are the in-plane stresses at the inner or outer surface, where they reach a maximum because of shell bending. For any one location on the hub, these are defined by three quantities at each surface: the in-plane direct stresses in two directions at right angles, and the in-plane shear stress. In general, these stresses will not vary in-phase with each other over time, so the principal stress directions will change, complicating the fatigue assessment. The following methods cater for one or more series of repeated stress cycles rather than the random stress fluctuations resulting from turbulent loading:

1. *Maximum shear method:* The fatigue evaluation is based on the maximum shear stress ranges, calculated from the $(\sigma_1 - \sigma_2)/2$, $\sigma_1/2$ or $\sigma_2/2$ time histories. The effect of mean stress is allowed for using the Goodman relationship:

$$\frac{\tau_a}{S_{SN}} + \frac{\tau_m}{S_{SU}} = \frac{1}{\gamma} \tag{5.104}$$

where τ_a is the alternating shear stress, τ_m is the mean shear stress, S_{SN} is the alternating shear stress for N loading cycles from the material S–N curve, S_{SU} is the ultimate shear strength, and γ is the safety factor. Having used Eq. (5.104) to determine S_{SN}, the permitted number of cycles for this loading range can be derived from the S–N curve, enabling the corresponding fatigue damage to be calculated.

2. *ASME boiler and pressure vessel code method:* This is similar to the maximum shear method, but the shear stress ranges are based on national principal stresses calculated from the changes in the values of σ_x, σ_y, σ_z, τ_{xy}, τ_{yz} and τ_{zx} from datum values occurring at one of the extremes of the stress cycle. Mean stress effects are not included.

3. *Distortion energy method:* The fatigue evaluation is based on the fluctuations of the effective or Von Mises stress. In the case of hub shell, the stress perpendicular to the hub surface (and hence the third principal stress) is zero, so the effective stress is given by:

$$\sigma' = \sqrt{\frac{(\sigma_1 - \sigma_2)^2 + \sigma_1^2 + \sigma_2^2}{2}} \tag{5.105}$$

As the effective stress is based on the distortion energy, it is a scalar quantity, so it needs to be assigned a sign corresponding to that of the dominant principal stress. The effect of mean stress is allowed for in the same way as for the maximum shear method, except that the stresses in Eq. (5.104) are now direct stresses instead of shear stresses.

REVIEW QUESTIONS

1. A wind machine with 60 m diameter rotor, cut-in wind speed, V_c, of 3 m/s and furling wind speed, V_F of 25 m/s is available with a wind machine manufacturing organization. The machine is to be marketed in the region with wind speed probability distribution at 20 m height given by Weibull parameters $k = 2.2$ and $c = 5$ m/s. The machine hub height is 60 m. The rated wind speed, V_R is to be optimized for this region and accordingly, the capacity of the generator is to be selected. The rotational speed of the wind machine is designed as 15 rpm and gear ratio of 100 resulting into typical efficiencies, η_{mech} and η_{elect} as 0.94 and 0.96, respectively. Assume air density of 1.2 kg/cu.m. The C_p–λ curve for the machine is given as,

$$C_p = 0.003\lambda^3 + 0.015\lambda^2 + 0.075\lambda$$

   (a) Find k and c for 60 m height, if

$$\left(\frac{c_2}{c_1}\right) = \left(\frac{Z_2}{Z_1}\right)^n \text{ and } \left(\frac{k_2}{k_1}\right) = \frac{1 - 0.0881 \ln\left(\frac{Z_1}{10}\right)}{1 - 0.0881 \ln\left(\frac{Z_2}{10}\right)}$$

where

$$n = \frac{0.37 - 0.0881 \ln(c_1)}{1 - 0.0881 \ln\left(\frac{Z_1}{10}\right)}$$

(b) Find an expression for $C_{p/Rated}$ as function of V_R.
(c) Calculate capacity factor CF for different values of V_R using the following equation:

$$CF = \frac{\exp(-(V_c/c)^k) - \exp(-(V_R/c)^k)}{(V_R/c)^k - (V_c/c)^k} - \exp(-(V_F/c)^k)$$

(d) Calculate P_N for different values of V_R using the above values. Find V_R for maximum P_N.
(e) Find rated electrical power output capacity, P_{eR} of the generator to be installed.
(f) Find its Annual Energy Output (AEO).
(g) Find the number of hours it will run in a year.
(h) Calculate energy content of the wind at 60 m height and compare it with AEO.

2. A wind machine has 3 linearly tapering blades. The radius at blade root is 0.75 m and chord at this point is 1.2 m. Radius and chord at tip of the blade are 15 m and 0.4 m, respectively.
   (a) Calculate the solidity of the rotor.
   (b) If each blade weighs 1.5 tons, then find linear density, ρ_{bl} of the blade. Assume uniform distribution of the mass along the length of the blade.
   (c) Derive an expression for centrifugal force of the blade element if the rotational speed of rotor is ω. Integrate it to yield an equation for centrifugal force acting at the root.
   (d) Calculate the centrifugal force acting at the blade root if the machine rotates at constant rotational speed of 42 rpm. Assume that the coning angle is zero.
   (e) If the machine develops 225 kW, calculate the driving torque on the rotor. Assume $\eta_{mech} = 0.95$ and $\eta_{elect} = 0.93$.
   (f) Draw velocity diagram at the blade cross section at 75 per cent radius and show, α, β and ϕ. Calculate ϕ.
   (g) Calculate the thrust on each blade if $V_f = 25$ m/s, $C_l = 0.5$ and $C_d = 0.01$ at furling all along the blade and that the aerodynamic configuration at 75 per cent radius represents the blade. Annual average air density = 1.155 kg/cu.m.
   (h) If the blades were free coning, what would be the coning angle at furling? Neglect changes in the values of thrust and centrifugal force due to change in angles.
   (i) If the wind machine is mounted upwind on a cylindrical steel tower of 30 m height and 0.5 m diameter midway, then what should be the overhang shaft length measured from the centreline of the tower if the clearance of 0.2 m has to be maintained between the blade tip and tower at furling?

3. λ_r, a and a' (using the usual notations) can be calculated from application of wake analysis. These values are given in Table 5.13. If $C_l = 2\pi a$ with α in radians for the blade airfoil and if the blade is to be designed for $C_l = 0.7$ for a 3-bladed propeller machine with rated wind speed of 8 m/s and rotor speed of 38.18 rpm, then based on blade element theory, calculate:
   (a) chord at radius of 4 m, 8 m and 12 m
   (b) blade angle at radius of 4 m, 8 m and 12 m.

Take air density = 1.165 kg/m³ and $C_d/C_l = 0.01$.

TABLE 5.13 Values of λ_r, a and a'

λ_r	a	a'
0.5	0.298	0.543
1.0	0.317	0.183
2.0	0.328	0.052
3.0	0.331	0.024
4.0	0.332	0.014
5.0	0.332	0.009
6.0	0.333	0.006
7.0	0.333	0.004
8.0	0.333	0.004
9.0	0.333	0.003
10.0	0.333	0.002
11.0	0.333	0.002
12.0	0.333	0.002

4. A horizontal axis wind machine with 30 m diameter rotor has 3 blades with blade-root at 0.5 m radius from the axis. At root, its chord is 0.75 m and tip the chord is 0.5 m. The lift coefficient, C_l for the blade airfoil can be taken as, $C_l = 2\pi\alpha$, where α is the angle of attack on the blade airfoil (radians) and ε for the blade airfoil can be taken as $\varepsilon = C_d/C_l = 1/200$. The rotor is designed to rotate at constant speed of 47 rpm. At the tip, the blade makes 4° w.r.t. the rotor plane.

   (a) Define different parameters of the rotor and write down the governing equations if the rotor is to be analysed based on the 'Blade Element Theory'.
   (b) Write down the equation for C_p.
   (c) Write down the procedure to solve the equations.
   (d) Is the given data sufficient to solve the equations? If not, then what additional information do you require?

5. Darrieus Machine of diameter, $D = 17$ m and height, $H = 17$ m has troposkien blades. If the swept area for this machine is given by $A_s = 2D^2/3$, then find power of the machine, if undisturbed wind speed = 10 mis, air density = 1.165 kg/m³ and $C_p = 0.3$.

6. The aerodynamic performance of this machine can be predicted on the basis of blade element momentum method. The machine is designed to have 60 m diameter and 13.37 rpm rotor speed. The root diameter is 2 m, number of blades is 3, and air density at the site is 1.16 kg/m³.

   (a) Explain the procedure followed in BEM method to calculate a, a' and ϕ.
   (b) Find $(1-a)$, a' and ϕ at 65.5 per cent of the blade length at rated wind speed. The design strategy is that the product of chord and lift coefficient is maintained at 0.32 m all along the blade using the following approximation for $\lambda > 2$.

For values of $\lambda > 2$, the relations between a, a' and λ_r can be given by

$$(1-a) = \frac{(1-A\lambda_r)^2}{(1-A\lambda_r)^2 + A\lambda_r}$$

$$a' = \frac{A}{\lambda_r - A}$$

with A = (No. of blades × Chord × Lift coefficient × Rotor speed)/(8π wind speed). ϕ can be calculated using trigonometric relation for velocity triangle at blade aerofoil section.

(c) Find the torque and thrust developed on the above rotor at rated wind speed assuming that the conditions at 65.5 per cent blade length represent the aerodynamic loading on the blade. The drag-to-lift ratio for the blade cross section is 0.007.

(d) The mechanical and electrical efficiencies are 0.95 and 0.97 at rated wind speed. Find the power developed by the above machine.

(e) The weight of each blade is 3 tons. Its weight per unit length is inversely proportional to its radius. Further, its weight per unit length at root is 4 times that at tip. Calculate the centrifugal force acting on the blades discussed in the above questions. Calculate the point of action of this force on the blade.

(f) Find the coning angle of the blade if moments due to thrust load and moments due to centrifugal force should balance each other at the rated wind speed.

7. A wind machine rotating at 17 rpm has 2 linearly tapering blades. Radius at root is 1.5 m and at tip of the blade is 47.3. The blade solidity is 0.01.

   (a) If each blade weighs 21 tons, then calculate the centrifugal force acting at the blade root. Assume that the coning angle is zero.

   (b) If the machine develops 1.2 MW, calculate the driving torque on the rotor. Assume $\eta_{mech} = 0.95$ and $\eta_{elect} = 0.93$.

   (c) Draw velocity diagram at the blade cross section at 66 per cent radius and show α, β and ϕ. Calculate ϕ.

   (d) Calculate the thrust on each blade if $V_{furling}$ = 42 m/s, C_l = 0.7 and C_d = 0.01 at furling all along the blade and that the aerodynamic configuration at 66 per cent radius represents the blade. Annual average air density = 1.12 kg/cu.m.

8. λ_r, a and a' (using the usual notations) can be calculated from application of wake analysis. These values are given in Table 5.14 for a propeller machine with rated wind speed of 8 m/s and rotor speed of 38.18 rpm. Then, based on wake analysis for actuator disc, with air density = 1.165 kg/m³, machine diameter = 30 m, hub diameter = 2 m and taking 2 m wide strips of the rotor disc from hub to tip for numerical calculations, calculate:

   (a) the thrust acting on the rotor at rated wind speed.
   (b) the driving torque on the rotor at rated speed.
   (c) The power developed by the rotor at rated wind speed.

TABLE 5.14 Values of λ_r, a and a'

λ_r	a	a'
0.5	0.298	0.543
1.0	0.317	0.183
2.0	0.328	0.052
3.0	0.331	0.024
4.0	0.332	0.014
5.0	0.332	0.009
6.0	0.333	0.006
7.0	0.333	0.004
8.0	0.333	0.004
9.0	0.333	0.003
10.0	0.333	0.002
11.0	0.333	0.002
12.0	0.333	0.002

9. A wind machine with 60 m diameter and 1 m hub radius has 3 blades weighing 10 tons each. The machine rotates at 17 rpm The blade mass per unit length is given as, $\rho_b = A/r$, where r is the radius and A is a constant. The blades are coned at an angle of 3°.
   (a) Find the value of A.
   (b) Calculate the centrifugal force acting on the blades as a function of r and on the hub.
   (c) Calculate the bending moment due to the centrifugal force acting on the hub.
   (d) If the coning effectively nullifies bending moment at the blade root due to thrust at rated speed, then find the thrust acting on the rotor. Assume that the thrust may be considered to be acting at 20 m radius at rated speed.
   (e) Calculate maximum driving torque acting on the rotor, if $P_{eR} = 1.2$ MW, $\eta_m = 0.95$ and $\eta_e = 0.97$.

10. A wind machine rotating at 17 rpm has 2 linearly tapering blades with the ratio of blade weight per unit length at root to that at tip as 5 : 1. Radius at blade root is 0.85 m and at tip of the blade is 47.3 m.
    (a) If each blade weighs 12 tons, then calculate the centrifugal force acting at the blade root. Assume that the coning angle is zero.
    (b) If the machine has $C_p = 0.378$ at 11.5 m/s, calculate driving torque on the rotor and the power developed at generator. Assume $\eta_{mech} = 0.95$ and $\eta_{elect} = 0.93$.
    (c) Draw velocity diagram at the blade cross-section at 66 per cent radius and show α, β and ϕ. Calculate ϕ with ideal values of a and a'.
    (d) Calculate thrust on each blade if $V_{furling} = 42$ m/s, $C_l = 0.7$ and $C_d = 0.01$ at furling at 66 per cent radius and that the aerodynamic configuration at 66 per cent radius represents the blade. Annual average air density = 1.12 kg/cu.m.
    (e) Calculate moment at the blade root due to thrust at furling speed if the thrust force can be considered to be acting at 66 per cent radius.

11. Define power coefficient, C_P and tip speed ratio λ, for a horizontal axis wind rotor of area, A in a wind speed of U.

(a) At an element of the blade of the rotor, the relative velocity of the wind to the blade undergoes an axial velocity reduction by a factor a and a transverse velocity increase by a factor a'. Explain the origins of these factors and state the conditions under which C_P attains its maximum value.

(b) With reference to a large wind turbine suitable for use in a wind farm, explain why aerofoil sections having very high lift/drag ratio are necessary.

(c) Describe, with the use of sketches, the general arrangement of gear box, electrical machinery and controls in the nacelle of a large wind turbine.

12. What are cut-in, rated and cut-out wind speeds with reference to power performance characteristics of a wind turbine? How it is chosen for optimal design of wind turbine?

13. How the strip theory is used for design of rotor blades of HAWT?

14. List all the design tasks involved in the design of a modern wind turbine.

15. For one blade 25 kW, 10 m diameter, pitch 0° and blade is divided into ten stations with wind speed 10 m/s and tip speed ratio $X = 6.11$ following sample output table is given as design example:

Blade element data for delta beta = 0.00

$$X = 6.11, \text{yaw} = 0.00$$

Refer Table 5.15, Figures 5.43, 5.44, 5.45, and 5.46.

TABLE 5.15 Design parameters of blade

Element	1	2	3	4	5	6	7	8	9	10
Theta	180	180	180	180	180	180	180	180	180	180
Vel	1.00	1.00	1.00	1.00	1.00	1.00	1.00	1.00	1.00	1.00
A	0.296	0.140	0.188	0.204	0.230	0.213	0.195	0.206	0.231	0.39
AP	0.073	0.021	0.016	0.012	0.010	0.008	0.006	0.006	0.006	0.39
C_l	0.813	1.005	1.160	1.206	1.334	1.311	1.168	1.037	0.918	0.770
C_d	0.014	0.098	0.053	0.043	0.020	0.019	0.016	0.014	0.013	0.010
Phi	49.92	42.48	27.54	20.14	15.45	13.03	11.35	9.74	8.34	6.71
ANG	7.92	19.18	15.74	14.84	13.35	12.93	11.35	9.74	8.34	6.71
TC	0.384	0.526	0.622	0.656	0.707	0.665	0.609	0.610	0.609	0.54
QC	0.040	0.059	0.073	0.075	0.083	0.079	0.074	0.073	0.0699	0.03
PC	0.243	0.363	0.443	0.459	0.508	0.485	0.453	0.443	0.421	0.34
TD, lb/ft	2.64	6.03	11.90	17.57	24.37	28.01	30.31	35.04	329.6	41.6
QD, ft-lb/ft	4.38	10.92	22.21	32.26	45.86	53.47	59.02	66.73	71.85	65.5
PD, kW	0.024	0.298	0.606	0.880	1.251	1.458	1.610	1.820	1.959	1.78
Rey, *10^6	0.92	0.862	0.922	0.931	0.910	0.868	0.890	1.004	1.132	1.13
Rotor 2 blades	pitch	X	TC	QC	PC	VO m/s	TD lb	MD ft-lb	QD ft-lb	PD kW
	0.0	6.1	0.614	0.070	0.427	10.0	752	3984	1372	23.3

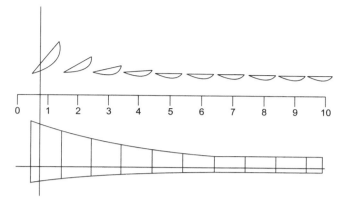

FIGURE 5.43 Variation of blade geometry along the length.

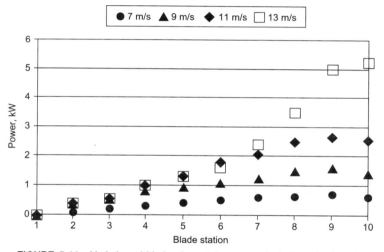

FIGURE 5.44 Variation of blade elemental power output along the length.

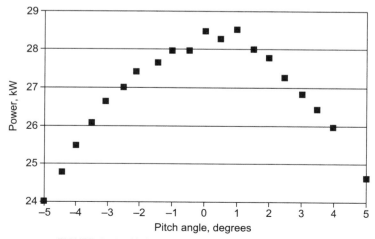

FIGURE 5.45 Variation of power output vs. pitch angle.

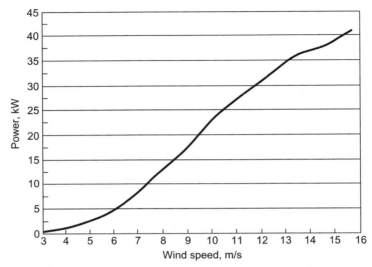

FIGURE 5.46 Power performance characteristics with varying wind speed.

Twist and plan form for 25 kW wind turbine blade. Blade is divided into ten sections for analysis, and the station is at the mid-point of the section.

Prediction of power output for one blade station for four wind speeds, tip speed ratio = 6.1.

Prediction of rotor power output for different pitch angles at 10 m/s. The turbine is a fixed pitch, constant-rpm machine.

Theoretical power curve for 25 kW wind turbine, tip speed ratio = 6.1.

16. A blade is 12 m long, weight is 500 kg, and the centre of mass is at 5 m. What is the torque if the force is 320 Nm?

17. Find the power loss for three struts on a HAWT. Struts are 4 m long, 2.5 cm in diameter. Rotor speed is 180 revolutions per minute. Use numerical approximation by dividing strut into 1 m sections and calculate at midpoint of section. Then add the value for each section, $C_d = 1$.

18. Calculate the power loss for the struts on VAWT. Centre tube, torque tube, diameter = 0.5 m. Struts are at top and bottom, 2 m long from torque tube to blades, diameter = 5 cm, rotor speed is 80 rpm. $C_d = 1$.

19. Find the value of u (speed of drag device) that produces the maximum C_P for a drag device. Use equation $F/A = 0.5\rho V^2 C_D$, wind speed is V and V_0 is the wind speed at infinity.

20. Aerodynamic efficiency can be maintained for different solidities of the rotor. If solidity increases, will you increase or decrease the tip speed ratio?

21. Explain the difference in performance of a wind turbine if it
    (a) Operates at a constant tip speed ratio
    (b) Operates at constant rpm.

22. What is the maximum theoretical efficiency for a wind turbine? What general principles were used to calculate this number?
23. If the solidity of the rotor is very small, for example, a one-bladed rotor, what is the value of the rpm for maximum C_P compared to the same size rotor with higher solidity?
24. Calculate the value of axial interference factor for which C_P is maximum for a lift device. Then show that this gives a maximum $C_P = 59$ per cent. Find the value of α that produces the maximum C_P by plotting the curve $\{P/A = 0.5\rho V_0^2\, 4\alpha(1-\alpha)^2\}$ for different values of α.
25. A rotor reaches maximum C_P at a tip speed ratio of 7. Calculate rotor rpm for four different wind turbines (diameters 5, 10, 50 and 100 m) at wind speeds of 10, 20 and 30 m/s.
26. A wind turbine that operates at constant rpm will reach maximum efficiency at only one wind speed. What wind speed should be chosen?

For Questions 27–33, specifications for a wind turbine are induction generator (rpm = 65), fixed-pitch, rated power = 300 kW, hub height = 50 m, rated wind speed = 18 m/s, tower head weight = 3,091 kg; rotor : three blades, mass of one blade = 500 kg, hub radius = 1.5 m, rotor radius = 12 m.

27. How fast is the tip of the blade moving?
28. How fast is the blade root (at hub radius) moving?
29. Put the mass at the midpoint and calculate the kinetic energy for one blade. Assume the mass of the blade is distributed evenly over ten sections. What is the kinetic energy for one blade?
30. At rated wind speed, calculate the torque.
31. At 10 m/s, what is the thrust (force) on the rotor trying to tip the unit over? Calculate for that wind speed over whole swept area.
32. If the unit produced 800,000 kWh/year, calculate output per rotor swept area.
33. Calculate the annual output per weight on top of tower, kWh/kg.

For Questions 34–41, specifications for a wind turbine are induction generator (rpm = 21), variable-pitch, rated power = 1,000 kW, hub height = 60 m, rated wind speed = 13 m/s, tower head weight = 20,000 kg, rotor: three blades, mass of one blade = 3,000 kg, hub radius = 1.5 m, rotor radius = 30 m.

34. How fast is the tip of the blade moving?
35. How fast is the blade root (at hub radius) moving?
36. Place the mass at the midpoint of the blade and calculate the kinetic energy for one blade. Assume the mass of the blade is distributed evenly over ten sections. Now, what is the kinetic energy for one blade?
37. Calculate the torque at the rated wind speed.
38. At 15 m/s, what is the thrust (force) on the rotor trying to tip the unit over? Calculate for that wind speed over the whole swept area.

39. If the unit produces 2,800,000 kWh/year, calculate the specific output, annual kWh/rotor area.
40. Calculate the annual output weight on top of tower, kWh/kg.
41. For a 12 m blade, centre of mass at 5 m, weight = 500 kg, calculate the angular momentum if the rotor is operating at 60 rpm.
42. For the blade in Question 41, the angular momentum is around 8×10^4 kg m²/s. Calculate the torque on the blade at that point if the angular momentum of the rotor is stopped in 5 s. Use $T = \Delta L/\Delta t$, where T = torque, ΔL = change of angular momentum and Δt = change in time. Then estimate the force trying to bend the blade.
43. Why are the blades for large wind turbines made from fibreglass-reinforced plastics?
44. Why are yaw rates limited on large wind turbines or yaw dampers installed on small wind turbines?
45. How does furling work on small wind turbines?
46. For loss of load on small wind turbines connected to the utility grid, how long can it take for overspeed shutdown?
47. For megawatt-size wind turbines, what is the most common configuration?
48. From internet search for blade design, make a paper model of the blade to see the principle for passive control.
49. List two methods with brief description of non-destructive testing of wind turbine blades.
50. By approximately what per cent will bugs on blades reduce the power?

Chapter 6

Siting, Wind Farm Design

6.1 INTRODUCTION

Looking at nature itself is usually an excellent guide to finding a suitable wind turbine site. If there are trees and shrubs in the area, it gives a good clue about the prevailing wind direction. Meteorology data, ideally in terms of a wind rose calculated over a period of 30 years, is probably the best guide, but these are rarely collected at the site, and here are many reasons to be careful about the use of meteorological data. If there are already wind turbines in the area, their production gives a guide to the local wind conditions. It is desirable to have as wide and open look as possible in the prevailing wind direction, and few obstacles and as low a roughness as possible in the same wind direction. A rounded hill is very good place for establishing a wind farm. Obviously, large wind turbines have to be connected to an electrical grid. For smaller projects, it is therefore essential to be reasonably close to a 10–30 kV power line if the costs of extending the electrical grid are not excessively high. Both the feasibility of building the foundation of the turbines, and road construction to reach the site with heavy trucks must be taken into account with any wind project.

The major goal in siting wind turbines is to maximize energy capture and thereby reduce the unit cost of generating electricity. Maximum energy capture can be difficult to achieve because of the wide variability of the wind over both space and time. A number of potential wind power sites are available all over the world. Of these, not all the sites have wind turbine generators installed.

Experiences with the existing wind farms have shown that some of the wind power plants have failed completely or performed poorly, especially in developing countries, because the installed wind turbine systems do not match the site. Hence, there arises a need for a systematic approach towards the problem of optimum siting of wind turbine generators.

It is very important for the wind industry to be able to describe the variation of wind speeds. Turbine designer needs the information to estimate their income from the electricity generation. If the wind speeds throughout a year, it is noticed that in most of area as strong gale force winds are rare, while moderate and fresh winds are common. The wind variation for a typical site is usually described using the so-called Weibull distribution. The shape of wind curve is determined by a so-called shape parameter.

Usually, wind turbines are designed to start running at wind speeds somewhere around 3 to 5 m/s. This is called cut-in wind speed. The wind turbine will be programmed to stop at high wind speeds above, say, 25 m/s, in order to avoid damaging the turbine or its surroundings. This speed at which turbine stops is called *furling wind speed*. There are many different models of wind turbine generators, commercially available, with same MW ratings. Each of these wind turbines has their own specifications and speed parameters. These speed parameters affect the capacity factor at a given specific site, and subsequently affect the choice of optimum wind turbine generator.

Small scale wind turbines (rotor diameters up to about 12 m) are typically used to generate power for consumption near the turbine by a private household or a small business. This restricts the location of the machine to a relatively small area. The siting approach under these circumstances must address two basic problems: finding the best (or at least the most acceptable) site for the turbine in a given area, and accurately estimating the wind characteristics at the selected site. Picking the initial location can be done using empirical guidelines formulated for evaluating the effects of the local topography and roughness elements of the wind, including the effects of trees and buildings.

Depending on the time and funds available, there are three approaches for estimating the wind characteristics at the site: first, using wind data from a nearby site, which is the quickest and least costly approach; second, correlating limited on-site wind measurements with data from nearby site; and third, collecting a representative data sample at the proposed site, which is the most accurate approach, but also the most time-consuming and most expensive. The success of first and second approaches depends very heavily on the complexity of the terrain between the proposed site and site where data were collected, and on whether the two locations are similarly exposed to the prevailing wind in the area.

In order to assess the economic viability of installing a wind energy conversion system (WECS) at a site, it is necessary to know the wind characteristics at that site. Since it is usually impractical to measure wind at all potential sites over a suitably long period of time, it is necessary to develop a methodology that can provide accurate estimates of wind economically at potential WECS sites from data that are already available. The wind energy potential is directly proportional to the air density and the third power of the wind speed averaged over a suitable time period. The wind speed and air density have temporal and spatial stochastic variability.

The meteorological aspects of siting medium and large-scale siting activities for utility applications include more than just the identification of a large windy site. Economic viability, compatibility with wind turbine design capabilities, and acceptabilities to the public are also important. The process combines the meteorological aspects of siting (e.g. annual average wind speed, frequency distribution, persistence, turbulence and wind shear) with the non-meteorological aspects (economic potential, safety and environment considerations) that are equally essential to a successful wind power project.

The spacing between turbines in a wind power station is an important factor in extracting maximum energy from the wind without significantly affecting turbine life. Optimal spacing requires knowledge of the size and intensity of wakes downwind of each type of turbine. The wake analysis problem is complicated by the terrain induced variability of the ambient wind over the station.

6.2 WIND FLOW MODELLING

The so-called field measurement method of determining the spatial variation of the wind resource across a wind power station is currently the procedure generally followed by private wind power developers. Wind flow modelling consists of subjectively interpolating between on-site measurement of the wind speed at several locations to estimate wind speeds at potential turbine sites. Two alternative approaches are wind tunnel physical models and computerized numerical models, but these have not yet been used extensively by private industry.

A basic assumption for determining spatial variations in the wind resource is that power producing winds are associated with a persistent set of characteristic meteorological conditions such that when these conditions occur, the wind field is spatially well correlated. It is operationally assumed that the wind flow pattern is fixed and that the temporal variation in speed can be scaled locally with speed, once the wind direction at a given reference point is known from the prevailing wind direction. Ratios of wind speed measured simultaneously at separate locations on the terrain are constants under this assumption.

The general procedure of the field measurement is to establish the wind speed ratios between the short-term atleast 12 months data from on-site anemometers and an existing long-term data base (several years) from a nearby reference station. Typically, two to three short-term anemometer stations are located per square kilometer, although the trend recently has been towards twice the density. The correlation of wind speeds between short- and long-term data records is obtained by linear regression analysis. For large-scale turbines, it is economically justified to utilize one anemometer per turbine site for at least one year.

Kite anemometer measurements can be made with commercially available equipment to provide additional definition of the spatial variation of the wind resource to supplement the fixed anemometer data. Kite anemometers typically add approximately 60–80 estimates of wind speed and direction to the data already available from fixed anemometers. About one kite measurement per 20,000–60,000 m^2 is recommended, depending on the steepness of the terrain. Data records are about 20 min long, and three or more flights are generally made per kite location on different days. Kite elevation is usually the same as that of the centre of the swept area of the prospective turbines.

6.2.1 Physical Modelling

Physical modelling consists of obtaining velocity, direction, and turbulence measurements over a scale model of selected terrain in a wind or water tunnel. Data from the tunnel model may be acquired in sufficient detail to determine the spatial variations in the wind flow field. Physical modelling of the flow field over a planned wind power station can be very useful in the final stages of wind turbine siting. It can be used to locate individual machine and wind monitoring equipment within a small area.

6.2.2 Numerical Modelling

Numerical flow modelling uses digitized and gridded terrain elevations as the lower boundary of the model and produces vector and contour plots of predicted local wind speeds and directions for specified boundary conditions. Complex terrain in 2-km square section of a wind power station was numerically modelled using nodal points a 40 by 40 grid.

The accuracy of numerical flow models is significantly improved when measurements from a few locations are used to adjust the direction of the initial flow and the model parameters that simulate the enhancement or suppression of vertical motion due to thermal stratification. Measurements for this purpose can be made over a very short time period when atmospheric conditions are climatologically representative of the power producing winds. The few measurements recommended to help determine the spatial variability of the wind over an area should not be confused with measurements made to estimate the long-term average of the wind speed. Under most conditions, a minimum of one year of data is required to estimate a long-term mean wind speed to an accuracy of 10 per cent with a confidence level of 90 per cent.

6.2.3 Modelling Goals

Recent advances in the ability to accurately model the wind resource provide a new tool to better understand the long-term spatial and temporal variability of the wind resource at a specific project site, or across an entire region, and over a range of time scales. The ultimate goal is to progressively increase the level of certainty throughout the development process. A thorough, scientifically-based programme during the development process should include the following components: the preliminary regional identification of the wind resource, site-specific identification, long-term resource variability analysis, and project specific performance characteristics.

The off-site correlation of data is used to understand how the short-term data collected compress to longer-term observation. Large distances between the project site and off-site measurement sites reduce the correlation, and complex terrain features can magnify the discrepancies even more. Additionally, there is no way of controlling the quality of the historical data. The simulation of the past 10, 25 or 30 years allows greater insight into the historical patterns of the wind source. Comparing a year of on-site measurement data with a long-term mesoscale numerical weather prediction (NWP), analytical model simulation for the site puts the measurement data in a more reliable long-term perspective, and can greatly increase the certainty around the wind resource. In so doing, it is also possible to determine if on-site measurements were taken during the average, high or low wind year, and the expected distribution of these variances over the lifetime of the project. From that foundation, it is possible to build a realistic financial model for the project, and make better, more informed decision.

The final step in completing a thorough wind resource assessment involves integrating all the component pieces: high-resolution maps of the project site, the long-term climatic variability analysis data, direct measurement data, off-site data, turbine placement and turbine-specific power curves.

Synthesizing this data gives a complete picture of the wind resource at a specific project site and should be used to develop a comprehensive energy analysis.

6.3 CAPACITY FACTOR

The production of electricity by a wind turbine generator at a specific site depends upon many factors. These factors include the mean wind speed of the site and, more significantly, the speed characteristics of the wind turbine itself, namely, cut-in (V_c), rated (V_r) and furling (V_f) wind speeds including the hub height. There are many different models of wind turbine generators, commercially available with same kW ratings. Each of these wind turbines has their own specifications and speed parameters. The speed parameters affect the capacity factor at a given specific site, and subsequently affect the choice of optimum WEG for the site. The capacity factor (CF) is the most effective parameter that indicates the power generation efficiency of the WEG. The problem of optimum siting based on the selection of site with higher wind power potential and matching WEG that gives higher CF is the critical issue. It is important to have an accurate assessment of wind power potential at all the study sites and matching WEG to these sites. By capacity factor of turbine, "it means the actual annual energy output divided by the theoretical maximum output, if the machine were running at its rated speed during all of the 8766 hours of the year." Thus, it is another way of stating the annual energy output from the turbine.

$$\text{Capacity Factor} = \frac{\text{Actual energy output}}{\text{Rated energy output}}$$

Although one would generally prefer to have large capacity factor, it may not always be an economic advantage. In a very windy region, it may be an advantage to use a larger generator with the same rotor diameter or a smaller rotor diameter for a given generator size. This would lower the capacity factor using less of the capacity of the larger generator, but it may mean a substantially larger annual production. Whether it is worthwhile to go for a lower capacity factor with a relatively large generator depends on both the wind conditions, and on the cost of the different turbine model. The choice can also be made between a relatively stable power output with high capacity factor or a high energy output with a low capacity factor.

The CF says nothing about the physical processes associated with the conversion of power carried by the wind into electric power. It is a description of the relationship between the power output of a wind turbine as a function of wind speed and the variation of wind speeds throughout a given period. Together, these functions describe the matching of the WEGs power generator characteristics to those of the wind regime in which the WEG is situated. These functions can be used to predict or estimate the WEGs energy production.

During the year, there are times when the wind does not blow, or blows at speeds below the cut-in wind speed of a WEG. Obviously, wind systems do not produce energy during all of the 8760 hours in a year. Even when a wind system does produce energy, it does not always do so at its full rated power. What is required is a measure of the energy productivity of the WEG. This measure is the CF, a descriptive parameter defined and used in the utility industry.

If the CF is considered at the planning and development stages of installation of wind power stations, it will enable the wind power developers or the utility engineer to make a judicious choice of a potential site and WEG from the available potential sites and WEGs, respectively.

The analytical approach for calculating CF based on the cubic mean of wind speeds and Weibull statistical model, the parameterised continuously changing wind speed distributions and its computational steps involved in optimum siting analysis are discussed in the following section.

6.3.1 Estimation of Capacity Factor

Step-I: The calculation of mean wind speeds:

The mean wind speed is calculated by the following equation:

$$\bar{V} = \left(\frac{\sum_{i=1}^{N} f_i V_i^n}{\sum_{i=1}^{N} f_i} \right)^{1/n} \tag{6.1}$$

The standard deviation of wind speed distributions are computed using the following equation:

$$\sigma = \sqrt{\frac{\sum_{i=1}^{N} f_i (V_i - \bar{V})^2}{\sum_{i=1}^{N} f_i}} \tag{6.2}$$

Step-II: The calculation of the Weibull scale and shape parameter:

The wind speed frequency curve is modelled by continuous mathematical function, called the *probability density function f(V)*. The mean value and the variance of the wind speed is given by:

$$\bar{V} = \int_0^\infty V f(V)\, dV \tag{6.3}$$

and

$$\sigma^2 = \int (V - \bar{V})^2 f(V)\, dV \tag{6.4}$$

There are several density functions used to describe the wind speed frequency curve. The Weibull probability density function has two parameters. The Rayleigh probability density function has an advantage over the Weibull density function since only one parameter is involved. This makes its use easier. The Weibull density function is more accurate than the Rayleigh density function because it uses two parameters. So, the Weibull density function is used. The Weibull density function is a special case of Pearson type-III or generalised Gamma distribution with two parameters. The wind speed V is distributed as the Weibull distribution, if its probability density function is given by:

$$f(V) = \frac{k}{C}\left(\frac{V}{C}\right)^{k-1} e^{(-(V/C))^k} \qquad (k > 0, V > 0, C > 1) \tag{6.5}$$

where k is the shape parameter and C is the scale parameter.

Substituting $f(V)$ from Eq. (6.5) in Eq. (6.3), the mean wind speed is computed as,

$$\bar{V} = \int_0^\infty V \frac{k}{C}\left(\frac{V}{C}\right)^{k-1} e^{\left(-\frac{V}{C}\right)^k} dV \tag{6.6}$$

The mean wind speed, on simplification of Eq. (6.6), is given by:

$$\overline{V} = C\Gamma\left(1 + \frac{1}{k}\right) \quad (6.7)$$

where Γ is the gamma function, which is given as,

$$\Gamma(n) = (n-1)! \quad (6.8)$$

On similar lines, the variance of the Weibull density function can be shown to be,

$$\sigma^2 = \frac{\overline{V}^2 \Gamma(1 + 2/k)}{\Gamma(1 + 1/k)} - 1 \quad (6.9)$$

There are several methods for determining the shape parameter k and scale parameter C. If the mean and variance of the wind speed for a site are known, then Eqs. (6.7) and (6.8) can be solved directly for C and k. An acceptable approximation to obtain the shape parameter k as given by Justus was as follows:

$$k = \left(\frac{\sigma}{\overline{V}}\right)^{-1.086} \quad (6.10)$$

Once k is determined, Eq. (6.7) can be solved for c as:

$$C = \frac{\overline{V}}{\Gamma(1 + 1/k)} \quad (6.11)$$

Step-III: Calculation of the Rayleigh probability density function:

The Rayleigh distribution is a subset of the Weibull distribution and has one parameter. This makes it simpler to use and implement. It is given by:

$$f(V) = \frac{\pi V}{2\overline{V}^2} e^{\left[-\frac{\pi}{4}\left(\frac{V}{C}\right)^2\right]} \quad (6.12)$$

Equation (6.12) can be simplified and written as

$$f(V) = \frac{\pi}{C^2} e^{\left(-\frac{V}{2C^2}\right)} \quad (6.13)$$

The approximate parameter c is given by Justus as:

$$C = \frac{\overline{V}}{1.253} \quad (6.14)$$

Step-IV: Estimation of capacity factor

The capacity factor is defined as the ratio of the average output power to the rated power output. It implies the percentage of electrical power which a wind turbine generator can generate from the available wind at a site. The capacity factor is computed as under:

$$CF = \frac{P_{e,\text{ave}}}{P_{er}} \quad (6.15)$$

Consider the basic wind electrical system as shown in Figure 6.1. P_w is the power in the wind, P_m is the turbine output power, P_t is the generator input power, and P_e is generator power output η_m is the transmission efficiency and η_g is the generator efficiency.

FIGURE 6.1 Block diagram of wind electric system.

From the block diagram shown in Figure 6.1, the electrical power output can be written as:

$$P_e = \frac{1}{2} \rho A C_p \eta_m \eta_g V^3 \qquad (6.16)$$

The average electrical power output of the generator is the electrical power produced at each wind speed multiplied by the probability of the wind speed experienced and integrated over all possible wind speeds. In integral form, it is given by,

$$P_{e,\text{ave}} = \int_0^\infty P_e(V) f(V) \, dV \qquad (6.17)$$

Equation (6.16) relating power and wind speed can be rewritten as:

$$\begin{cases} P_e(V) = 0 : V < V_c \\ P_e(V) = \frac{1}{2} \rho A C_p \eta_m \eta_g V^3 : V_c \leq V < V_r \\ P_e(V) = \frac{1}{2} \rho A C_p \eta_m \eta_g V_r^3 : V_r \leq V < V_f \\ P_e(V) = 0 : V > V_f \end{cases} \qquad (6.18)$$

where V_c, V_r and V_f are cut-in, rated and furling wind speeds of the wind turbine generator. The average electrical power output can now be written as:

$$P_{e,\text{ave}} = \int_{V_c}^{V_r} \frac{1}{2} \rho A C_p \eta_m \eta_g V^3 f(V) \, dV + \int_{V_r}^{V_f} \frac{1}{2} \rho A C_p \eta_m \eta_g V_r^3 f(V) \, dV \qquad (6.19)$$

The rated electrical power output of the generator is the electrical power produced at rated wind speed of the turbine and is given by,

$$P_{er} = \frac{1}{2} \rho A C_p \eta_m \eta_g V_r^3 \qquad (6.20)$$

From Eqs. (6.15), (6.19) and (6.20), the capacity factor is obtained as:

$$CF = \frac{1}{V_r^3} \int_{V_c}^{V_r} V^3 f(V) \, dV + \int_{V_r}^{V_f} f(V) \, dV \qquad (6.21)$$

where $f(V)$ is the Weibull density function given as Eq. (6.6) or Rayleigh probability density function Eq. (6.13).

The calculation of *CF* is shown in Figure 6.2 in the form of a logic for developing a computer programme.

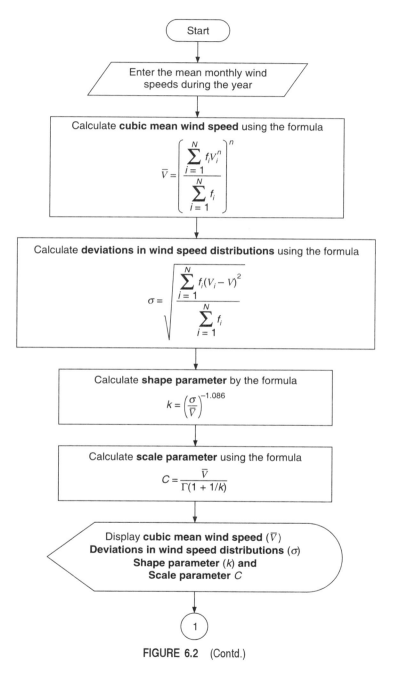

FIGURE 6.2 (Contd.)

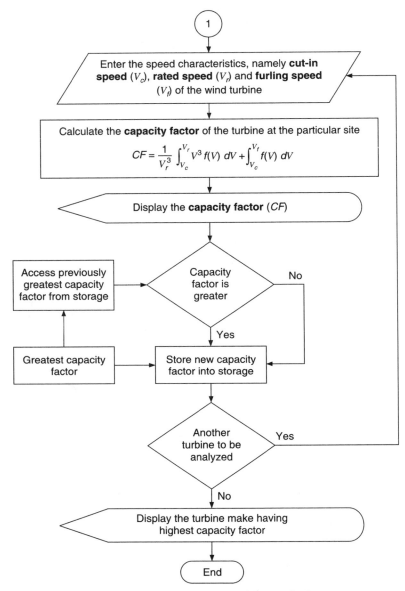

FIGURE 6.2 Flowchart for calculating capacity factor.

6.4 PLANNING OF WIND FARM

Once a site has been identified for further investigation, the developer will conduct a preliminary site characterization to determine the initial suitability of that site. The preliminary site characterization can also be a useful tool for performing an alternative analysis of multiple potential sites. The major steps involved in this initial stage include:

- *Analyze the wind resource:* When existing site data is not available, this process typically takes one to three years.
- *Conduct an initial site visit:* To determine any obvious constructability and/or environmental constraints.
- *Establish the economics of the project:* The criteria for economic success and/or the business model used to develop the wind project.
- *Conduct critical environmental issues analysis and identify regulatory framework:* It is a must to identify early in the development process the government–state and local regulatory issues that will influence the project.
- *Conduct transmission capacity analysis:* To determine if the existing transmission system will be able to support the proposed project, work with utility to conduct a transmission capacity analysis.
- *Assess public acceptability:* Reaching out to the community and understanding the level of public acceptability within the project area is a critical component of successfully developing a wind project. The public out reach process is best initiated in the early stages of development and maintained throughout the entire development process and operations.
- Selected site(s) should go with a master plan approach, i.e. the machine deployment should be in such a manner that machine that may come in future are taken into account.

The questions frequently asked by wind developers, utilities and financial players include:

Where is the best place to build a project?
What is the long-term wind resource variability?
Is the wind resource available when it is needed?
What are the long-term performance characteristics of the wind resource?

All too often, wind developers make the mistake of collecting measurement data over a period of 12–18 months, and then falsely assume that they have captured a representative sample of the wind resource at the project site. In truth, the data captured is merely a randomly selected snapshot of the resource without any long-term historical context. Assuming that the year in which you took measurements is an average year can have significant financial consequences. The management and engineering inputs are illustrated in Figure 6.3 for development of wind farm.

The steps involved in buildilng a wind farm are as follows:

1. **Understand your wind resource:** The most important factor to be considered in the construction of a wind energy facility is the site's wind resource. A site must have a minimum annual average wind speed in the neighbourhood of 5–6 m/s to even be considered. Local weather data available from airports and meteorological stations may provide some insight regarding averages. In time, a monitoring device can be installed to record the site's wind characteristics.

2. **Determine proximity to existing transmission lines:** A critical issue in keeping costs down in building a wind farm is minimizing the amount of transmission infrastructure that has to be installed. High voltage lines can cost thousands of dollars per mile. Whenever possible, availability and access to existing lines should be considered in selecting a site.

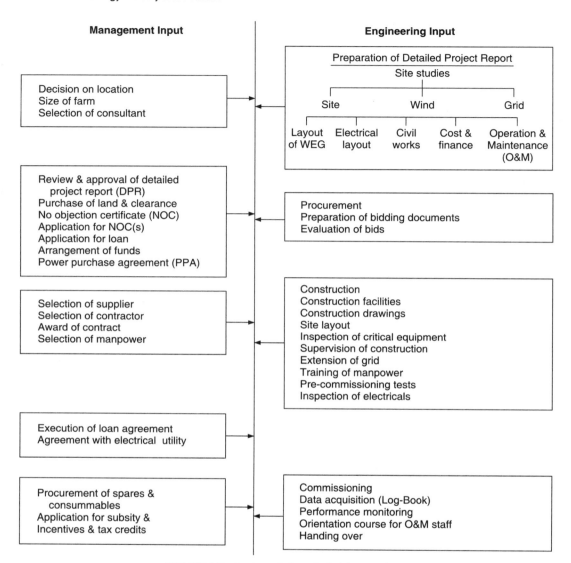

FIGURE 6.3 Implementation of wind farm project.

3. **Secure access to land:** Landowners, both private and public, will expect to be compensated for any wind energy development that occurs on their land. Royalty or lease agreements will need to be discussed with all parties involved. Roads, transmission equipment, maintenance infrastructure, turbine, and the like all need to be considered. Moreover, the construction of a wind farm necessitates the use of heavy industrial equipment. Developers will need to invest in roads capable of accommodating significant weight.

4. **Establish access to capital:** Building a wind farm is not cheap. On an average, wind power development costs around $ 1 million per megawatt of generating capacity installed.

To take advantage of economies of scale, wind power facilities should be in excess of 20 MW.

5. **Identify reliable power purchaser or market:** To date, wind energy is the most cost competitive renewable energy option on the market. In fact, wind energy's cost has declined so much that it rivals many traditional power generation technologies. Before investing into wind resource assessments, permitting, and pre-construction activities, a developer will secure tentative commitments from one or more buyers for the wind plants output over 10 to 30 years of its operational lifetime.

6. **Address siting and project feasibility considerations:** The fact that a site is windy does not mean it is suitable for wind power development. A developer needs to consider many factors in siting a project. Are there endangered or protected species that could be jeopardized by the presence of the facility? Is the site's geology suitable and appropriate for industrial development? Will noise and aesthetics be issues for the local community? Will the turbine obstruct the flight path of local air traffic? There are quite a few environmental and social issues that will need to be addressed in the siting of a wind power facility.

7. **Understand wind energy's economics:** There are many factors contributing to the cost and productivity of a wind plant. Financing methods can make a major difference in the project economics as well. Securing significant investment capital or joint ownership of a project can cut costs significantly. Furthermore, there are government incentives for which a project may qualify and which could reduce costs and encourage more favourable investment.

8. **Obtain zoning and permitting expertise:** Siting any power project can be a daunting task due to the dizzying array of social and enviromental factors at play. A wind power developer would be well served to obtain the services of a professional familiar with the regulatory environment surrounding wind power development. Additionally, legal counsel familiar with the local political climate may be able to help nagate the permitting process.

9. **Establish dialogue with turbine manufacturers and project developers:** Every turbine is different despite seemingly similar power ratings. Some machines are designed to operate more efficiently at lower wind speeds while others are intended for more robust wind regimes. A prospective wind power developer would be wise to investigate the various considerations and compare the performance to existing machines. Moreover, anecdotal information and even the professional services of wind power developers may prove helpful.

10. **Secure agreement to meet O and M needs:** Wind turbine technology has made great strides in recent years. Todays's machines are more efficient and cost-effective than ever. However, they are also more complex. Turbine availability (reliability) is a major factor in project success, and the services of professionals familiar with the operation and maintenance of wind turbines can prove to be invaluable. Also, turbine manufacturers may offer more favourable product gurantees knowing that qualified project operators will be on site to maintain the equipment.

6.4.1 Environmental Issues

- Identifying state-listed threatened or endangered species or habitat within the project area. Identifying bird and bat species, habitat (year-round and seasonal) and migration pathways.
- Identifying wet land and protected areas.
- Identifying land development constraints. Some of these are:
  - Noise limits
  - Setback requirements
  - Flood plain issues
  - Height restrictions
  - Zoning constraints
- Identifying the interference with telecommunications transmissions and microwave paths.
- Identification of known airports, landing strips and other aviation considerations.

The planning application will require the preparation of an environmental statement and the scope of this is generally agreed, in writing, with the civic authorities during the Project Feasibility Assessment. The purpose of a wind farm environmental statement is summarized below:

- To describe the physical characteristics of the wind turbines and their land-use requirements.
- To establish the environmental character of the proposed site and the surrounding area.
- To predict the environmental impact of the wind farm.
- To describe measures which will be taken to mitigate any adverse impact.
- To explain the need for the wind farm and provide details to allow the planning authority and general public to make a decision on the application.
- Typically, it includes topics like: policy framework, site selection, designated areas, visual and landscape areas, noise assessment, ecological assessment, archaeological and historical assessment, hydrological assessment, interference with telecommunication systems, aircraft safety, general safety, traffic management and construction, electrical connection economic effects on the local economy, global environmental benefits, decommissioning, mitigating measures, non-technical summary, etc.

It is recommended that wind projects should be lit at night. Now the lights can be upto 0.8 km apart and be placed only around the project perimeter, reducing the number of lights needed overall. Red lights are recommended which are less annoying than white lights. No day lighting is necessary if the turbines and blades are painted white or off-white.

6.4.2 Constraints Map

Constraints map is a useful tool for graphically depicting the environmental and land use constraints that limit the desirable area for development at a site. The set back constraints may include:
Set back from:

- Sensitive buildings such as residence, schools, hospitals, etc.
- Roads
- Electric transmission lines, oil and gas facilities, mining areas, etc.
- Wet lands, surface waters, drinking water supplies, etc.
- National parks and bird sanctuaries

- Areas of known geotechnical instability
- Communication/radar-related constraints
- Areas impacted by air traffic
- Any other environmental and land use constraints identified for the site.

High resolution maps, which provide insight into the geographical dispersion of the wind over the site and illustrate how the wind interacts with terrain features, should be used to ensure proper, representative meteorological tower placement. A typical constraint map is shown in Figure 6.4.

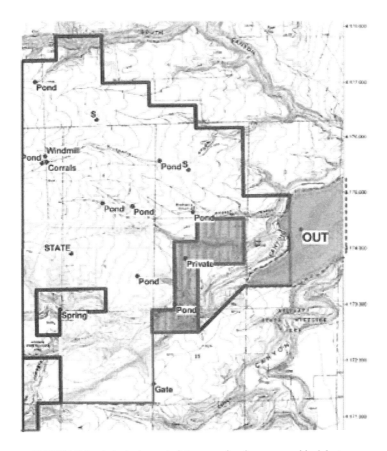

FIGURE 6.4 A typical constraints map showing geographical features.

6.5 SITING OF WIND TURBINES

6.5.1 Siting in Flat Terrain

Choosing a site in flat terrain needs two primary questions to be considered:

- What surface roughness might affect the wind profile in the area?
- What barriers might affect the free flow of the wind?

Terrain can be considered flat if it meets the following conditions (shown in Figure 6.5): the elevation difference between the site and the surrounding terrain is less than 100 m for 5 to 8 km in any direction and the ratio of *h/l* in Figure 6.5 is less than 0.03. The potential user can determine if the site in question meets these conditions either by inspecting it or by consulting topographical maps.

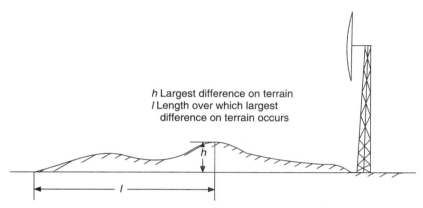

FIGURE 6.5 Determination of flat terrain.

The conditions given for determining flat terrain are very conservative. If there are no large hills, mountains, cliffs, etc. within 1.5 km or so of the proposed WECS site, the data in this section can be used for siting.

Wind rose information can also guide the user in determining the influence of nearby terrain. For example, suppose a 15 m high hill lies 1.0 km north-east of the proposed site (this classifies the terrain as non-flat); also assumes the wind rose indicates that winds blow from the north-east quadrant only 5.0 per cent of the time with an average speed of 3.0 m/s. Obviously, so litle power is associated with winds blowing from the hill to the site that the hill can be disregarded. If there are no terrain features upwind of the site along the principal wind power direction(s), the terrain can be considered flat.

6.5.2 Uniform Roughness

Surface roughness describes the texture of the terrain. The rougher the suface, the more the wind flowing through it is impeded. Flat terrain with uniform surface roughness is the simplest type of terrain for a WECS site. A large area of flat, open land is a good example of uniform terrain. Provided there are no obstacles (i.e., buildings, trees or hills), the wind speed at a given height is nearly the same over the entire area.

The only way to increase the available power in uniform terrain is to raise the machine higher above the ground. A measurement or estimate of the average wind speed at one level can be used to estimate wind speed (thus available power) at other levels. Table 6.1 provides estimates of wind speed changes for several roughnesses at various tower heights. The numbers in the table are based on wind speeds measured at 10 m because Weather Department wind data are frequently measured near that height. To estimate the wind speed at another level, multiply the 10 m speed by the factor for the appropriate surface roughness and height.

TABLE 6.1 Extrapolation of wind speed from 10 m to higher heights over flat terrain of uniform roughness

Roughness characteristics	20 m	30 m	40 m	50 m	60 m	70 m
Smooth surface ocean, sand	1.10	1.18	1.21	1.26	1.29	1.31
Low grass or fallow land	1.12	1.21	1.25	1.31	1.33	1.37
High grass or low row crop	1.13	1.24	1.28	1.35	1.38	1.43
Tall row crop or low trees	1.16	1.29	1.34	1.42	1.46	1.52
High trees	1.21	1.40	1.47	1.60	1.65	1.73
Suburbs small town	1.39	1.78	1.95	2.23	2.36	2.51

If the height of the known wind speed is not 10 m, wind speed can be estimated using the following equation:

$$\text{Estimated wind speed} = \frac{E}{K} \times S$$

where
E = table value for the height of the estimated wind
K = table value for the height of the known wind
S = known wind speed.

Over areas of dense vegetation (such as a crop field or a forest), a new "effective ground level" is established at approximately the height where the branches of adjacent trees touch. Below this level, there is little wind; consequently, it is called the *level of zero wind*. In a dense wheat field, the level of zero wind would be the average height of the wheat plants. The height at which this level occurs is called the *zero displacement height*, and is labeled d in Figure 6.6. If d is less than 3 m, it can usually be disregarded in estimating speed changes.

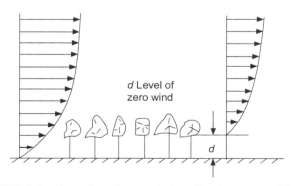

FIGURE 6.6 Formation of new wind profile above ground level.

6.5.3 Changes in Roughness

Often, roughness varies upwind of the wind turbine. Figure 6.7 shows how a sharp change in roughness affects the wind profile. If a wind turbine were sited at the first level in Part A of Figure 6.7, the user would be greatly underutilizing wind energy, since roughness changes cause a sharp increase in wind speed slightly above the first level. Part B of Figure 6.7 shows that in smooth terrain, little, if anything, would be gained by increasing tower height from the first level to even as high as the third. One principle stands out: the user will gain more in terms of available power by increasing the height of a wind turbine located in rough terrain than it will be by increasing the height in smoother terrain.

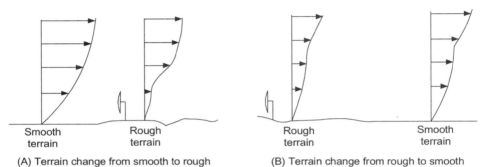

(A) Terrain change from smooth to rough
(B) Terrain change from rough to smooth

FIGURE 6.7 Wind speed profiles near change in terrain.

Which upwind surface roughness is influencing the wind profile at the height of the wind turbine? It is crucial to know which wind directions are associated with the most power. Roughness changes along the most powerful wind directions will have the greatest effect on power availability at the site.

Barriers in flat terrain

Barriers produce disturbed area of airflow downwind, called *wakes,* in which wind speed is reduced and turbulence increased, because exposure to turbulence may greatly shorten the life span of wind turbine rotors. Barrier wakes should be avoided whenever possible, not only to maximize power but also to minimize turbulence.

Buildings

Since it is likely that buildings will be located near a wind turbine, it is important to know how they affect airflow and available power. A general rule for avoiding most of the adverse effects of building wakes is to site a wind turbine according to one of the following guidelines:

- Upwind,* a distance of more than two times the height of the building
- Downwind,* a minimum distance of ten times the height of the building
- At least twice the height of the building above ground if the wind turbine is to be mounted on the building.

---

* Upwind and downwind indicate directions along the principal power directions.

Figure 6.8 illustrates this rule with a cross sectional view of the flow wake of a small building. The above rule is not fool proof, because the size of the wake also depends upon the building's shape and orientation to the wind.

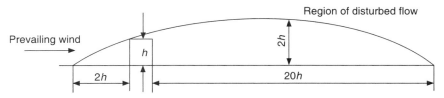

FIGURE 6.8 Zone of disturbed flow over a small building.

Table 6.2 summarizes the effects of building shape on wind speed, available power and turbulence, for buildings oriented per pendicular to the wind flow. Building shape is given by the ratio of "width divided by height." As might be expected, power reduction is felt farther downstream for wider buildings. The speed, power and turbulence changes reflected in Table 6.2 occur only when the wind turbine lies in the building wake. Wind rose information will indicate how often this actually occurs. Annual percentage time of occurrence multiplied by the percentage power decrease in the table will give the net power loss.

TABLE 6.2 Wake behaviour of variously shaped buildings

	Downwind distances (in terms of building heights)								
	5H			10 H			20 H		
Building shape (Width ÷ Height)	% speed decrease	% power decrease	% turbulence increase	% speed decrease	% power decrease	% turbulence increase	% speed decrease	% power decrease	% turbulence increase
4	36	74	25	14	36	7	5	14	1
3	24	56	15	11	29	5	4	12	0.5
1	11	29	4	5	14	1	2	6	
0.33	2.5	7.3	2.5	1.3	4	0.75			
0.25	2	6	2.5	1	3	0.50			
Height of the wake flow region (in building heights)		1.5			2.0			3.0	

If a tower is located on the roof of a building, the turbulence near the roof should be considered. A slanted roof produces less turbulence than does a flat roof and may actually increase the wind speed over the building. The zone of speed increase may extend upto twice the building height if the building is wider than it is tall and is perpendicular to the prevailing wind. It is generally wiser to raise wind trubine as high as is economically practical, taking afvantage of the fact that winds usually increase and turbulence decreases with height.

6.5.4 Shelterbelts

Shelterbelts are windbreaks, usually consisting of a row of trees. When selecting a site near a shelterbelts, the user should:

- Choose a site far enough upwind/downwind to avoid the disturbed flow.
- Use sufficient high tower to avoid the disturbed flow.
- Minimize power loss and turbulence by examining the nature of the wind flow near the shelterbelt and choose a site accordingly, if the disturbed flow at the shelterbelt cannot be entirely avoided.

The degree to which the wind flow is disturbed depends on the height, length and porosity of the shelterbelt. Porosity is the ratio of the open area in a windbreak to the total area, expressed here as the percentage of open area.

Figure 6.9 locates the region of greatest turbulence and wind speed reduction near a thick break. How far upwind and downwind this area of disturbed flow extends? It varies with the height of the wind break. Generally, the taller the wind break, the farther the region upwind and doownwind that will experience a disturbed airflow.

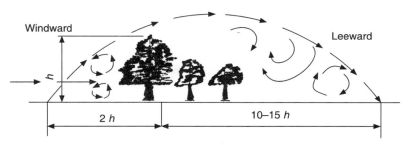

FIGURE 6.9 Airflow near a shelterbelt.

6.5.5 Trees and Scattered Barriers

The trees near a prospective WECS site may not be organized into a shelter belt. The wake of disturbed airflow behind individual trees grows larger (but weaker) with distance, much like a building wake. However, the highly disturbed portion of a tree wake extends farther downstream than does that of a solid object. If available, wind rose information can be used to estimate the percentage of time a site will be in the tree wake. The WECS should be raised above the most highly disturbed airflow. To avoid most of the undesirable effects of trees and other barriers, the rotor disc should be situated on the tower at a minimum height of three times that of the tallest barrier in the vicinity.

6.5.6 Siting in Non-flat Terrain

Any terrain that does not meet the criteria listed in Figure 6.10 is considered to be non-flat or complex. A site that features significant variations in topography and presence of obstacles like buildings, hills, forests, etc. that may cause flow distortions is called complex terrain. To select candidate sites in such terrain, the potential user should identify the terrain features, i.e. hills, ridges, cliffs and valleys located in or near the siting area.

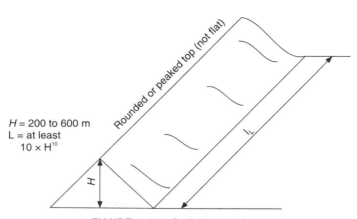

FIGURE 6.10 Definition of ridge.

The terrain which is composed of randomly steep slopes is called complex terrain. The roughness value of such terrain is high. If such terrain has too many steep slopes, ruggedness index (RIX) has been indicated. RIX has been developed by Mortensen et al. to quantify the roughness. The wind flow over such complex terrain is further modified due to different temperature fields. In hilly terrain, during day time, up-slope flow of wind dominates as top of the hills are heated much by sun than lower slopes and valleys. The flow reverses in night time and becomes down-slope in hills and valleys due to rapid cooling. Such flow exhibits sharp gradient in all directions with unsteadiness. The detailed flow structure will be governed by number of factors such as solar input, steepness of slope, pressure gradient and orientation of slope across the landscape. In such terrain, the meteorological measurements must be carried out at several points combined with modelling to represent an acceptable flow visualization. The period of one year is the longest and complete seasonal cycle to understand the climate for estimating wind resource in terms of wind speed and direction distributions at wind turbine hub height. Several years of wind data is useful to account for inter-annual variability. For turbine rotor load estimation, the extreme values of turbulence intensity, wind shear and wind speed values at a location are to be recorded.

Wave height and wave direction are to be known to establish wave climate in offshore sites for determining wave loads on turbine tower and foundation. Such records are needed in return of one to several years. Gust wind loads on wind turbine blades is design concern. For dynamic load calculations on a wind turbine components, wind speed and three directions spatially are to be correlated for gust wind by sweeping blades.

In complex terrain, land forms affect the airflow to some height above the ground in many of the same ways as surface roughness does. However, topographical features affect airflow on a much larger scale, overshadowing the effects of roughness. When weighing various siting factors by their effects on wind power, topographical features should be considered first, barriers second, and roughness third.

6.5.7 Ridges

Ridges are defined as elongated hills rising above 200 m to 700 m above the surrounding terrain and having little or no flat area on the summit (see Figure 6.10). There are many advantages of locating a WECS on a ridge:

- The ridge acts as a huge tower.

206 Wind Energy: Theory and Practice

- The undesirable effects of cooling near the ground are avoidable.
- The ridge may accelerate the airflow over it, thereby increasing the available power.

Figure 6.11 shows the cross-sectional shapes of several ridges and ranks them by the amount of acceleration they produce. Notice that a triangular-shaped ridge causes the greatest acceleration,

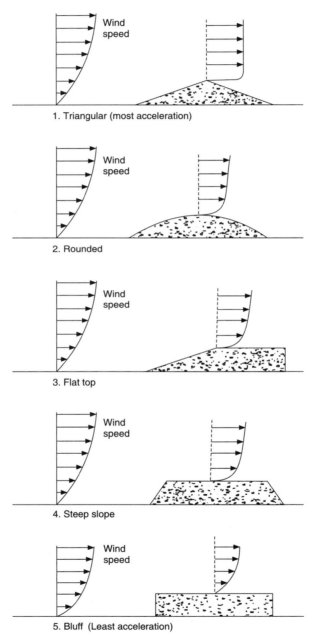

FIGURE 6.11 Ranking of ridge shape by amount of wind acceleration.

and that the rounded ridge is a close second. It indicates that certain slopes, primarily in the nearest few hundred metres to the summit increase the wind more effectively than do others. The portion of the ridge just below the summit has the greatest infuence on the wind profile immediately above the summit. Table 6.3 classifieds smooth, regular ridge slopes according to their values as wind power sites. Figure 6.12 gives approximate percentage variations in wind speed for an ideally-shaped ridge. The user should expect similar wind speed patterns along the path of the flow. Generally, wind speed decreases significantly at the foot of the ridge, then accelerates to a maximum at the ridge crest. It only exceeds the upwind speed on the upper half of the ridge.

TABLE 6.3 WECS site suitability based upon slope of the ridge

	Slope of the hill near the summit	
WECS site suitability	Percentage grade*	Slope angle
Ideal	10	16°
Very good	6	10°
Good	3	6°
Fair	2	3°
Avoid	Less than 2 Greater than 16	Less than 3° Greater than 30°

* Per cent grade as used above is the number of metre of raise per 30 metre of horizontal distance.

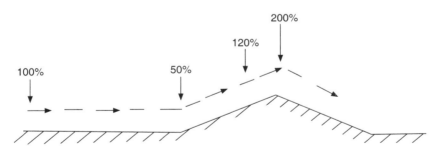

FIGURE 6.12 Percentage variation in wind speed over an idealized ridge.

Flat-topped ridges present special problems because they can actually create hazardous wind shear at low levels, as Figure 6.13 illustrates. The hatched area at the top of the flat ridge indicates a region of reduced wind speed due to the "separation" of the flow from the surface. Immediately above the separation zone is a zone of high wind shear. This shear zone is located just at the top of the shaded area in the figure. Siting a WECS in this region will cause unequal loads on the blade as it rotates through areas of different wind speeds and could decrease performance and the life of the blade. The wind shear problem can be avoided by increasing tower height to allow the blade to clear the shear zone or by moving WECS towards the windward slope.

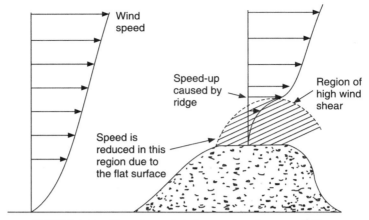

FIGURE 6.13 Hazardous wind shear over a flat-topped ridge.

The effects of barriers and roughness should not be overlooked. Figure 6.14 shows how a rough surface upwind of a ridge can greatly decrease the wind speed. After selecting the best

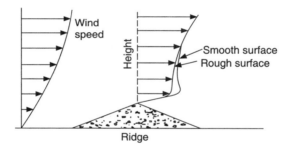

FIGURE 6.14 Effect of surface roughness on wind flow over a low sharp-crested ridge.

section of a ridge, based upon its geometry, the potential user should consider the barriers, then the upwind surface roughness. The important points for siting WECS are as follows:

- The best ridges or sections of a single ridge are most nearly perpendicular to the prevailing wind.
- Ridges or sections of a single ridge having the most ideal slopes within several hundred metres of the crest should be selected (use Table 6.3).
- Sites where turbulence or excessive wind shear cannot be avoided should not be considered.
- Roughness and barriers must be considered.
- If siting on the ridge crest is not possible, then site should be either on the ends or as high as possible on the windward slope of the ridge.
- The foot of the ridge should be avoided.
- Vegetation may indicate the ridge section having the strongest winds.

6.5.8 Isolated Hills and Mountains

An isolated hill is 200 m to 700 m high, is detached from any ridges and has a length of less than 10 times its height. Hills greater than 700 m high will be referred to as *mountains*. Hills, like ridges, may accelerate the wind flowing over them but not as much as ridges will, since air tends to flow around the hill as shown in Figure 6.15. Table 6.3 can be used to rank hills according to their slopes. Benefits are gained by siting on hills:

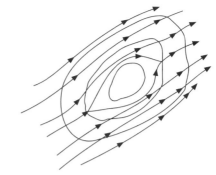

FIGURE 6.15 Airflow around an isolated hill (top-view).

- Airflow can be accelerated.
- The hill acts as a huge tower, raising the WECS into a stronger airflow.

The best WECS sites on an isolated hill may be along the sides of the hill tangent to the prevailing wind. For comparing hill sides and hill top sites, the surest method currently is simultaneous wind recordings and analysing to find out the best.

6.5.9 Passes and Saddles

Passes and saddles are low spots or notches in montain barriers. Such sites offer the following advantages to WECS siting:

- Since they are often the lowest spots in a mountain chain, they are more accessible than other mountain locations.
- They are flanked by much higher terrain, the air is funneled as it is forced through the passes.
- Depending upon the steepness of the slope near the summit, wind may accelerate over the crest as it does over a ridge.

Factors affecting airflow through passes are orientation to the prevailing wind; width and length of the pass; elevation differences between the pass and adjacent mountains; the slope of the pass near the crest; and the surrounding roughness. Some desirable characteristics of passes are as follows:

- The pass should be open to the prevailing wind, preferably parallel to the prevailing wind.
- The pass should have high hills or mountains on both sides—the higher, the better.
- The slope of the pass near summit should be sufficient to further accelerate the wind.
- The surface should be smooth—the smoother, the better. If pass is very narrow, the user should consider the roughness of the sides of the pass.

Figure 6.16 shows that maximum wind is located in the centre of the pass. The WECS should be sited near the centre of the pass. Since the location of maximum wind speed will vary from pass to pass, wind measurements are recommended before a final decision on WECS placement is made.

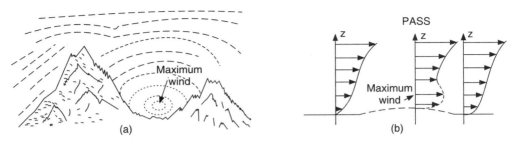

FIGURE 6.16 A schematic of (a) the wind pattern and (b) velocity profile through a mountain pass.

6.5.10 Gaps and Gorges

Gaps and gorges are generally deeper than passes and can significantly enhance even relatively light wind. A river gorge can augment mountain-valley or land-sea breezes, providing a reliable source of power. Gaps and gorges are also usually more accessible than mountain passes. The major drawback to sites in gaps and gorges is that, because they are narrow, there is often much turbulence and wind shear. Figure 6.17 shows gap through mountain barriers, a low level path and much air is forced through it.

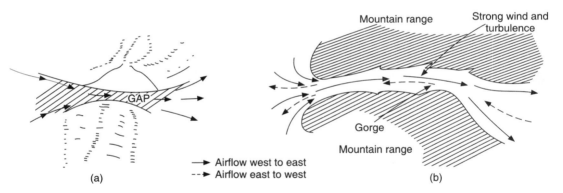

FIGURE 6.17 A schematic illustration of flow patterns that may be through gaps and gorges.

Valleys and canyons

The airflow pattern is a pattern in a particular valley or canyon depends on such factors as the orientation of the valley, to the prevailing wind; the slope of the valley floor; the height, length and width of the surrounding ridges; irregularities in the width; and the surface roughness of the valley. Funneling occurs only if the valley or canyon is constricted, at some point. Unless the valley is constricted, the surrounding ridges will provide better WECS sites than will the valley floor.

The day time wind blowing up the valley tends to be more sensitive to factors such as heating of the sun, the sun, the driving force for this wind and the winds blowing high overhead. As a result, the valley winds are more variable, and often weaker, than the mountain winds. Unlike the mountain wind, which is strongest near the centre of the valley, valley winds are normally greatest along the side most directly facing the sun. Figure 6.18 shows how to take advantage of mountain and valley winds.

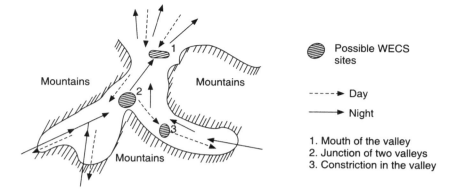

FIGURE 6.18 Possible WECS sites in sloping valleys and canyons.

Figure 6.19 shows possible wind sites, where valley channeling enhances the wind flow. In Figure 6.19(a), a funnel-shaped valley on the windward side of a mountain range is shown. The constriction (or narrowing) near the mouth produces a zone of accelerated flow. In Figure 6.19(b), a narrow valley in the lee of a mountain range is illustrated. It is parallel to the prevailing wind and is constricted slightly near its mouth.

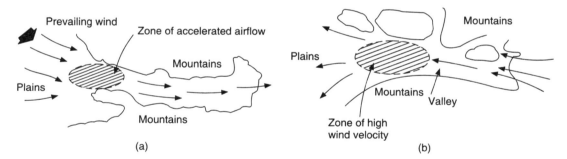

FIGURE 6.19 Possible WECS sites where prevailing winds are channeled by valleys.

To site WECS in valleys and canyons, the potential user should:
- Select wide valleys parallel to the prevailing wind or long valleys extending down from mountain ranges.
- Choose sites in possible constrictions in the valley or canyon, where the wind flow might be enhanced.
- Avoid extremely short and/or narrow valleys and canyons, as well as those perpendicular to the prevailing winds.
- Choose sites near the mouth of the valley, where mountain valley winds occur.
- Ensure that the tower is high enough to place the WECS as near to the level of maximum wind as is practical.
- Consider nearby topographical features, barriers and surface roughness.

6.5.11 Basins

Basins are depressions surrounded by higher terrain. The flow into and out of a basin is similar to the mountain-valley cycle. Valleys sloping down into basins may provide sufficient channeling to warrant consideration as WECS site. The following points are helpful when siting WECS in basins.

- Consider only large, shallow inland basins.
- Use vegetation indicators of wind to locate areas of enhanced wind in basins.
- Consider all topographical features, barriers and surface roughness effects.

6.5.12 Cliffs

A cliff is a topographical feature of sufficient length, 10 or more times the height to force the airflow over rather than around its face. For such long cliffs, the factors affecting the airflow are the slope both on windward and lee sides, the height of the cliff, the curvature along the face and the surface roughness upwind.

Figure 6.20 shows how the air flows over cliffs or different slopes. The swirls in the flow near the base and downwind from the cliff edge are turbulent regions, which must be avoided.

Turbulent swirls or flow separation becomes larger as the face of the cliff leans more into the wind. When the cliff slopes downward on the lee side, as in Figure 6.20(c), the zone of turbulence moves more downwind from the face. Part of the turbulence can be avoided by siting a WECS very close to the face of such hill-shaped cliffs. Selecting a section of the cliff having a more gradual slope (a) is sometimes advantageous because the tower height required to clear the turbulent zone is reduced.

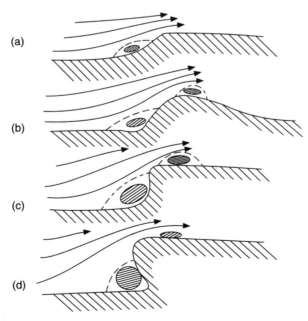

FIGURE 6.20 Airflow over cliffs having differently sloped faces. (hatched area is turbulent region)

Any curvature along the face of a cliff should also be considered. Figure 6.21 illustrates a top view of a curved cliff section. The curvature of the face channels the winds into the concave portions. Figure 6.22 shows the vertical wind profile of air flowing over a cliff. Wind speed rapidly increases near the top of the flow separation. This region of shear should be avoided, either by choosing a new site or by raising the WECS so that the rotor disc is above the shear zone. For maximum enhancement of the wind speed, the prevailing wind direction should be perpendicular to the cliff section on which the WECS will be located.

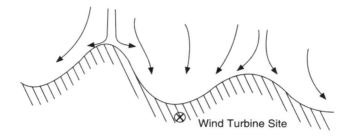

FIGURE 6.21 Top view of airflow over concave and convex portions of cliff face.

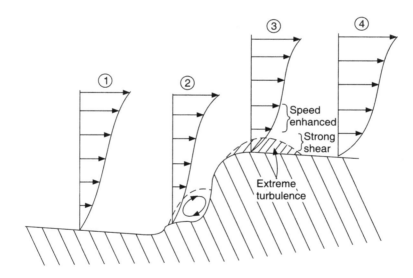

FIGURE 6.22 Vertical profiles of air flowing over a cliff.

When chooing a site on a cliff, the following points should be considered:
- The best cliffs are well exposed to the wind.
- The best cliffs are oriented perpendicular to the prevailing winds.
- The shape and slope of the cliff which causes the least turbulence should be selected.
- General wind patterns near cliffs may be revealed by the deformation of trees and vegetation.
- The entire rotor disc should clear the zone of separation.

6.5.13 Mesas and Buttes

Smaller buttes those less than 600 m in height, and less than about five times as long as they are high can be considered flat topped hills. The best WECS sites on such buttes appear to be along the windward edge. Figure 6.23 shows some flow patterns over and around mesas and buttes. In Figure 6.23(a), the wind accelerates over the top, although not as much as over triangular or rounded hills. When a mesa or butte is located in an area where the winds are already enhanced by valley funneling or other effects, additional power benefits may be gained.

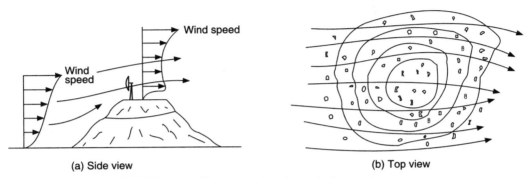

(a) Side view   (b) Top view

FIGURE 6.23 Flow around and over buttes and mesas.

6.6 ECOLOGICAL INDICATORS

Vegetation deformed by average winds can be used both to estimate the average speed and to compare candidate sites. This technique works in regions:

- Along coasts
- In river valleys and gorges exhibiting strong channeling of the wind
- In mountainous terrain.

The most easily observed deformities of trees illustrated in Figure 6.24 are listed and elaborated below:

Brushing   Branches and twigs bend downwind like the hair of a pelt which has been brushed in one direction only. This deformity can be obseved in decidous trees after their leaves have fallen. It is the most sensitive indicator of light winds.

Flagging   Branches stream downwind, and the upwind branches are short or have been stripped away.

Throwing   A tree is wind-thrown when the main trunk and the branches lean away from the prevailing wind.

Clipping   Because strong winds prevent the leader branches from extending up to their normal height, the tree tops are held to an abnormally low level.

Carpeting   The winds are so strong that every twig reaching more than several cms above the ground is killed, allowing the carpet to extend far downwind.

Figure 6.24 uses the Griggs-Putnam classification of tree deformities described by indices from 0 to VII. Following points summarizes the ecological indicators:

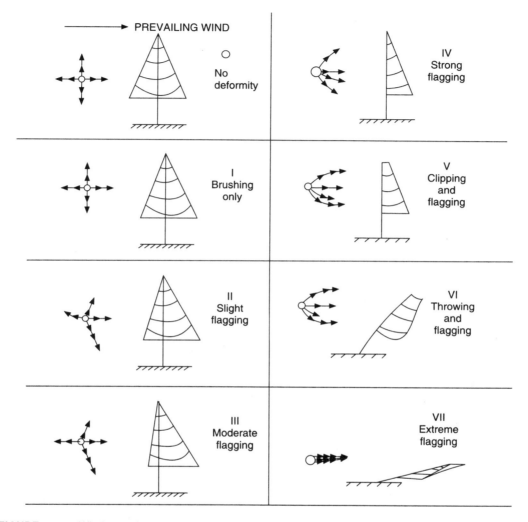

FIGURE 6.24 Wind speed rating scale based on the shape of the crown and degree twigs, and trunk are bent.

- Detect ecological indicators of strong wind.
- Compare isolated trees within the strong wind areas to select candidate sites.
- Consider flow patterns over barriers, terrain features and suface roughness in the final selection.
- Measure the wind to ensure that the best site in complex terrain is selected.
- Base selection of a particular WECS and estimation of its power output on wind measurements, not on ecological indicators alone.

6.7 SITE ANALYSIS METHODOLOGY

Table 6.4 presents three general approaches to site analysis and the respective advantages and disadvantages of each.

TABLE 6.4 Various approaches to site analysis

Method	Approach	Advantages	Disadvantages
1.	Use wind data from a nearby station: Determine power output characteristics	Little time or expense required for collecting and analysing data. If used properly, can be acceptably accurate.	Only works well in large areas of flat terrain where average annual wind speeds are 4.5 m/s or greater
2.	Make limited on-site measurements estimate rough correlation with nearby stations; compute power output using adjusted wind data.	If there is a high correlation between the site and the station, this method should be more accurate than the first method.	Of questionable accuracy, particularly where there is reasonal modulation variation in correlation between the WECS site and the nearby station.
3.	Collect wind data for the site and analyze it to obtain power output characteristics.	Most reliable method. Works in all types of terrain.	Requires at least a year of data collection. Added costs of wind recorders. Data period should represent typical wind conditions.

- **Use of available data:** This method uses only wind data collected at a representative weather station, which is a station that can be expected to have wind characteristics similar to the WECS site because of similar exposure to prevailing winds. Determining whether a nearby weather station is representative is not simple, wind conditions can vary significantly over short distances. The relationship of the site and the weather station to local terrain is very important when using data from a nearby weather station.

    In rugged, hilly or mountainous terrain, however, the winds from a nearby station are usually not applicable for a site analysis.

- **Limited on-site data collection:** In this method, site analysis might be considered whenever nearby weather stations may not adequately represent the WECS site. Weather stations may not be representative for the following reasons:
    (i) If they have slightly different exposure to the prevailing winds than the WECS site; or
    (ii) If they have the same exposure but may be too far away to adequately represent the WECS site.

    Ideally, the anemometer would be placed at the same location and at the same height as the planned WECS. The wind monitoring station is installed and data is collected over a specified time interval. Then, the wind speed at the site is averaged and that average is divided by the average wind speed for the same period (say for three months) at a nearby weather station. The result of dividing the site's average wind speed by that of the weather station is the correction factor, which is then applied to the weather station's long-term wind speed averages to make those averages more representative of the WECS site.

- **Extended on-site data collection:** In this method of site analysis, extended on-site wind measurement is involved, usually for a full year or more. Although this method is more reliable, it is also more expensive and time-consuming. The following points are to be considered for extended on-site data collection:
  - A listing of the information needed to evaluate WECS economics and performancs.
  - An estimate of the time and money available for data analysis.
  - The actual siting of the wind monitoring station.

6.8 LAYOUT OF WIND FARM

The wind turbulence caused by upstream turbines interferes with the downstream turbines which leads to higher structural loads and fatigue damage on downstream turbines. It results in increased maintenance costs and adds to overall generation costs. The simulation by computational fluid dynamic technique is available nowadays to assess wind farm array efficiency.

Wake effects of one machine on another decide the spacing between the wind machines in a farm. Typical spacing between the machines in a wind farm is shown in Figure 6.25 and effects of spacing on energy loss is shown in Figure 6.26. Wind farming is a serious financial and technological challenge. A small miscalculation can lead one to a disastrous condition.

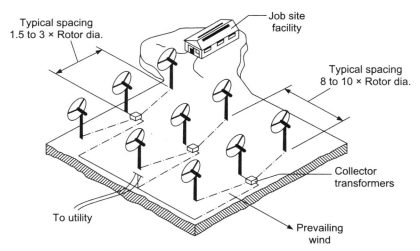

FIGURE 6.25 Typical spacing for a wind farm.

Wind development requires large areas of land. The actual footprint of wind development projects range from 2 per cent to 5 per cent for the turbines and related infrastructure. An important factor to note is that, wind energy projects use the same land area each year, coal and uranium must be mined from successive areas, with total disturbed area increasing each year. In agricultural areas, land used for wind generation projects has the potential to be compatible with some uses because only a few hectares are taken out of production, and no mining or drilling is needed to extract the fuel.

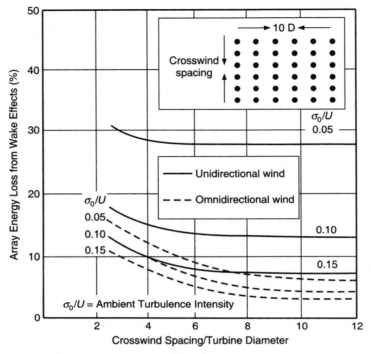

FIGURE 6.26 Effect of spacing on energy loss.

Estimates of temporary impacts range from 0.2 to 1.0 hectare per turbine; estimates of permanent habitat spatial displacement range from 0.3 to 0.4 hectare per turbine. Some studies found bird species displacement of 180 m to 250 m from the turbine. The wake behind a wind turbine rotor is illustrated in Figure 6.27. The array of wind turbines and their influence downstream is illustrated in Figure 6.28. The layout of turbines in a wind form typically follows a configuration shown in Figure 6.29 with spacing in the prevailing wind direction and across, in terms of multiples of diameter. A virtual turbine layout topography of turbines configuration is illustrated in Figure 6.30.

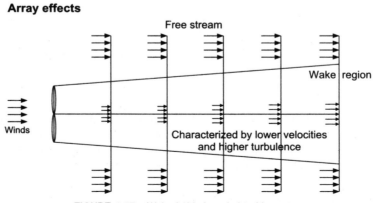

FIGURE 6.27 Wake behind a wind turbine rotor.

Siting, Wind Farm Design 219

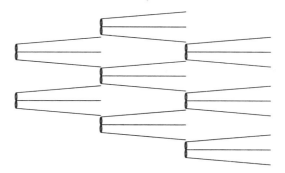

FIGURE 6.28 Wind turbines in an array.

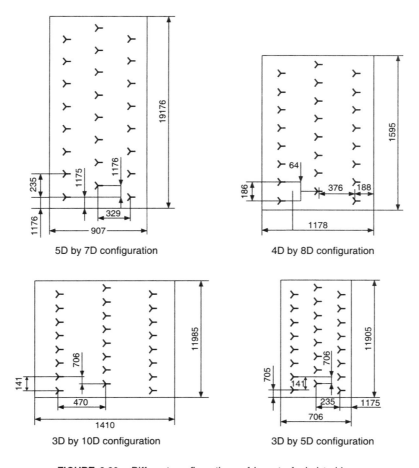

FIGURE 6.29 Different configurations of layout of wind turbines.

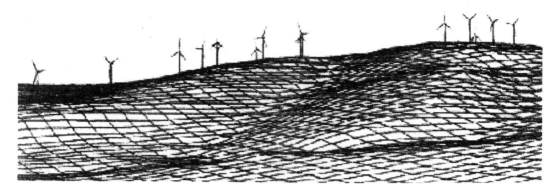

FIGURE 6.30 The appearance from defined viewpoints: 'wireframe' representations of the topography.

6.9 INITIAL SITE SELECTION

To locate a suitable site and to confirm it as a potential candidate for the location of a wind farm, the power production for a wind turbine (assuming 100 per cent availability) is given by

$$E = T \int P(U) f(U) \, dU \qquad (6.22)$$

where $P(U)$ is the power curve of the wind turbine, $f(U)$ is the probability density function (PDF) of the wind speed, and T is the time period.

The PDF is based on Weibull distribution and takes account of regional climatology, roughness of the surrounding terrain, local obstacles and topology. PDFs are calculated for 12.30° sectors and integrated with the power curve usually using numerical evaluation techniques.

On-site measurements of wind characteristics (speed and direction) are required for an accurate potential assessment. If measured site wind speed data is available, then the energy yield of a wind turbine can be estimated as shown in Figure 6.31 by combining the binned wind speed distribution with the power curve:

$$\text{Energy} = \sum_{i=1}^{i=n} H(U_i) P(U_i) \qquad (6.23)$$

where $H(U_i)$ is the number of hours in wind speed bin U_i, $P(U_i)$ is the power output at that wind speed and there are n wind speed bins.

It is also necessary to confirm that road access is available. For large wind farms the heaviest piece of equipment is likely to be the main transformer. The local electricity utility should be able to provide information on the amount of generation which the distribution network can accept, although for a first approximation it may be useful to consider rules-of-thumb as indicated in Table 6.5.

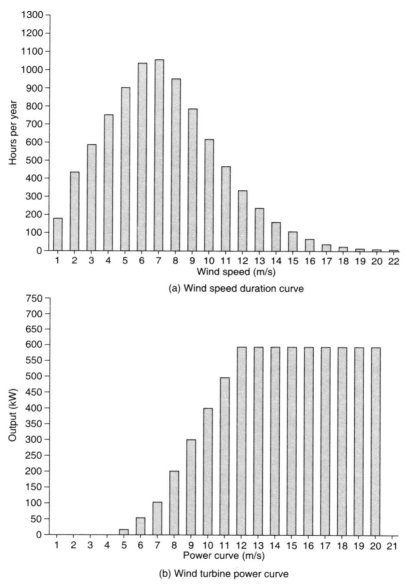

FIGURE 6.31 Annual energy calculations of a wind turbine by combining (a) and (b).

TABLE 6.5 Indication of possible connection of wind farms

Location of connection	Maximum capacity of wind farms (MW)
Out on 11 kV network	1–2
At 11 kV bus bars	8–10
Out on 33 kV network	12–15
At 33 kV bus bars	25–30
On 132 kV network	30–60

6.10 MEASURE-CORRELATE-PREDICT (MCP) TECHNIQUE

The MCP approach is based on taking a series of measurements of wind speed at the wind farm site and correlating them with simultaneous wind speed measurements made at a meteorological station. The averaging period of the site-measured data is chosen to be the same as that of the meteorological station data. In its simplest implementation, linear regression is used to establish a relationship between the measured site wind speed and the long-term meteorological wind speed data of the form:

$$U_{site} = a + bU_{long\text{-}term} \tag{6.24}$$

Coefficients are calculated for the 12.30° directional sector and the correlation for the site applied to the long-term data record of the meteorological station. This allows the long-term wind speed record held by the meteorological station to be used to estimate what the wind speed at the wind-farm site would have been over the last, say, 20 years. It is then assumed that the site long-term wind speed is represented by this estimate—which is used as a prediction of the wind speed during the life of the project. Thus, MCP requires the installation of a mast at the wind farm site on which are mounted anemometers and a wind vane. There are many difficulties in using MCP including:

- With modern wind turbines, high site meteorological masts are necessary and these may themselves require planning permission.
- There may not be a suitable meteorological station nearby (within say 50–100 km) or with a similar exposure and wind climate.
- The data obtained from the meteorological station may not always be of good quality and may include gaps. So, it may be time-consuming to ensure that it is properly correlated with the site data.
- It is based on the assumption that the previous long-term record provides a good estimate of the wind resource over the lifetime of the wind farm.

Conventional MCP techniques assume that the wind direction distribution at the site is the same as that of the meteorological station. Investigations by Addison et al. suggest that this assumption is a source of significant error and a correlation technique based on artificial neural network is proposed. Using this approach, an improvement in predictive accuracy of 5–12 per cent was obtained.

6.11 MICROSITING

The conventional MCP technique is well established, and specially designed site data loggers, temporary meteorological masts and software programs for data processing are commercially available. The estimate of the long-term wind speeds obtained from MCP may then be used together with a wind-farm design package to investigate the performance of a number of turbine layouts. These programmes take the wind distribution data and combine them with topographic wind-speed variation and the effect of the wakes of the other wind turbines to generate the energy yield of any particular turbine layout. The constraints such as turbine separation, terrain slope, wind turbine noise and land boundaries may also be applied. Optimization techniques are then used to optimize the layout of the turbines for maximum energy yield of the site, taking into account local wind speeds and wake effects. The programmes have also visualization facilities to generate zones of visual impact, views of the wind farm either as wire frames or as photomontages.

For more detailed investigation of the proposed site, careful assessment of existing land use and how best the wind farm may be integrated with, for example, agricultural operations should be made. The ground conditions at the site also need to be investigated to ensure that the turbine foundations, access roads and construction areas can be provided at reasonable cost. Local ground conditions may influence the position of turbines in order to reduce foundation costs. It may also be important to undertake a hydrological study to determine whether spring water supplies are taken from the wind farm site and if the proposed foundations or cable trenches will cause disruption of the ground water flow. More detailed investigation of the site access requirements will include assessment of bend radii, width gradient, and any weight restrictions on approach roads.

As a rough guide, the installed capacity is likely to be of the order of 12 MW/km^2, unless there are major restrictions that affect the efficient use of available land. A key element in the layout design is the spacing used between the turbines. The appropriate spacing for turbine is strongly dependent on the nature of the terrain and the wind rose at a site. For areas with predominantly uni-directional wind roses greater distances between turbines in the prevailing wind direction will prove to be more productive.

The detailed design of the wind farm is facilitated by the use of commercially available wind farm design tools. Once an appropriate analysis of the wind regime at the site has been undertaken, a model is set up, which can be used to fine-tune the layout, predict the energy production of the wind farm and address a number of economic and planning related issues. Some of the research findings and thumb rules are summarised below:

- The distance that the turbine should be from each other for minimum wind interference is three rotor diameters (D) when aligned perpendicular to the wind, and ten rotor diameter (D) when aligned parallel to the wind direction.
- The minimum area required around each turbine in a single line where wind is generally from one direction would be $13D \times 3D$.
- For stand-alone turbine, the area required would be $13D \times 6D$ oval at a site where the wind is generally from one direction and to a $20D$, wind circle to use the wind from any direction.
- The GE 1.5 MW turbine, with a 70.5 m rotor span requires at least 48 acres per tower in a single line perpendicular to wind, 32 acre/MW and 123 acre per tower and 82 acre/MW in an array.
- Each Vestas V90, 1.8 MW turbine with a 90 m rotor requires 78 acres per tower in a single line perpendicular to the wind and 200 acre in an array. For single, it comes out 43 acre/MW and in the array it would be 111 acre/MW.
- According to Tom Gray of AWEA, the rule of thumb is 60 acre per MW for wind farm on land.
- The capacity is different from actual output. Typical average is only 25 per cent of the capacity, so the area required for a MW of actual output is 4 times the area listed for a MW capacity. And because 2/3 of the time wind turbines produce power at a rate far below average, even more X3 perhaps for a total of X12, dispersed across a wide geographical area–would be needed for any hope of a steady supply.
- The minimum area in an array would be 10D × 10D.

- If turbines are spaced closer than 5D, it is likely that unacceptably high wake losses will result.
- Tighter spacing means that turbines are more affected by turbulence from the wake of upstream turbines. This will create high mechanical loads and will require approval from the turbine supplier.
- Turbine loads are also affected by natural turbulence caused by obstructions, topography surface roughness, thermal effects extreme winds, etc.

6.12 WAKE MODELS

The calculation of wake effects employs a systematic approach where each turbine is considered in turn in order of increasing axial displacement downstream. The first turbine considered is not subjected to wake effects. The first turbine's wind speed, the thrust coefficient and the tip-speed ratio are calculated. There are two wake models described here:
- Park model and modified park model
- Eddy viscosity model

Due to the complexity of the wake directly behind the rotor, all models are initiated from two diameters downstream. This is assumed to be the distance where pressure gradients no longer dominate the flow.

6.12.1 Park and Modified Park Model

Park and modified park model, a two-dimensional model, uses the momentum theory to predict the initial profile assuming a rectangular wind speed profile and that the wake expands linearly behind the rotor. Figure 6.32 outlines the flow field used by this model.

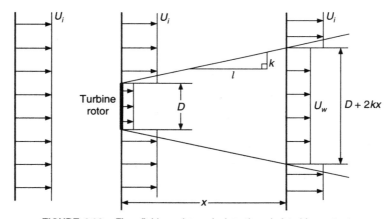

FIGURE 6.32 Flow field used to calculate the wind turbine output.

The downstream wind speed is calculated by

$$U_w = U_i \left[1 - (1 - \sqrt{1 - C_t}) \left(\frac{D}{D + 2kx} \right)^2 \right] \qquad (6.25)$$

where U_w = downstream wind speed
U_i = unperturbed upstream wind speed
D = rotor diameter
C_t = turbine thrust coefficient

Here k is the wake decay constant that is defined by the following expression:

$$k = \frac{A}{\ln\left(\frac{h}{Z_o}\right)} \quad (6.26)$$

where A is a constant which is if equal to 0.5 means that k is set at the same value for all wind directions. This assumes that there is no significant variation in surface roughness length over the site and the surrounding area.

If the turbine under consideration is in the wake of another turbine, the initial wake velocity deficit is corrected from the incident rotor wind speed to the force stream wind speed. This correction is necessary in order to ensure that at distances far downstream, the wake wind speed will recover to the free stream value rather than that incident on the rotor. Therefore, the initial central line velocity U_{wi} is scaled by the ratio of the average influx velocity U_i and the free upstream wind velocity according to the following equation:

$$U_w = \left(\frac{U_i}{U_o}\right) U_{wi} \quad (6.27)$$

To combine the wakes of two wind turbines onto a third turbine, the deficit of kinetic energy originating from each turbine is added and averaged over the incident rotor area of the third turbine.

The modified park model incorporates that if a turbine is in more than one wake, the overall wake effect is taken as the largest wind speed deficit, and other smaller wake effects are neglected.

6.12.2 Eddy Viscosity Model

The eddy viscosity wake model is a calculation of the velocity deficit field using a finite-difference solution of the thin shear layer equation of the Navier-Stokes equations in axi-symmetric co-ordinates. The eddy viscosity model observes the conservation of mass and momentum in the wake. An eddy viscosity, averaged across each downstream wake section, is used to relate the shear stress to gradients of velocity deficit. The mean field can be obtained by a linear superposition of the wake deficit field and the incident wind flow. Figure 6.33 shows wake profile.

The Navier-Stokes equations with Reynolds stresses and the viscous terms dropped give

$$U\frac{\partial U}{\partial x} + V\frac{\partial U}{\partial r} = -\frac{1}{r}\frac{\partial (\overline{ruv})}{\partial r} \quad (6.28)$$

The turbulent viscosity concept is used to describe the shear stresses with an eddy viscosity defined as:

$$\varepsilon(x) = L_m(x) \cdot U_m(x) \quad (6.29)$$

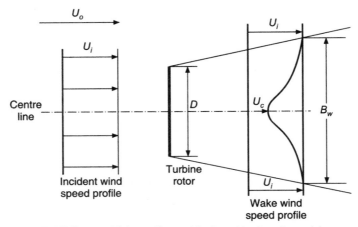

FIGURE 6.33 Wake profile used in the eddy viscosity model.

and
$$-\overline{uv} = \varepsilon \frac{\partial U}{\partial r} \tag{6.30}$$

L_m and U_m are suitable length and velocity scales of the turbulence as a function of the downstream distance x, but independent of r. The length scale is taken as proportional to the wake width B_w and the velocity scale is proportional to the difference $U_i - U_c$ across the shear layer.

Equation (6.30) permits the shear stress term $\overline{uv}$ to be expressed in terms of the eddy viscosity. The governing differential equation to be solved becomes:

$$U \frac{\partial U}{\partial x} + V \frac{\partial U}{\partial r} = \frac{\varepsilon}{r} \frac{\partial (r \partial U / \partial r)}{\partial r} \tag{6.31}$$

When considering the ambient wind flow for a wind farm, it must be considered as turbulent. Therefore, the viscosity in the wake cannot be wholly described by the shear contribution alone and thus an ambient term is included. Hence, the overall eddy viscosity is given by:

$$\varepsilon = F K_1 B_w (U_i - U_c) + \varepsilon_{amb} \tag{6.32}$$

where the filter function F is a factor applied for near wake conditions. This filter can be introduced to allow for the build up of turbulence on wake mixing. The dimensionless constant K_1 is a constant value over the whole flow field and the deficit value is 0.015. The ambient eddy viscosity term is calculated by the equation proposed by Ainslie:

$$\varepsilon_{amb} = \frac{F \cdot K_k^2 \cdot I_{amb}}{100} \tag{6.33}$$

K_k is the Von Karman constant with a value of 0.4. As a result of comparison between the model and measurements reported by Taylor, the filter function F is fixed at unity.

6.12.3 Initialization of the Model

The centre line velocity deficit D_{mi} can be calculated at the start of the wake model (two diameters downstream) using the following empirical equation proposed by Ainslie:

$$D_{mi} = 1 - \frac{U_c}{U_i} = C_t - 0.05 - [(16 C_t - 0.5) I_{amb} / 100] \tag{6.34}$$

Assuming a Gaussian wind speed profile and momentum conservation, an expression for the wake width is obtained:

$$B_w = \sqrt{\frac{3.65 C_t}{8 D_m (1 - 0.5 D_m)}} \qquad (6.35)$$

The wake width B_w used is defined as 1.89 times the ball-width of the Gaussian profile.

Using the above equations, the average eddy viscosity at a distance. $2D$ downstream of the turbine can be calculated.

However, if the turbine under consideration is in the wake of another turbine, this initial wake velocity deficit is corrected from the incident rotor wind speed to the free stream wind speed. This correction is necessary in order to ensure that at distance far downstream, the wake wind speed will recover to the free stream value rather than that indicent on the rotor. Therefore, the initial centre line velocity D_{mi} is scaled by the ratio of the average influx velocity U_i and the free upstream wind velocity according to the following formula:

$$D_m = \left(1 - \frac{U_i}{U_o}\right) + \left(\frac{U_i}{U_o}\right) D_{mi} \qquad (6.36)$$

For each downstream turbine that falls inside the wake, the incident wind speed needs to be calculated. The velocity profile across the turbine affected by wake is calculated by assuming a Gaussian profile based on the centre line velocity at that distance downstream. If some of the rotor is outside the wake, then the wind speed for that portion of the rotor is set as the incident wind speed of the turbine creating the wake. The velocity profile across the turbine rotor at the hub height is integrated to produce a mean wind speed incident across the rotor at the hub height.

6.12.4 Turbulence Intensity

The eddy viscosity model relies on a value of incident ambient turbulence intensity for Eq. (6.33) and Eq. (6.34). For a turbine in the free wind stream, the calculation must be initiated using the ambient turbulence level. For a turbine within a wind farm, it is necessary to calculate the increased turbulence level which results from the presence of upstream turbines.

Wind farm turbulence levels are calculated using an empirical characterization developed by Quarton and Ainslie. This characterization enables the added turbulence in the wake to be defined as a function of ambient turbulence, the turbine thrust coefficient, the distance downstream from the rotor plane and the length of the near wake:

$$I_{add} = 4.8 C_t^{0.7} I_{amb}^{0.68} \left(\frac{X}{X_n}\right)^{-0.57} \qquad (6.37)$$

where X_n is the calculated length of the near wake. The characterization was subsequently amended slightly to improve the prediction,

$$I_{add} = 5.7 C_t^{0.7} I_{amb}^{0.68} \left(\frac{X}{X_n}\right)^{-0.96} \qquad (6.38)$$

Using the value of added turbulence and the incident ambient turbulence, the turbulence intensity at any turbine position in the wake can be calculated. The model also accounts for the turbine not being completely in the wake. The ambient turbulence intensity is best derived from measurements. The turbulence intensity can be calculated from an input surface roughness length, which is representative of the site,

$$I_{amb} = \frac{1}{\ln(h/z_o)} \qquad (6.39)$$

The turbulence intensity is defined here as the quotient of standard deviation σ and mean wind speed $\bar{U}$ at high wind speeds.

$$I = \frac{\sigma}{\bar{U}} \qquad (6.40)$$

6.13 RE-POWERING

Re-powering in wind energy means replacement of installed old wind turbines of lower capacity by modern turbines of higher capacity normally in lesser numbers. It also refers to replace first generation turbines installed for more than fifteen years back. It has generally been accomplished by installing fewer, larger capacity turbines. The modern commercial grid connected wind turbines are multi-megawatt machines equipped with advanced operation and control systems. As per a study carrried out at *Risoe Laboratory, Denmark,* the process of re-powering will double the capacity, triple the energy yield with half the infrastructure. Another report of *Leonardo Energy* outlines the re-powering programme for following factors:

- More annual energy production from the same site and land area as capacity is multiplied without additional land.
- Fewer wind turbines than earlier which improves the appearance of landscape. The hub height is also increased after re-powering.
- Higher efficiency of individual turbines and higher array efficiency with usually reduced cost per unit energy generation.
- Better visual appearance as modern wind turbines rotate with lower rotational speeds.
- Possibly better grid integration due to improved power electronics and power quality from multi-megawatt modern turbines.
- Wind characteristics are known in terms of speed, direction, turbulence, therefore, better prediction of annual energy generation of an existing site.

In one of the Examples of re-powering in Germany (Bundesverband Windenergie), the investment was increased by 3 fold, annual energy yield increased by 4 fold, installed capacity was increased by 3.5 fold for replacing 20 old turbines by 7 new turbines.

Wind farms in India were developed in the areas where the wind regime is often very good. Therefore, the best windy locations are occupied by old or first generation wind turbines. These sites could benefit by replacing old turbines by modern machines of higher hub heights. New locations for wind farms are becoming less and less available due to scarcity of land, environmental protection, green belt requirements and sometime resistance from local people. It offers a huge opportunity and great technical and economical challege to replace old turbines. Re-powering priority is to be given for wind farms with low plant load factor (PLF).

Issues and challenges

Normally wind turbines are designed for a service life of about 25 years. Replaced turbines due to re-powering are usually not installed on the same site, these can be sold for installation at elsewhere for their remaining service life or sold for scrap for recycling of different parts. In a wind farm, turbines are placed in rows, typically perpendicular to prevailing wind direction by micro siting process involving flow modelling, micro surveys and wind monitoring, and determining array efficiency and turbulens in the downstream. The challenges for re-powering are many and some are identified as follows:

Turbine ownership: Issue of ownership is to be resolved in cases where more number of turbines are replaced by few and one to one replacement is not possible.

Land ownership: Multiple ownership of wind farm land is to be resolved.

Power purchase agreement (PPA): PPA might have been signed for long duration and before end of that period re-powering may pose difficulties.

Electricity evacuation: The grid is designed to handle current power supply, but enhanced power output due to re-powering may require modification or replacement of equipment and systems.

Additional cost: The decomisioning cost of existing turbine is to be estimated.

Disposal of existing turbines: Many options are to be analysed like scrap value, buy-back by manufacturer, relocation, etc.

Re-powering approach

To determine the re-powering potential of an existing wind farm site, following technical aspects are to be considered:

- Wind resource of the site in terms of speed frequency distribution curve of past several years, Weibull parameters, wind power density, turbulence intensity, power law index, wind rose, prevailing wind direction and other characteristics.
- Existing wind turbines, numbers of turbines, capacity, power curve, cut-in, rated and cut-out speed, rotational speed per minute, hub height, type of generator, rating, thrust and power coefficient, etc.
- Proposed wind turbine, capacity and detailed specifications.
- Available area with necessary off-set from approach roads and dwellings.
- Estimation of annual energy production, gross and net energy production and array efficiency from proposed re-powering turbines at new hub height.
- Plan(s) of proposed wind turbines with micro-siting and turbine array spacing(s).
- Re-powering ratio of proposed turbines to existing turbines.
- Energy yield ratio from proposed and existing turbines for same site or area.
- Economic analysis of unit cost of energy after re-powering, dismantling cost of existing wind turbines, cost of new turbines installation, and commissioning and cost of other modified facilities.
- Master plan approach to make the re-powering a dynamic exercise.

High capacity wind turbines are particularly well adapted for sites with higher average wind speeds. Report of *Leonardo Energy* outlines:

For utilities trying to scale-up their capacity to achieve set target put preference for larger wind turbines.

The practical reason is that power output of a wind turbine depends on the square of rotor diameter, therefore, larger wind turbine is better than two smaller wind turbines of same capacity. The larger wind turbines are preferred for off-shore applications as it minimises installation cost per MW in terms of foundation cost which takes significant proportion of total cost of installation.

Techno-economic analysis

Flow modelling requires three essential sets of technical data related to wind, site and turbine:

- Wind characteristics
- Site characteristics, and
- Wind turbine and farm characteristics

Wind characteristics: For the development of wind energy, the site characterization has to include following major parameters and information: annual average wind speed (WAsP requires the wind data measured at 10 minutes intervals by a meteorological mast in the same region), wind power density, wind rose, wind resource map, prevailing wind direction, speed frequency distribution and persistence, vertical wind speed profile, wind shear exponent, Weibull shape parameter (k), scale parameter (c), turbulence intensity, wind density and its variation vertically and seasonally, historical wind data (including frequency and intensity of past storms) and so on.

Site characteristics: Location (latitude, longitude and mean sea level), topographic maps provide the analyst with a preliminary look at site attributes, including available land area, contour map, roughness class of the site, grid-related data, transmission line map, positions of existing roads and dwellings, land cover (e.g. forests), political and administrative boundaries, parks, national parks, forest reserves, restricted areas, proximity to transmission lines, location of significant obstructions, potential impact on local aesthetics, cellular phone service reliability for data transfers, and other infrastructural facilities.

Wind turbine and farm characteristics: Wind farm layout of existing and after re-powering, power and thrust curve of the turbine(s), hub height, rotor diameter, pitch mechanism, braking mechanism, generator type, gear or gearless machine, type of power electronics, cut-in, rated and cut-out speeds, capacity factor, etc.

In addition to above, the parameters related to machine availabilty, grid availabilty, transmission losses and WAsP prediction error are to be considered for arriving at annual energy production (AEP) value. The annual energy production of the wind farm after re-powering can be predicted by using industry standard WAsP (Wind Atlas Analysis and Application Programme) and simulation can be performed by using MATLAB. Following parameters are to be determined after analysis for re-powering:

- Improvement in plant load factor (PLF)
- Increase in annual energy production (AEP) or energy yield
- Increase in installed capacity of wind farm
- Change in reactive power consumption

The replacement of old wind turbines with modern versions can be done in several ways. According to studies undertaken by *Grontmij* for replacement of existing wind turbines, following alternatives are suggested:

- One to one replacement of wind turbines with similar capacity, but with newer machines;
- One to one up-scalling of solitary wind turbines;
- Replacement of two smaller wind turbines by one large wind turbine;
- Clustering of solitary wind turbines into farms, for example, 20 solitary wind turbines by clustering 6 to 10 wind turbines at one location; and
- One to one up-scaling of wind farms.

For each alternative, there is positive impact on landscape and increase in annual energy production. The best alternative is the fourth one of replacing larger cluster with small cluster. The capacity after re-powering will increase, and quality of landscape will also improve. The benefits of re-powering of existing wind farms are as follows:

- Annual energy production will increase as taller turbines access the increased wind speeds at higher altitude, and thereby have better power-wind velocity curves;
- The revenue generation will increase so business model will become more profitable in majority of cases;
- Landscape appearance will improve due to newer turbines;
- Number of turbines will decrease and more land will be available;
- Possible reduction in visual interference;
- Possible reduction in noise level;
- Possibly the new turbines will be placed at more acceptable locations.

REVIEW QUESTIONS

1. List various geographical features of land and their signficance for consideration of wind turbine siting.
2. How the slopes of hills affect the siting of wind turbines?
3. What are the factors and criteria for siting of individual wind turbine (stand alone systems) and a cluster of wind turbines (wind farm)?
4. Explain the various factors influencing the wind farm design.
5. Explain wind rose and its significance in wind turbine siting.
6. A constant speed pitch regulated wind machine with 60 m diameter, 14.32 rpm and 1.0 MW rated generator capacity is installed at a site with annual values of Weibull parameters C and k given by 7.2 m/s and 1.8, respectively. It has cut-in speed of 4 m/s and cut-off speed of 27 m/s. It has 3 blades and the mechanical and electrical efficiencies for all wind speeds are given as 0.95 and 0.96, respectively. Its coefficient of performance, C_p as function of tip speed ratio, λ is given by,

$$C_p = 0.089 + 0.0123\lambda^2 - 0.00091\lambda^3$$

Find the electrical power developed by the machine as function of speed from cut-in speed to cut-off speed in the step of 1 m/s. Take air density as 1.165 kg/m³.
(a) Identify the speed at which regulation should start.
(b) Find its Annual Energy Output (AEO) at the site.
(c) Find for how many hours this machine will run in a year?
(d) What is the capacity factor of the machine?
(e) What is the probability of the wind speed exceeding the furling velocity at this site?
(f) Find the most frequent wind speed.

7. A 3-bladed propeller wind machine fixed with 220 kW generator has diameter = 30 m and tower height = 30 m. It is designed to have C_p = 0.39 at wind speed of 11 m/s when regulation starts. The cut-in and furling speeds for the machine are 3.5 m/s and 25 m/s, respectively. Mechanical and electrical efficiencies of the machine are 0.96 and 0.97, respectively.

(a) What is the power developed by the rotor at the wind speed = 11 m/s? When the regulation starts? If it is calculated for mean air density at the surface = 1226 gm/cu.m.
(b) If tip speed ratio at this wind speed is 4.2, then what is the speed of rotation of the machine?

8. It has been decided to instal a small wind turbine to supply some of the electricity and domestic hot water requirements of a house.

(a) Discuss the factors which determine the size and location of such a wind turbine.
(b) Outline possible systems which use the wind power and comment on those which are likely to be most cost-effective.
(c) What back-up or additional energy storage might be necessary?

9. From Table 6.6, calculate annual specific output, kWh/kW, for two different wind turbines.
10. From Table 6.6, calculate capacity factor for Fayette, Vestas 23 and Bonus 120.

TABLE 6.6 Specific output, kWh/m² for wind turbines

Turbine	Diameter (m)	Rated (kW)	No. units	Capacity (MW)	Per turbine (kWh)	(kWh/m²)
Fayette	10	90	1,363	123	41,000	522
Bonus 65	15	65	644	42	113,000	640
Vestas 15	15	65	1,330	86	53,000	300
Micon 60	16	60	531	32	95,000	473
Nordtank 60	16	60	152	9	170,000	846
Micon 65	16	65	126	8	184,000	916
Nordtank 150	16	65	375	24	100,000	498
Vestas 17	17	100	1,071	107	145,000	639
U.S. Windpower	18	100	3,419	342	220,000	865

(Contd.)

Siting, Wind Farm Design **233**

TABLE 6.6 Specific output, kWh/m² for wind turbines (*contd.*)

Turbine	Diameter (m)	Rated (kW)	No. units	Capacity (MW)	Per turbine (kWh)	(kWh/m²)
Micon 108	20	108	967	104	230,000	732
Bonus 120	20	120	316	38	276,000	879
Carter 250	21	250	24	6	250,000	722
Nordtank 150	21	150	164	25	240,000	693
Flowind 19	21	250	200	50	142,000	410
Danwin 23	23	160	151	24	390,000	939
Vestas 23	25	200	20	4	434,000	885
WEG MS2	25	250	20	5	560,000	1,141
Mitsubishi	25	250	360	90	486,000	991
DWT 400[a]	35	400	35	14	1,000,000	1,040
[a]*Estimated kilowatt-hour*					Average	756

11. From Table 6.7, calculate the average capacity factor for 1989 through 1996.

TABLE 6.7 Enertech 44 wind turbine, fixed pitch, induction generator

Year	Operating time (h)	Connect time %	Energy (kWh)	Capacity factor %	Availability %	Wind speed (m/s)	Rated power (kW)
82	3,218	63.0	48,092	40	99.9	5.7	25
83	5,567	63.6	63,710	29	92.6	6.0	
84	4,611	52.6	72,295		86.3	5.9	40
85	4,662	55.5	91,732	17	94.9	5.6	60
86	4,121	47.1	77,522	15	82.1	5.7	
87	3,850	44.0	65,638	12	81.0	5.6	
88	3,971	45.3	71,643		77.0	5.6	40[a]
89	5,893	67.3	83,452	19	99.4	5.3	
90	5,831	66.6	86,592	20	97.5	5.6	
91	5,705	65.1	82,390	19	96.6	5.9	
92	5,641	64.6	73,510	17	98.0	5.4	
93	5,754	65.9	88,363	17	96.4	5.7	
94	5,769	66.4	79,392	18	95.7	5.6	
95	4,099	46.8	51,931	12	72.8	5.7	
96	4,991	56.8	76,470	17	86.8	5.8	
97	4,608	52.6	56,958	13	75.4	5.5	Hybrid
98	4,944	56.4	68,885	16	93.2	5.5	
99	4,487	51.2	65,147	15	93.3	5.7	
00	4,241	48.3	66.589	15	85.3	5.7	
					Average	5.7	

Note: Data for 1982 is not a full year. Wind speed at 10 m height.
[a]*60 kW generator, 40 kW gearbox.*

12. From Table 6.8, calculate the specific output, kWh/m² in (a) 1985 for Enertech 44/60 and (b) 1990 for Enertech 44/40.

13. From Table 6.8, calculate for 7 months for Enertech 44/25: (a) kWh/m² and (b) capacity factor.
14. From Table 6.8, calculate for 1985: (a) kWh/m² and (b) capacity factor.
15. From Table 6.8, calculate kWh/m² for May and August 1984. Does specific output depend on wind?

TABLE 6.8 Performance, Enertech 44/40 kW, 44/60 kW, Bushland, Texas, April 1984–September 1986 (Anemometer at 10 m)

Date	No. days	Operating time (h)	Connect time %	Energy (kWh)	Availability %	Average speed (m/s)
3/20/84–4/01/84					Shakedown	
4/02/84–4/30/84	29	571	82	11,148	100	7.4
May	31	568	76	9,078	99.7	6.4
June	30	511	71	8,281	100	6.3
July	31	430	58	5,017	100	5.0
August	31	302	41	2,443	99.7	4.1
September	30	461	64	7,240	100	5.8
October	31	412	55	6,260	100	5.3
Summary	213	3254	64	49,467	100	5.8
44/60						
11/17/84–11/30/84	17				Shakedown	
December	31	366	49	7,877	87.3	5.6
1985	365	4,897	56	91,732	94.9	5.7
January–September 1986	273	3,824	58	72,905	100	5.8
Summary	686	9,087	57	172,514	97	5.7

Information for Question 16–18: (Initial Cost = IC)

Specification	Turbine-I	Turbine-II
Diameter, m	24	27
No. of blades	2	3
Rated power, kN	300	225
Tower height, m	50	31.5
Installed cost (IC)	$100,000	$225,000
Estimated annual energy kWh	600,000	500,000
Weight specs, kg		
Rotor	1,340	2,900
Tower head (nacelle)	2,091	7,900
Tower	8,023	12,000
Guy cables, winch	1,336	
Control box and panel	155	
Total	**14,250**	**22,800**

16. For the turbine-I, calculate (a) kWh/m², (b) $IC/kW, (c) kWh/kg, and (d) kWh/$IC.
17. For the turbine-II, calculate (a) kWh/m², (b) $IC/kW, (c) kWh/kg, and (d) kWh/$IC.
18. Estimate the annual capacity factor for the turbine-I and turbine-II.
19. Go to the Vestas website, www.vestas.com (a) For the Vestas V82 (1.65 MW), estimate the annual kWh/m² for a good wind regime. (b) For the Vestas V90 (3 MW), estimate the annual kWh/m² for a good wind regime.
20. For an annual average wind speed of 6 m/s, compare the predicted annual energy production for the Enertech 44 for the 25 kW and 60 kW wind generators. Use Figure 6.34 for power curves and use Rayleigh distribution (1 m/s bin width).

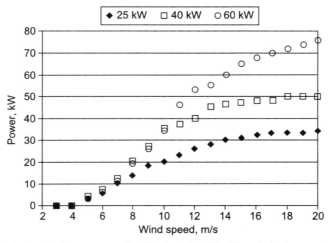

FIGURE 6.34 Power curves for Enertech 44 with different sized generators.

21. Select a wind farm that is close to your town or city. What is the installed capacity? How much electricity did it produce last year? If values are not available, estimate from installed capacity and capacity factor.
22. A building is 20 m by 15 m and 15 m tall. You want to install a 10 kW wind turbine. How tall a tower and how far away from the building would you place it?
23. There are a number of trees (20 to 30 m in height) close to a house. You want to install a 10 kW wind turbine. What is the minimum height of the tower? What is the approximate cost of that tower?
24. Refer Figure 6.7 and Table 6.2. The building is 15 m tall. What is the power production at 15 m height at a distance of 75 m downwind? At 150 m downwind? Would it be cheaper to use a taller tower or to move the location farther away from the building? Show all cost estimates.
25. Use equation,

$$\frac{P}{P_a} = \frac{\left[\ln\left[\dfrac{H_h + E}{Z_0}\right]\right]^3}{\left[\ln\left[\dfrac{H_h}{Z_0}\right]\right]^3}$$

P_a = average power/area from normalized wind map
H_h = hub height, 50 m
E = exposure, m
Z_0 = roughness length; crop land 0.03 m, crop land/mixed woodland 0.1–0.3 m, forest 0.8–1.0 m

Care must be taken in use of P_a. Do you use the bottom or the middle of the wind class? Do you limit the number of wind class changes to one, especially for mountains terrain?

Calculate the corrected power for a class 3 wind area if the terrain exposure is 80 m and area is grassland. Use the bottom and middle value for class 3.

26. Estimate the annual energy production for a 50 MW wind plant where the average wind power potential is 500 W/m² at 50 m height. Select the size of turbine from commercial turbines available today.

27. Do Question 26. However, now the land is high priced, so select close spacing and estimate array losses.

28. What size of land area do you need to lease for a 50 MW wind farm? Select the size of turbine from commercial turbines available today and the spacing. Remember, if your spacing between turbines is too close, you will have array losses. How many megawatts can you install per square kilometres?

29. The array spacing is 4D by 8D, for 3 MW wind turbines, 90 m diameter. How many can be placed in a square kilometre?

30. The row spacing is 2D for 3 MW wind turbines, 90 m diameter. How many wind turbines can be placed per linear kilometre on a ridge?

31. Assume you have complex terrain. What size of land area do you need to lease for a 50 MW wind farm?

32. What are some advantages and disadvantages of using vector or raster based GIs in determining wind energy potential?

33. Go to Flash Earth, www.flashearth.com and search for White Deer, Texas (latitude, N 35° 27′; longitude, W 101° 10′). The wind farm is just to the northwest of the town. Zoom in to see the layout of the wind farm. Estimate approximate number of wind turbines per square km for the wind farm. Remember, not all the land will have wind turbines on it within the area of wind farm.

34. Go to Flash Earth, www.flashearth.com, and search for the wind farms in San Gorginia Pass, California, just northwest of Palm springs. Estimate the spacing for one of the densely packed wind farms.

35. Go to www.topozone.com. Find quadrangle map that shows Mesa Redonda, New Mexico. It is in Quay County. What is the elevation of Mesa? You can see all of Mesa in 1:200,000 view. You will need 1:50,000 view to read elevation. Or go to www.newmexico.org/map/ or www.awstruewind.com (Wind Navigator) and use terrain map.

36. What is the general rule for MW/km² in plains and rolling hills? For MW/km for ridges and narrow mesas?

37. How many met stations, at what height, and at what time period are needed for determining the wind potential for a 50 MW wind farm or larger? In general, terrain will not be completely flat. Also remember, wind turbine are getting larger, which means hub heights are larger. For your selection of number, height, instrumentation, and time period, estimate the costs.

38. Wind velocity measured on a day at 10 min interval from a prospective wind farm site and meteorological station (10 years daily data averaged) are given in Excel file. Transform the meteorological data to that of wind farm site data using the "measurecorrelate and predict technique".

39. A wind farm has 5 wind turbines of 1.8 MW size arranged as shown in Figure 6.35. The cut-in, rated and cut-out velocities of the turbine are 4, 15 and 25 m/s respectively. The diameter of the turbine is 80 m and the spacing between the turbines in the wind direction is 400 m. Find out the velocity deficit at T_2 due to the wake effect from T_1, taking shadowing ratio 0 g 0.3. Use the WAKE calculator.

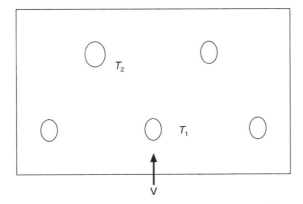

FIGURE 6.35 Wind farm showing locations of wind turbines.

40. Estimate the wind energy potential of a site for which the wind velocity data are given in the Excel file 2 using the WERA model.

41. A 2 MW wind turbine with cut-in, rated and cut-out velocities 3.5 m/s, 13.5 and 25 m/s respecctively is to be installed at a site. Estimate the performance of the wind turbine using the Weibull model.

42. Specification of four commercial wind turbines are given below. Select the turbine best suited for the wind regime described in Table 6.9.

Specification	*Wind turbine*			
	1	2	3	4
Rated power	250	250	250	250
Rated velocity	13.5	13	15	15
Cut-in velocity	3.5	4	3	4
Cut-out velocity	25	25	25	25

TABLE 6.9

Wind farm site velocity, m/s				Meteorological station velocity, m/s			
9.6	9.4	9.4	8.9	8.7	6.5	11.3	3.1
9.7	8.7	9.7	9.3	8.2	5.3	10.3	3.3
8.9	8.5	9.4	10.6	8.1	5.1	9.4	4.2
9.3	8.4	8.6	11.6	8.6	3.9	9.3	6.9
9.9	7.9	9.4	11.6	8.1	3.5	9	8.1
9.9	7.9	9.1	12.6	8.3	3.3	9.4	8.3
11.2	6.9	7.7	12.4	9.9	3.4	9.1	9.1
11.3	7.2	7.4	12.4	9.2	3.2	9.2	9.2
11.7	8.4	6.3	11.2	9	2.9	8.8	10.1
13.8	6.8	5.7	10.6	10.1	3.4	8.9	8.5
13.6	6.3	4.4	10.1	11.2	3.9	7.2	9
14.3	7.8	4.2	9.4	13.1	4.4	7.9	8.9
13.9	10	4.1	10.1	13	5.4	7.4	7.5
13.3	10.8	3.6	10.4	12.7	7.1	6.9	7.9
13.6	10.6	4.5	10.4	12.4	8.4	7.1	8.1
13.1	10.7	3.8	9.4	11.6	8.2	6.1	6.9
11.1	9.7	4.4		10.5	7.9	6.9	
11.1	10.4	4.6		10.4	8.1	7.4	
10.6	10.3	5.6		8.4	10.1	5.9	
10.8	11.2	6.6		9.7	11.1	5.5	
10.7	10.5	8.1		8.1	11.3	6.4	
11.2	10.4	10.2		9.2	9.7	9.1	
8.5	9.4	9.4		7.7	6.9	9.4	
11.6	9.4	10.1		9.5	5.5	9.3	
9.6	9.7	10.7		7.9	5.7	9.4	
8	8.7	11.8		7.1	6.1	8.4	
8.7	8.4	12.4		7.3	6	9.5	
9.1	6.4	12.4		7.9	5.9	9.4	
9.4	5.9	10.7		8.4	7.5	10.3	
8.8	5.4	8.4		8.4	7.5	9.3	
10.6	4.6	6.3		9.1	5.7	9.1	
10.5	4.1	6.6		9.2	6.9	9.6	
10.9	4.5	7		9	6.6	8.9	
12.1	4.1	6.8		10.7	8.6	8.4	
10.8	4.6	6.7		8.9	7.3	7.7	
11.5	5.5	8.4		10.1	7.4	7.4	
11.1	7.2	8.3		10.6	8.2	6.1	
9.8	9.1	6.6		8.1	9.7	5	
10.5	9.6	7.3		9.2	9.9	4.9	
9.1	11	7.8		8.4	10.1	4.1	
8.4	10.5	9.1		7.6	11.1	3.7	
8.9	11.2	8		7.4	10.5	3.8	

CHAPTER 7

Wind Energy Economics

7.1 INTRODUCTION

The economic appraisal of wind energy involves a number of factors. They include:
- Annual energy production from the wind turbine installation
- Capital cost of the installation
- Annual capital charge rate, which is calculated by converting the capital cost plus any interest payable into an equivalent annual cost
- Length of the contract with the purchaser of the electricity produced
- Number of years over which the investment in the project is to be recovered (or any loan repaid), which may be the same as the length of the contract
- Operation and maintenance cost, including maintenance of the wind turbines, insurance, land leasing, etc.

The annual energy produced by a wind turbine installation depends on the wind speed-power curve of the turbine, the wind speed frequency distribution at the site, capacity factor and availability of the turbine. As the cost of wind energy does not include the cost of fuel, it is relatively straightforward to determine as zero. Wind turbines are very quick to install, so they can be generating before they incur significant levels of interest during construction.

A number of factors determine the economics of utility scale wind energy and its competitiveness in the energy market place. These are as follows:

- The cost of wind energy varies widely depending upon the wind speed at a given project site. The energy that can be tapped from the wind is proportional to the cube of the wind speed, so a slight increase in wind speed results in large increase in electricity gneration. Consider two sites, one with an average wind speed of 6.22 m/s and the other with an average wind of 7.11 m/s. All other things being equal, a wind turbine at the second site will generate nearly 50 per cent more electricity than it would at the first location.
- Improvement in turbine design brings down costs: The taller the turbine tower and the larger the area swept by the blades, the more powerful and productive the turbine. Advances in electronic monitoring and controls, blade design, and other features have also contributed to drop in cost.
- A large wind farm is more economical than a smaller one. A larger project has lower operation and maintenance costs per kilowatt-hour because of the efficiencies of managing a larger wind farm.
- Optimal configuration of the turbines to take the best advantage of micro-features on the terrain will also improve a project's productivity.
- Transmission, tax, environmental, and other policies also affect the economics of wind energy. Transmisson and market access constraints can significantly affect the cost of wind energy. Since wind speeds vary, wind plant operators cannot perfectly predict the amount of electricity they will be delivering to transmission lines in a given hour. Deviations from schedule are often penalised without regard to whether they increase or decrease system costs. Interconnection procedures are not standardized, and utilities have on occasions imposed such difficult and burdensome requirement on wind plants for connection to transmission lines that wind companies have chosen to build their own lines instead. As electricity markets are restructured and long-term power purchase agreements give way to trading on power exchanges, transmission and market access condition will play an increasingly important role in the economics of a wind project.
- Stricter environmental regulations enhance wind energy's competitiveness. Wind power's environmental impact per unit of electricity generated is much lower than that of mainstream forms of electricity generation, as wind energy neither emits pollutants, wastes, or greenhouse gases, nor damages the environment through resource extraction.
- The project viability exclusively based on financial engineering principles is to be avoided.

7.1.1 Cost Calculation

The cost per unit of electricity generated, g, by a wind farm can be estimated using the following formula:

$$g = \frac{CR}{E} + M \tag{7.1}$$

where
C = capital cost of the wind farm
R = capital recovery factor or the annual capital charge rate expressed as a fraction
E = wind farm annual energy output
M = cost of the operating and maintaining the wind farm per unit of energy output

The capital recovery factor, R is defined as:

$$R = \frac{x}{1-(1+x)^{-n}} \tag{7.2}$$

where
  x = required rate of return net of inflation expressed as a fraction
  n = number of years over which the investment in the wind farm is to be recovered.

An estimate of the energy, E (in kilowatt-hours), can be made using the following formula:

$$E = h * P_r * CF * T \tag{7.3}$$

where
  h = number of hours in a year, 8760
  P_r = rated power of the wind turbine in kilowatts
  CF = net annual capacity factor for the turbines at the site
  T = number of wind turbines

The operating and maintenance cost, M is defined by:

$$M = \frac{KC}{E} \tag{7.4}$$

where K is a factor representing the annual operating costs of a wind farm. The European Wind Energy Association (EWEA) has estimated this to be 2.5 per cent of capital cost.

7.2 ANNUAL ENERGY OUTPUT (AEO)

For the site of interest, the total energy output of the machine considering the power output at any wind speed and frequency of occurrence of that wind in a year is AEO. Combining electrical power developed by wind machine as a function of wind speed and wind speed frequency distribution described by Weibull parameters for the site, the annual average power for the machine is given by,

$$P_{e,\text{ave}} = \left[\frac{e^{(-Q_c)} - e^{(-Q_r)}}{(Q_r - Q_c) - e^{(-Q_f)}} \right] \tag{7.5}$$

where
  $Q_c = (U_c/C)^k$
  $Q_r = (U_R/C)^k$
  $Q_f = (U_F/C)^k$
  U_C = cut-in wind speed
  U_R = rated wind speed
  U_F = furling or cut-out wind speed
  k = Weibull shape factor
  C = scale factor
  $P_{e,\text{ave}}$ = average power developed by wind machine
  P_{eR} = rated power developed by wind machine.

Then, the Capacity Factor (CF) for the wind machine is given by,

$$CF = \frac{P_{e,\text{ave}}}{P_{eR}} \tag{7.6}$$

Wind machine speed range for a given site is selected such that Capacity Factor (CF) or Annual Energy Output (AEO) is maximum.

The effect of rated wind speed for the machine on its CF or AEO can be studied by studying the behaviour of normalized power, P_N. Normalized power P_N is defined as:

$$P_N = CF \left(\frac{U_R}{C} \right)^3 \tag{7.7}$$

Figure 7.1(a) and (b) shows P_N plotted vs (U_R/C) for different values of Weibull shape parameter k. Plot in Figure 7.1(a) is for $(U_C/U_R) = 0.5$ and $(U_F/U_R) = 2$ and plot in Figure 7.1(b) is for $(U_C/U_R) = 0.4$ and $(U_F/U_R) = 2$. These plots can be used for selecting U_C, U_R and U_F for maximum

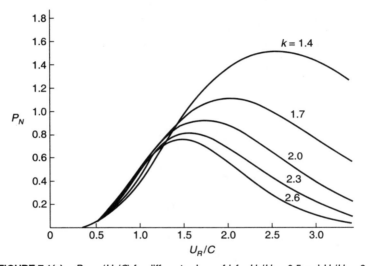

FIGURE 7.1(a) P_N vs (U_R/C) for different values of k for $U_C/U_R = 0.5$ and $U_F/U_R = 2$.

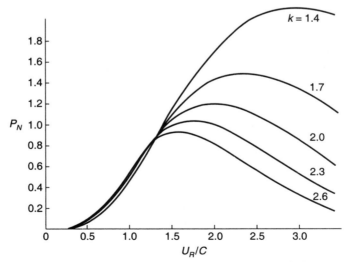

FIGURE 7.1(b) P_N vs (U_R/C) for different values of k for $U_C/U_R = 0.4$ and $U_F/U_R = 2$.

capacity factor or annual energy output for C and k of the site. Selecting of the ratio (U_C/U_R) depends upon the starting characteristics of the wind machine. This analysis can also be applied to see how much power can be generated for a wind machine of given parameter.

7.3 TIME VALUE OF MONEY

The main issue in assessing the viability of an investment is comparing the investment made today with the benefits which are expected in the future. Money now is valued more by individuals or companies than the same amount of money in the future. In general, the money today can be used productively to create goods and services in the economy and can hence grow to a larger amount in the future. This effect is usually represented by an index called the *discount rate*.

The discount rate represents how money is worth more now than in the future. The discount rate determines how any future cash flow is discounted or reduced to make it correspond to an equivalent amount today. If the discount rate is d, the value of unit cash flows in different years and as given in Table 7.1.

TABLE 7.1 Value of unit cash flow in different years

	2006	2007	2006 + k
Value in year	1	1	1
Present value	1	$1/(1+d)$	$1/(1+d)^k$

The choice of the discount rate is critical in the evaluation of the project. The availability and value of capital varies depending on the individual or organization making the investment so that there is no theoretically correct value of the discount rate. The discount rate would always be higher than the bank interest rate as the bank interest rate represents the minimum returns available by placing the money in a bank account. A discount rate can be calculated from the manner in which people and companies make investments among alternatives that have costs and benefits spread in time. If a consumer's discount rate is 20 per cent, what does it imply? This means that the consumer will make an investment of ₹ 100 today only if annual returns of ₹ 20 (or more) are expected every year in the future.

7.3.1 Present Worth Approach

The future value of an investment C made today is give by

$$A_1 = C(1+i)$$
$$A_2 = C(1+i)^2$$
$$A_3 = C(1+i)^3$$
$$\dots\dots\dots\dots\dots$$
$$\dots\dots\dots\dots\dots$$
$$A_n = C(1+i)^n \tag{7.8}$$

where $A_1, A_2, A_3, ...,$ and A_n indicate the value in the 1st, 2nd, 3rd,...and nth year, respectively.

'i' is the interest rate or, as more commonly termed, the discounting rate.

In other words, the present value of a receipt after n year (A_n) is given by

$$pV(A) = \frac{A_n}{(1+i)^n} \tag{7.9}$$

Now, let us consider a uniform cash flow for n years as shown in Figure 7.2.

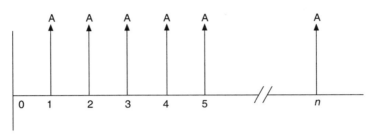

FIGURE 7.2 Cash flow diagram.

Present values of payment A in different years, discounted to the initial year (year 0) is,

$$pV(A)_1 = \frac{A}{(1+i)}$$

$$pV(A)_2 = \frac{A}{(1+i)^2}$$

$$pV(A)_3 = \frac{A}{(1+i)^3}$$

and
$$pV(A)_n = \frac{A}{(1+i)^n} \tag{7.10}$$

Thus, the accumulated present value of all the payments put together is,

$$pV(A)_{1-n} = A\left\{\frac{1}{(1+i)} + \frac{1}{(1+i)^2} + \frac{1}{(1+i)^3} + \cdots + \frac{1}{(1+i)^n}\right\} \tag{7.11}$$

Equation (7.11) can be brought to standard geometric series by taking $1/(1+i)$ common.

$$pV(A)_{1-n} = \frac{A}{(1+i)}\left\{1 + \frac{1}{(1+i)} + \frac{1}{(1+i)^2} + \cdots + \frac{1}{(1+i)^{n-1}}\right\} \tag{7.12}$$

which can be further reduced to

$$pV(A)_{1-n} = \frac{A}{(1+i)} \cdot \frac{1-\left\{\frac{1}{(1+i)}\right\}^n}{1-\left\{\frac{1}{(1+i)}\right\}} \tag{7.13}$$

This can be simplified as

$$pV(A)_{1-n} = A\left\{\frac{(1+i)^n - 1}{i(1+i)^n}\right\} \tag{7.14}$$

EXAMPLE 7.1 A wind turbine generates 15,76,800 kWh electricity in a year. The generated electricity is sold to the utility at a rate of 5 cents/kWh. The discount rate is 5 per cent. Calculate the present worth of electricity generated by the turbine throughout its life period of 20 years.

Solution: Yearly revenue from the project is

$$15,76,800 \times \frac{5}{100} = \$78,840$$

The cash flow during 20 years is shown in Figure 7.3.

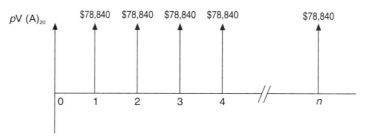

FIGURE 7.3 Cash flow from electricity sales.

Accumulated present value of electricity generated is given by

$$pV(A)_{1-20} = 78,840 \times \left\{ \frac{(1+0.05)^{20} - 1}{0.05(1+0.05)} \right\} = \$98,2521$$

EXAMPLE 7.2 An amount of $10,000 was borrowed at a discount rate of 7 per cent and invested in a project. The loan has to be settled in 10 years through uniform annual repayment. Calculate the amount to be paid annually.

Solution: The cash flow is shown in Figure 7.4

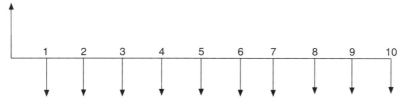

FIGURE 7.4 Cash flow of loan repayment.

The annual payment is given by Eq. (7.14)

$$A = pV(A)_{1-n} \left\{ \frac{i(1+i)^n}{(1+i)^n - 1} \right\}$$

That is,

$$A = 10,000 \left\{ \frac{0.07(1+0.07)^{10}}{(1+0.07)^{10} - 1} \right\} = \$1,424$$

7.3.2 Inflation

The economic evaluation of a project depends on the general level of prices. Over a time period, there is often a change in the purchasing price of the rupee. This effect is denoted as inflation. A rise in general price levels is known as *inflation* and a fall as *deflation*. Inflation is usually measured by the change in a price index—either the Consumer Price Index (CPI) or the Wholesale Price Index (WPI). The price index is the ratio of the price of a stipulated basket of goods and services at a particular year to the price in the base year (denoted as 100). Figure 7.5 shows the variation of the Consumer Price Index for India during the period 1970–1997. It is clear that the CPI increased from 100 in 1970 to 942 in 1997. (The average annual inflation rate during this period was 8.7 per cent). Except one year when prices declined (1976), in all years there has been a steady increase in prices. The normal way of dealing with inflation is to convert all cash flows into units of constant purchasing power.

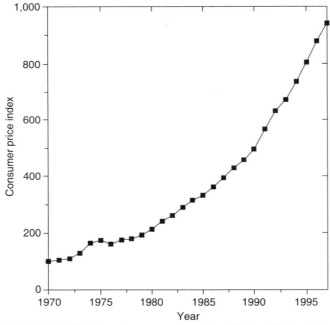

FIGURE 7.5 Variation of consumer price index (1970 = 100) for India.

The discount rate corrected for inflation is termed the real rate of discount (interest). The real discount rate (I) can be roughly taken as the difference between the nominal interest rate and inflation. I is given by

$$1 + I = \frac{1+i}{1+r} \tag{7.15}$$

where r is the rate of inflation.

The increase in the cost of a commodity, in comparison with the general inflation, is termed the escalation e. When the escalation rate e is combined with the inflation r, it is termed the apparent escalation rate e_a, which is given by

$$e_a = \{(1+e)(1+r)\} - 1 \tag{7.16}$$

The real rate of discount, adjusted for both the inflation and escalation, is then given by

$$I = \frac{1+i}{1+e_a} - 1 \qquad (7.17)$$

EXAMPLE 7.3 Calculate the annual repayment in Example 7.2, if the rate of inflation is 3 per cent.

Solution: The real discount rate, corrected for inflation is

$$I = \left\{\frac{1+0.07}{1+0.03}\right\} - 1 = 0.039$$

Therefore, the annual repayment is

$$A = 10,000 \left\{\frac{0.039(1+0.039)^{10}}{(1+0.039)^{10} - 1}\right\} = \$1,227$$

EXAMPLE 7.4 *Inflation ratio.* In a state the CPI in 1995 was 140 (with 1990 as the base year). In 1990, an investment was made in a fixed deposit account which had an interest rate of 10 per cent. What is the real interest rate obtained on the investment?

Solution: The inflation rate i can be calculated as

$$(1+i)^5 = \frac{140}{100}$$

Hence, $i = 7.0$ per cent.

The true interest rate obtained on the investment (r) is given by

$$(1+r)(1+0.07) = (1+0.1)$$

Hence, $r = 2.8$ per cent.

In economic evaluation of projects, it is necessary to distinguish between the real and nominal discount rates. If cash flows are adjusted for inflation and expressed in constant money terms, the discount rate used is the real discount rate (d_r). If the cash flows are not corrected for inflation, the discount rate used is termed the *nominal discount rate* (d_n). The real and nominal discount rates are related by the equation:

$$1 + d_n = (1+d_r)(1+i) \qquad (7.18)$$

Hence,

$$d_r \sim d_n - i \qquad (7.19)$$

As the inflation in a country varies, so does the nominal discount rate. The real discount rate tends to be more stable and reflects the real scarcity of capital. Normally, in calculating the viability of a project, it is better to use the real discount rate. The annual savings computed are based on present prices. An assumption is made that the annual savings escalate the inflation rate. Using a nominal discount rate would involve explicitly assuming the inflation rate and increase in the energy prices (inflation rate for the annual savings). Unless otherwise specified by a discount rate, we will refer to the real discount rate. Real discount rates for private companies may range between 15–20 per cent while the public sector (especially the infrastructure sector) may have lower discount rates. There is no theoretically correct discount rate for a company. The discount rate depends on the availability of capital. Depending on the total capital available to a company, it will choose the investments with the highest rates of return.

7.4 CAPITAL RECOVERY FACTOR

The problem in economic assessment is usually the comparison of future annual cash flows with initial investment. Consider annual cash flows A_k in the kth year. The present value P of these cash flows is,

$$P = \sum_{k=1}^{n} A_k/(1+d)^k \tag{7.20}$$

where n is the life of the equipment or the number of years during which annual cash flow streams exist. If the annual cash flow streams are constant (A), then the present value P of the annual cash flows can be obtained as the sum of a geometric progression with the first term being $A/(1+d)$ and the common ratio (multiplier) for successive terms being $1/(1+d)$. On simplification, this gives:

$$P = \frac{A[(1+d)^n - 1]}{\left[\dfrac{d}{(1+d)^n}\right]} \tag{7.21}$$

$$P = A \text{ [Uniform Present Value Factor]}$$

The uniform present value factor is defined as the ratio of the present value of the constant annual cash flows to the annual cash flow. For an ECO, the uniform present value factor can be multiplied by the annual cash flow to get the equivalent present value of the benefits (annual cash flows usually represent savings). The capital recovery factor (CRF) is the inverse of the uniform present value factor and is defined as,

$$\text{CRF} = \frac{A}{P} = \frac{[d(1+d)^n]}{[(1+d)^n - 1]} \tag{7.22}$$

The capital recovery factor enables the determination of the annualised value equivalent to the initial investment. The CRF is dependent on the equipment life n and the discount rate d. Consider an investment in an equipment with a life of 10 years and a real discount rate of 12 per cent. Substituting in Eq. (7.22), we get,

$$\text{CRF } (d = 12\%, n = 10 \text{ years}) = 0.177$$

This implies that an investment of ₹ 1,000 today is equivalent to annual investments of ₹ 177 over the lifetime of the equipment. When will this investment be worthwhile? This will be viable if the expected annual savings are greater than ₹ 177.

The main criteria used with discounted cash flows are the net present value (NPV), benefit/cost (B/C) ratio and the internal rate of return (IRR). The net present value of a project is the present value of the savings (benefits) minus the costs and is obtained as,

$$\text{NPV} = \frac{\sum_{k=1}^{n} A_k}{(1+d)^k} - C_0 \tag{7.23}$$

where C_0 is the initial investment. For uniform cash flows, this simplifies to

$$\text{NPV} = \frac{A}{(CRF(d,n))} - C_0 \tag{7.24}$$

For an ECO to be viable, the NPV should be greater than zero. While comparing two projects, the project with the higher NPV can be selected. At times, the NPVs of two projects may be similar but one may involve a higher initial investment. An alternative criterion which may be preferable in this case is the ratio of the net present value of benefits to cost. The B/C ratio greater than 1 is essential for a viable project. In a comparison between two projects, the project with a higher B/C ratio would be preferred. Another criterion which is commonly used is the internal rate of return. To compute the IRR, the net present value is equated to zero and the following equation is solved:

$$C_0 = \frac{\sum_{k=1}^{n} A_k}{(1 + \text{IPR})^k} \tag{7.25}$$

This is a non-linear equation and can be solved iteratively (using the bisection, secant or Newton–Raphson method) to obtain the IRR. The company usually has a hurdle rate or minimum acceptable rate of return for an investment.

7.5 DEPRECIATION

Depreciation is an accounting concept. This attempts to distribute the cost or basic value of a tangible asset, less salvage value, over the estimated useful life of the asset in a systematic manner. For most assets, the market value of the asset decreases with time. For accounting purposes, depreciation is subtracted as an expense from the profits to obtain the net profits which is taxable. In the indices considered, the initial investment is considered as a cost. Hence, depreciation should not be separately accounted for in the annual cash flows. Otherwise, this would lead to double counting. For many ECOs, the government provides for accelerated depreciation, as an incentive. In the evaluation of an ECO, the tax benefits from depreciation should be considered. A commonly used method of depreciation is the straight line method. In this method, the annual depreciation (A_D) is computed as,

$$A_D = \frac{C_0 - S}{n} \tag{7.26}$$

where S is the salvage value at the end of the life of the equipment. If the salvage value is zero, the annual depreciation is,

$$A_D = \frac{C_0}{n} \tag{7.27}$$

If the company is making profits and the corporate tax rate is t, there is an annual reduction in tax equal to $t\,A_D$. This may be included in the annual cash flows, especially for large projects, but is often negligible. For options where accelerated depreciation is permitted, this needs to be considered.

7.6 LIFE CYCLE COSTING

The life cycle cost (LCC) is the present value of all expenses related to a specific option during its lifetime. To compute this, the present value of the annual expenditures is added to the initial investment. The annual expenditures will include the energy, maintenance, labour and other costs. The LCC can be written as,

$$\text{LCC} = C_0 + \frac{\sum_{k=1}^{n} AC_k}{(1+d)^k} \quad (7.28)$$

where AC_k is the annual cost or expenditure in the kth year. For constant annual costs, this reduces to,

$$\text{LCC} = C_0 + \frac{AC}{\text{CRF}(d,n)} \quad (7.29)$$

To use this index, it is necessary to compare the LCCs of different options that perform the same function. The option with the lower LCC is preferable. It is difficult to apply this index to options with different lives. An alternative that is preferable for such cases is the annualised life cycle cost (ALCC). Normally, the annual operating expenses are estimated at the start of the project. Hence, we deal with constant annual costs. The initial investment is added to the annual operating cost to obtain the ALCC.

$$\text{ALCC} = C_0 \times \text{CRF}(d,n) + AC \quad (7.30)$$

The ALCC is related to the LCC as,

$$\text{ALCC} = \text{LCC} \times \text{CRF}(d,n) \quad (7.31)$$

The ALCC represents the annual cost of owning and operating the equipment. In pieces of equipment which have components with different lives, the ALCC can be easily computed as,

$$\text{ALCC} = AC + \sum_{m=1}^{p} C_{0m} \times \text{CRF}(d, n_m) \quad (7.32)$$

where there are p sub-components with initial investments $C_{01}, C_{02}, ..., C_{0p}$ and lives $n_1, n_2, ..., n_p$.

It should be understood that the economic viability of an ECO is sensitive to a number of factors—energy prices, equipment life, number of hours of operation. The choice of a discount rate affects the viability of projects, especially those with long lives. If energy prices are expected to escalate at rates higher than the general inflation rate, the future inflation rates can be explicitly assumed and the economic criteria computed using a nominal discount rate. The criteria discussed can also be used when the company takes a loan to pay for the project. The payments for the principal and the interest can be included in the yearly cash flows. Tax deductions for loan repayments can also be added as a benefit in the cash flow stream. In such cases, it is better to explicitly put down annual cash flows before discounting and calculating the NPV, B/C ratio, IRR or LCC. For dealing with uncertainty in energy prices (or any other parameter which affects the economic viability), discrete future scenarios can be constructed to check the robustness of the decision.

7.6.1 Cost of Wind Energy

To ensure sustained and healthy growth of wind power sector, it is necessary to rationally evaluate the cost of generation, and to determine a selling rate which should be acceptable to consumers and attractive to investors. The selling/purchase rate would, however, vary for three mode of use/sale of wind energy:

- Captive consumption
- Third party sale
- Sale to utility

The cost of wind energy generation varies from site to site, depending on the wind resource and also on year-to-year basis due to yearly variation in wind speed. The cost of energy should be decided based on long-term average generation. The cost can be calculated accurately provided the assumptions are realistic corresponding to wind resource and market conditions.

The parameters which would determine the cost of wind energy are given in Table 7.2.

TABLE 7.2 Cost of wind energy

Parameter	Assumption
Capital cost	The capital cost varies between $1.0 million to $1.5 million per MW depending on the technology of wind turbine.
Capacity Utilization Factor (CUF) (Annual Generation)	It varies from 18 per cent to 30 per cent (average of few years) depending on wind resource at particular site and the technology of wind turbine. The CUF should be the net after deducting internal consumption.
Operation and Maintenance (O&M)	This cost includes manpower, consumables, spares, breakdown maintenance, insurance and also all other statutory duties and expenses.
Interest on debt	This varies depending on the credit worthiness of the borrower.
Loan repayment period	Wind energy project being an infrastructure project, repayment period of ten years may be considered. The repayment schedule can be structured to match with revenue inflow.
Debt equity ratio	Depending on financial institution and credit rating of borrower, this may be taken as average figure of 70:30.
Return on equity	It may be around 16 per cent, though a higher return may be justified in view of uncertainties involved in wind flow pattern.
Rate of depreciation	For the sake of simplicity, the common system of straight line method (SLM) @5.28 per cent can be considered. While 5.28 per cent depreciation under SLM method is suitable for calculation of cost, the rate should be higher for calculating selling rate in first 10 years if loan is to be repaid in 10 years.
Wheeling and banking charges	These are not applicable against sale to utility.
Income tax liability and other benefit	Depends upon the policies of the present government.

The selling rate, if based on the actual cost of generation, would be quite front-loaded and may have a significant impact on the existing electricity tariff.

For wind farms to be profitable, not only the initial cost, but the reliability, the operation and maintenance costs, the effectiveness of siting each turbine, and the interference effects between adjacent turbines are all important factors to be considered. The evaluation of the economy of purchase of an individual wind turbine involves the following steps:

1. Estimating the output-power characteristics of the machine to be used versus wind speed.
2. Estimating the wind spectrum at the proposed site at hub level.
3. Estimating all costs associated with the purchase, installation, and operation of the machine.
4. Determining the worth of the utilized and 'exported' electrical power produced.
5. Evaluating financial implications, including all applicable tax credits, depreciation costs of borrowing capital, and the effect of applicable current income taxes.
6. Applying basic principles of engineering economy to evaluate the economic performance of the investment.

Standard measures of economic performance include the effective cost of the power produced, the net present value, the payback period, and the internal rate of return. In calculating the net effective cost of electrical power, it is common to 'levelize' expenses or incomes over the year of a wind turbine into an equivalent yearly amount. This means taking an expense that occurs during a particular year and converting it to an equivalent amount for each year.

There are three different ways in which the cost of a wind energy system is commonly determined. They are:

- Cost per rated power of the turbine.
- Cost per unit rotor size
- Cost per kWh of electricity generated.

If C_T is the cost of the turbine with a rated power P_R, then the cost per kW is given by

$$C_{PR} = \frac{C_T}{P_R} \tag{7.33}$$

If A_T is the rotor area, then cost per unit size of the turbine C_A is given as

$$C_A = \frac{C_T}{A_T} \tag{7.34}$$

Two turbines with the same rotor area may have different rated wind speeds. In a wind regime with stronger winds, the system with higher P_R will deliver more power and hence will have economic advantage over the other.

The cost/kWh is a better economic indicator. The characteristic wind regime is a critical factor in deciding the cost of wind-generated electricity. If C_A is the cost of operation and E_I is the kWh generated, both expressed in annual basis, then the unit cost of wind-generated electricity is

$$C_E = \frac{C_A}{E_I} = \frac{C_A}{8,760 \, C_F P_R} \tag{7.35}$$

where C_F is the capacity factor.

The annual cost of a project would essentially have two components—fixed cost and variable cost. Fixed cost refers to the cost that is to be incurred by virtue of mere existence of the project. That is, fixed costs are to be met irrespective of whether the project is functional and how much power is being generated. On the other hand variable costs vary in proportion to the quantity of project output.

As illustrated in Figure 7.6, the total annual cost of operation C_A is the sum of annual fixed costs F_C and variable costs V_C.

$$C_A = F_C + V_C \qquad (7.36)$$

The fixed component of annual cost is contributed by the initial investment and variable costs consist of the expenditure on the operation and maintenance of the system.

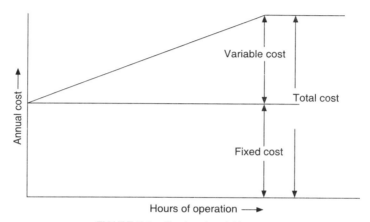

FIGURE 7.6 Fixed and variable costs.

The initial investments include cost of land, power conditioning unit, civil works, electrical infrastructure and installation charges. Break-up of the capital cost shows that 70 per cent investment is for wind turbine, 10 per cent for electrical infrastructure, 8 per sent and 7 per cent for civil work and power conditioning, respectively. Installation and other miscellaneous charges account for 5 per cent. It may be noted that the land costs are not included and figures are only indicative and vary from project to project.

7.6.2 Present Value of Annual Costs

Annual costs involved in a wind energy project over its life span of n years are shown in Figure 7.7. Let C_I be the initial investment of the project and C_{OM} be the operation and maintenance cost. Expressing C_{OM} as a percentage m of C_I

$$C_{OM} = m\, C_I \qquad (7.37)$$

Now, discounting the operation and maintenance costs for n years to the initial year,

$$pV\,(C_{OM})_{1-n} = m\, C_I \left[\frac{(1+I)^n - 1}{I(1+I)^n} \right] \qquad (7.38)$$

Including the initial investment C_I, the accumulated net present value of all the costs is represented as

$$\text{NPV}\,(C_A)_{1-n} = C_I \left[1 + m \left\{ \frac{(1+I)^n - 1}{I(1+I)^n} \right\} \right] \qquad (7.39)$$

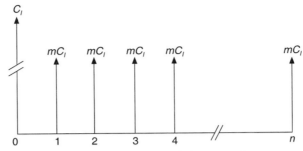

FIGURE 7.7 Costs of the wind energy project.

Hence, the yearly cost of operation of the project is

$$\text{NPV}(C_A) = \frac{\text{NPV}(C_A)_{1-n}}{n} = \frac{C_I}{n}\left[1 + m\left\{\frac{(1+I)^n - 1}{I(1+I)}\right\}\right] \qquad (7.40)$$

If P_R is the rated power of the turbine and C_F is the capacity factor, the energy generated by the turbine in a year is

$$E_I = 8{,}760\, P_R\, C_F \qquad (7.41)$$

Thus, the cost of kWh wind-generated electricity is given by

$$C = \frac{\text{NPV}(C_A)}{E_I} = \frac{C_I}{8{,}760\, n}\left(\frac{1}{P_R\, C_F}\right)\left[1 + m\left\{\frac{(1+I)^n - 1}{I(1+I)^n}\right\}\right] \qquad (7.42)$$

EXAMPLE 7.5 The cost of a 600 kW wind turbine is $5,50,000 other initial costs including that for installation and grid integration are 30 per cent of the turbine cost. Useful life of the system is 20 years. Annual operation and maintenance costs plus the land rent come to 3.5 per cent of the turbine cost. Calculate the cost of generating electricity from the turbine when it is installed at a site having a capacity factor of 0.25. The real rate of interest may be taken as 5 per cent.

Solution: The installation cost of the turbine is

$$5{,}50{,}000 \times \frac{30}{100} = \$1{,}65{,}000$$

So, the total initial investment for the project is

$$5{,}50{,}000 + 1{,}65{,}000 = \$7{,}15{,}000$$

Hence, the cost of one kWh of electricity is

$$C = \frac{7{,}15{,}000}{8{,}760 \times 20}\left(\frac{1}{600 \times 0.25}\right)\left[1 + 0.035\left\{\frac{(1+0.05)^{20} - 1}{0.05(1+0.05)^{20}}\right\}\right]$$

$$= \$0.04/\text{kWh}$$

EXAMPLE 7.6 Illustrate the effect of capacity factor on the cost/kWh in Example 7.5. If a utility company buys the generated electricity at a rate of $0.03/kWh, find out the break-even capacity factor.

Solution: As per Eq. (7.42)

$$C = \frac{C_I}{8,760\, n}\left(\frac{1}{P_R\, C_F}\right)\left[1 + m\left\{\frac{(1+I)^n - 1}{I(1+I)^n}\right\}\right]$$

$$0.03 = \frac{7,15,000}{8,760 \times 2}\left(\frac{1}{600 \times C_F}\right)\left[1 + 0.035\left\{\frac{(1+0.05)^{20} - 1}{0.05(1+0.05)^{20}}\right\}\right]$$

$$C_F = 0.33$$

7.6.3 Value of Wind Generated Electricity

Consider the retail energy price as the value of wind generated electricity. If the project delivers a benefit of B_A annually through electricity sales, then the accumulated present value of all benefits over the life of the project is

$$\text{NPV}\,(B_A)_{1-n} = B_A\left[\frac{(1+I)^n - 1}{I(1+I)^n}\right] \tag{7.43}$$

EXAMPLE 7.7 Calculate the net present worth of electricity sales from the wind turbine described in Example 7.6. Electricity price is $ 0.045 per kWh. Take the interest rate as 7 per cent, inflation as 3 per cent and escalation as 2 per cent.

Solution: The apparent rate of escalation, corrected for inflation, is given by Eq. (7.16)

$$e_a = \{(1+e)(1+r)\} - 1$$
$$= \{(1+0.02)(1+0.03)\} - 1$$
$$= 0.05$$

Therefore, the real rate of discount, adjusted for inflation and escalation is given by Eq. (7.17)

$$I = \frac{1+i}{1+e_a} - 1$$
$$= \frac{1+0.07}{1+0.05} - 1 = 0.02$$

With the rated capacity 600 kW and the capacity factor 0.25, annual energy production is given by Eq. (7.41).

$$E_I = 8,760\, P_R\, C_F$$
$$= 8,760 \times 600 \times 0.25 = 13,14,000 \text{ kWh}$$

Annual revenue from the sale of electricity at a rate of $0.045/kWh is

$$= 13,14,000 \times 0.045$$
$$= \$59,130$$

The cash flow is shown in Figure 7.8.

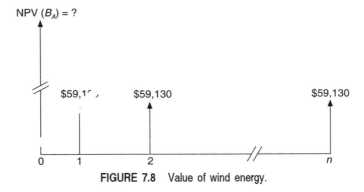

FIGURE 7.8 Value of wind energy.

Discounting the benefits of all 20 years to the project beginning at real discount rate and adding together, from Eq. (7.43)

$$\text{NPV }(B_A)_{1-n} = B_A \left[\frac{(1+I)^n - 1}{I(1+I)^n} \right]$$

$$B_A = 59{,}130 \left[\frac{(1+0.02)^{20} - 1}{I(1+I)^n} \right]$$

$$= \$9{,}66{,}860$$

7.6.4 Economic Merit

From the cash flow diagram shown in Figure 7.8, apart from initial investments, there are annual cash flows and outflow throughout its life span. The merit of the project can be seen in terms of the following indices:

- **Net present value (NPV):** The present worth of the entire project.
- **Benefit cost ratio:** The benefits from the project are in proportion with the costs involved.
- **Payback period:** How many years will it take to get our investment back from the project?
- **Internal rate of return (IRR):** What is the real return of the project or the maximum rate of interest at which we can arrange capital for the project?

7.6.5 Net Present Value

The net present value (NPV) is the net value of all benefits (cash inflows) minus costs (cash outflows) of the project, discounted back to the beginning of the investment. The benefits include the income from the sale of electricity generated. The cash outflow includes capital investment and annual operation and maintenance costs. Thus, the NPV is

$$\text{NPV} = \text{NPV }(B_A)_{1-n} - \{C_I + \text{NPV }(C_A)_{1-n}\} \quad (7.44)$$

Substituting for NPV $(B_A)_{1-n}$ and NPV $(C_A)_{1-n}$, we have

$$\text{NPV} = B_A \left[\frac{(1+I)^n - 1}{I(1+I)^n} \right] - \left[C_I \left\{ 1 + m \left(\frac{(1+I)^n - I}{I(1+I)} \right) \right\} \right] \quad (7.45)$$

If the NPV is greater than 0, the project is economically viable. For comparision, the project with higher NPV should be selected.

7.6.6 Benefit Cost Ratio

The project involving higher investment may tend to show an impressive NPV than the one requiring lower capital. In such a situation, benefit cost ratio (BCR) is a better tool to judge the economic viability.

Benefit cost ratio is the ratio of the accumulated present value of all the benefits to the accumulated present value of all costs, including the initial investment.

$$\text{BCR} = \frac{\text{NPV}(B_A)_{1-n}}{C_I + \text{NPV}(C_A)_{1-n}} \tag{7.46}$$

Thus, we have

$$\text{BCR} = \frac{B_A \left[\frac{(1+I)-1}{I(1+I)^n} \right]}{C_I \left[1 + m \left(\frac{(1+I)^n - 1}{I(1+I)^n} \right) \right]} \tag{7.47}$$

A project is acceptable if BCR is greater than 1.

7.6.7 Pay Back Period

Pay back period (PBP) is the year in which the net present value of all costs equals with the net present value of all benefits. Hence, PBP indicates the minimum period over which the investment for the project is recovered. At PBP,

$$\text{NPV}(B_A)_{1-n} = C_I + \text{NPV}(C_A)_{1-n} \tag{7.48}$$

That is,

$$B_A \left[\frac{(1+I)^n - 1}{I(1+I)} \right] = C_I \left[1 + m \left(\frac{(1+I)^n - 1}{I(1+I)^n} \right) \right] \tag{7.49}$$

The pay back period is computed by solving the above equation for n. The above equation can be rewritten as

$$\frac{C_I}{B_A - m C_I} = \left[\frac{(1+I)^n - 1}{I(1+I)^n} \right] \tag{7.50}$$

$$(1+I)^n = \left(1 - \frac{I C_I}{B_A - m C_I} \right)$$

Taking logarithm on both sides

$$n \ln(1+I) = -\ln \left(1 - \frac{I C_I}{B_A - m C_I} \right)$$

Thus,

$$n = -\frac{\ln\left(1 - \dfrac{I\,C_I}{B_A - m\,C_I}\right)}{\ln(1+I)} \tag{7.51}$$

The accumulated net present value of costs and benefits for a typical wind energy project is shown in Figure 7.9. The cost and benefit curves are meeting at a point A corresponding to the pay back period. Here, the project recovers its capital on 14th year of operation. The project with lower pay back period is preferred.

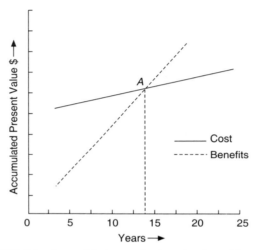

FIGURE 7.9 Accumulated present value of costs and benefits.

7.6.8 Internal Rate of Return

Internal rate of return (IRR) is defined as the discount rate at which the accumulated present value of all the costs becomes equal to that of the benefits. In other words, with IRR as the discount rate, the net present value of a project is zero.

IRR is the maximum rate of interest (in real terms) that the investment can earn. In physical sense, it is the interest rate upto which we can afford to arrange the capital for the project. The net present value of the project is discounted at zero IRR, the PBP at IRR is the projects' life period with IRR as the discounting rate,

$$\text{NPV}\,(B_A)_{1-n} = C_I + \text{NPV}\,(C_A)_{1-n} \tag{7.52}$$

That is,

$$B_A\left[\frac{(1+IRR)^n - 1}{IRR\,(1+IRR)^n}\right] = C_I\left[1 + m\left(\frac{(1+IRR)^n - 1}{IRR\,(1+IRR)^n}\right)\right] \tag{7.53}$$

From the above equation, IRR can be determined by trial and error method or, more precisely, using numerical techniques like the Newton–Raphson method.

The accumulated net present values of costs and benefits of a typical project are shown in Figure 7.10. The intersecting point of the cost and benefits curve corresponds to the IRR of the project. In the present case it is 13.33 per cent, which indicates that the project is acceptable up to a real interest rate of 13.33 per cent.

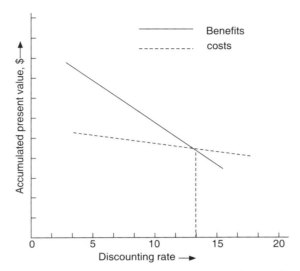

FIGURE 7.10 Accumulated present value of costs and benefits discounted at different rates.

EXAMPLE 7.8 A wind energy project of 2.4 MW installed capacity requires a capital investment of $22,00,000. The site's capacity factor is 0.35. Annual operation and maintenance costs are 2.0 per cent of the initial investment with 5 per cent real rate of discount, calculate (i) net present value, (ii) benefit cost ratio, (iii) pay back period and (iv) internal rate of return. The useful project life is 25 years and the local electricity price is $0.05/kWh.

Solution: Annual electricity production expected from the project is, Eq. (7.41)

$$E_I = 8{,}760 \ P_R \ C_F$$
$$= 8{,}760 \times 2{,}400 \times 0.35 = 73{,}58{,}400 \ \text{kWh}$$

At the rate of $0.05/kWh, the annual return from the electricity sates is B_A, benefit delivered annually

$$B_A = E_I \times r$$
$$= 73{,}58{,}400 \times 0.05$$
$$= \$3{,}67{,}920$$

Net present value of the benefits, Eq. (7.43)

$$\text{NPV} \ (B_A)_{1-n} = B_A \left[\frac{(1+I)^n - 1}{I(1+I)^n} \right]$$

$$\text{NPV }(B_A)_{1-25} = 3,67,920 \left[\frac{(1+0.05)^{25} - 1}{0.05\,(1+0.05)^{25}} \right]$$

$$= \$51,85,444$$

Annual operation and maintenance cost C_A, Eq. (7.37)

$$C_{OM} = C_A = m\,C_I$$
$$= 0.02 \times 22,00,000, = \$44,000$$

Net present value operation and maintenance cost, Eq. (7.38)

$$\text{NPV }(C_{OM})_{1-n} = m\,C_I \left[\frac{(1+I)^n - 1}{I\,(1+I)^n} \right]$$

$$= 44,000 \left[\frac{(1+0.05)^{25} - 1}{0.05\,(1+0.05)^{25}} \right]$$

$$= \$6,20,134$$

Therefore, the net present value of the project is, Eq. (7.44)

$$\text{NPV} = \text{NPV }(B_A)_{1-n} - \{C_I + \text{NPV }(C_{OM})_{1-n}\}$$

(i) NPV = 51,85,444 − (6,20,134 + 22,00,000)
   = \$23,65,311

(ii) Benefit cost ratio is Eq. (7.46)

$$\text{BCR} = \frac{\text{NPV }(B_A)_{1-n}}{C_I + \text{NPV }(C_{OM})_{1-n}}$$

$$= \frac{51,85,444}{22,00,000 + 6,20,134} = 1.84$$

(iii) Pay back period of investment is, Eq. (7.51)

$$n = -\frac{\ln\left(1 - \frac{I\,C_I}{B_A - m\,C_I}\right)}{\ln(1+I)}$$

$$= -\frac{\ln\left(1 - \frac{0.05 \times 22,00,000}{3,67,920 - 0.02 \times 22,00,000}\right)}{\ln(1+0.05)}$$

$$= 8.5 \text{ years}$$

(iv) Internal rate of return is, Eq. (7.53)

$$B_A \left[\frac{(1+IRR)^n - 1}{IRR\,(1+IRR)^n} \right] = C_I \left[1 + m\left(\frac{(1+IRR)^n - 1}{IRR\,(1+IRR)^n} \right) \right]$$

$$3,67,920 \left[\frac{(1+\text{IRR})^{25}-1}{\text{IRR}(1+\text{IRR})^{25}} \right] = 22,00,000 \left[1 + 0.02 \left(\frac{(1+\text{IRR})^{25}-1}{\text{IRR}(1+\text{IRR})^{25}} \right) \right]$$

Solving the above relationship following the Newton–Raphson method (see Appendix C) IRR is found to be 13.7 per cent.

7.6.9 Tax Deduction due to Investment Depreciation

Depreciation refers to the reduction in value of an item over its years of use. The value of the project at the end of its useful life period is known as the junk or salvage value. The depreciation on capital investment on a wind turbine is shown in Figure 7.11.

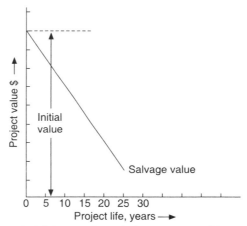

FIGURE 7.11 Depreciation of a wind turbine.

There are several ways to compute depreciation. Some extensively used methods are:
(i) straight line method, (ii) declining balance method, and (iii) sum of the years digit method.

7.6.10 Straight Line Depreciation

In straight line depreciation, the value of an item gets reduced at a uniform rate throughout its useful life period. Thus, C_I is the capital investment for a wind turbine and S is its salvage value after its life of n years. Then, the annual depreciation is given by

$$D_A = \frac{C_I - S}{n} \qquad (7.54)$$

For wind energy projects, the salvage value is taken as 10 per cent of the initial cost.

7.6.11 Declining Balance Depreciation

In declining balance depreciation, the value of an asset is written off at an accelerated rate. It allows higher rate of depreciation in the early years of the project. The depreciation at the end of a year is considered as a constant fraction of the book value of the project at the end of the previous year. The depreciation is given by

$$D_t = p\, C_I\, (1-p)^{t-1} \qquad (7.55)$$

where D_t is the depreciation at the tth year and p is the rate of depreciation. The rate of depreciation in this method may be taken as $2/n$. For example, if the project life is 30 years, p is taken as 0.15. Here, the salvage value of the project is not considered.

7.6.12 Sum of the Years' Digit Depreciation

In sum of the years' digit method the calculation of depreciation is at an accelerated rate in the initial years. The depreciation at t^{th} year is given by

$$D_t = \frac{n-(t-1)}{n(n+1)/2}(C_I - S) \tag{7.56}$$

EXAMPLE 7.9 The useful life of a 600 kW wind turbine costing $5,25,000 is 20 years. Compute its depreciation in the 5th, 10th and 15th year by (i) straight line depreciation, (ii) declining balance depreciation, and (iii) sum of the years digit depreciation.

Solution: Taking the salvage value as 10 per cent of the capital, the depreciation at the 5th year is given by straight line method, Eq. (7.54)

$$D_A = \frac{C_I - S}{n}$$

$$D_5 = \frac{5,25,000 - 52,500}{20} = \$23,625$$

In this method, depreciation for all other years is same.
Declining method, p is as $2/n = 2/20 = 0.1$
For 5th year, Eq. (7.55)

$$D_t = p\, C_I (1-p)^{t-1}$$

$$D_5 = 0.1 \times 5,25,000 (1-0.1)^{5-1} = \$34,445$$

Similarly, the depreciation during 10th and 15th years are calculated as $20,340 and $12,010, respectively.

Sum of the years' approach for 5th year of the project, depreciation is given by Eq. (7.56)

$$D_t = \frac{n-(t-1)}{n(n+1)/2}(C_I - S)$$

$$D_5 = \frac{20-(5-1)}{20(20+1)/2}(5,25,000 - 52,500)$$

$$= \$36,000$$

Similarly, for 10th and 15th years' the depreciation is $24,750 and $13,500, respectively.

The worth of the project at the end of each year can be computed by deducing the yearly depreciation from the value of the project at the beginning of the year. This is known as the book value of the project at that year. The book value of the turbine considered in the example, calculated using different methods of depreciation, are compared in Figure 7.12.

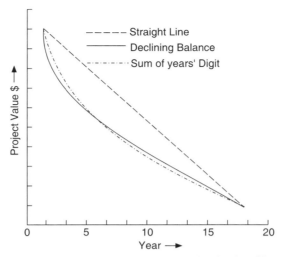

FIGURE 7.12 Depreciation of the wind turbine calculated using different methods.

EXAMPLE 7.10 *Levelized equivalent amount.* Suppose it was necessary to replace an alternator at a cost of $1,000 during the eleventh year of operation for a machine having an assumed 25-year life, and the interest rate as 15 per cent. Calculate the worth of $1,000 cost at the beginning (year zero) and after 25-year life.

Solution: The worth of $1,000 cost at year zero

$$= \$1,000 \, \frac{1}{(1+d)^k}$$

$$= \$1,000 \, \frac{1}{(1.15)^{11}} = \$214.94$$

The levelized equivalent amount over a 25-year life would be

$$= \$214.94 \, \frac{d(1+d)^k}{(1+d)^k - 1}$$

$$= \$214.94 \, \frac{0.15 \, (1.15)^{25}}{1.15^{25} - 1}$$

$$= \$214.94 \, [0.1547]$$

$$= \$33.25$$

EXAMPLE 7.11 *Single wind turbine purchase–analysis.* A wind turbine complete with tower, wiring and controls costs $4,000, installation costs another $1,000. Maintenance and insurance is estimated at $200 per year, interest at 10 per cent, and machine life at 20 years. Salvage value is estimated at $500 at the end of 20 years. The machine is estimated to produce 3,000 kWh of power at the proposed site per year. What is the estimated cost of the electricity produced?

Solution: Taking capital cost as an equivalent yearly cost, the capital recovery with return, and add to that the yearly operating costs. This yields a total equivalent yearly cost. Dividing this cost by the kilowatt hours produced yields the equivalent cost per kWh.

The capital recovery is

$$CR(i) = (P - F)(A/P\ i, n) + F_i$$

where P represent the present expense; F the future salvage value, i the interest rate ($i = 0.1$); and n, the number of years of machine operating life.

Capital recovery factor

$$A/P\ i, n = \frac{i(1+i)^n}{(1+i)^n - 1}$$

$$CR(0.1) = (5{,}000 - 500)(0.1175) + 50$$
$$= \$578.75$$

This figure represent the equivalent yearly cost of this investment, including recovery of both the capital and the interest on it. Adding the yearly maintenance and insurance of \$200, we obtain a total equivalent yearly cost of \$778.75. If this is the cost of 3,000 kWh of power, we have a unit cost of \$778.75/3,000, or 25.95 cents per kWh.

7.6.13 Wind Energy Economics Worksheet

A. Preliminary costs per machine

1. *Site analysis and site control*

   Wind resource assessment
   Environmental impact

2. *Preliminary/detailed engineering*

   Soil analysis
   Technical support
   Others

3. *Permitting and approval costs, financing, legal, etc.*

B. Purchasing, shipping and installation

1. *Factory price*

   Machine cost
   Tower cost
   Special electrical equipment costs
   Other costs

2. *Shipping expenses*

   Export packing
   Factory to dock charges
   Dock receiving charges
   Export declaration papers/brokers' fee

Wharfage
Ocean or air freight
Destination handling charges, storage
Trucking to site
Other expenses

3. *Site preparation*

   Access for equipment
   Fencing and security
   Others

4. *Installation*

   Footing excavation
   Reinforcement rods
   Anchor rods and plates
   Concrete/delivered to site/leasing/purchase
   Erection equipment rental/leasing/purchase
   Labour
   Electrical hookup
   Special electrical equipment such as power substation, transformer, telephone poles, etc.
   Transmission line or energy storage costs
   Others
   TOTAL CAPITAL or FIRST COSTS (1 to 4)

C. Tax benefits

NET CAPITAL COSTS

D. Annual costs and benefits

1. *Revenue factors*

   (a) Estimated total annual kWh
   (b) Value of kWh used by owner at the utility rate
   (c) Value of kWh sold to the utility
   GROSS REVENUE or ANNUAL VALUE

   Less down time (force outage)

2. *Operating expenses*

   (a) Operating and maintenance costs routine maintenance
       (servicing montly or every 6 months, painting, corrosion protection, etc.)
       extraordinary maintenance or parts replacement
   (b) Land rent
   (c) Insurance
   (d) Taxes–Income and general excise
   NET ANNUAL OPERATING EXPENSES
   NET ANNUAL INCOME

7.7 PROJECT APPRAISAL

When evaluating power projects, it has been the convention to use techniques based on discount cash flow (DCF) analysis. This is based on recognition that to receive money today not next year and would rather pay out money next year rather than today. The use of DCF analysis, with a high discount rate, tends to favour projects with short construction time, low capital costs and high operating costs.

The mechanics of DCF analysis are simple and the calculation may be implemented easily on a spreadsheet and the main functions are often included in commercially available packages. Given a discount rate of r, the value of a sum in n years time is given by:

$$V_n = V_p(1 + r)^n \tag{7.57}$$

and so the present value of a sum received or paid in the future is given by:

$$V_p = \frac{V_n}{(1+r)^n} \tag{7.58}$$

where V_n is the value of a sum in year n, and V_p is the present value of the sum.

Hence, a payment stream lasting m of years

$$V_p = \sum_{n=1}^{n=m} \frac{V_n}{(1+r)^n} \tag{7.59}$$

This is a geometric series which, for equal payments (A), sums to

$$V_p = \frac{1-(1+r)^{-n}}{r} A \tag{7.60}$$

This allows the calculation of the present value of any sum of money, which is either paid or received in the future. The net present value (NPV) is simply the summation of all the present values of future income and expenditures. Using identical discounting techniques, it is possible to calculate other financial indicators:

- The benefit/cost ratio is the present value of all benefits divided by the present value of all costs.
- The internal rate of return (IRR) is the value of the discount rate that gives a net present value of zero.

These should not be confused with indicators not based on discounted values such as:

- Payback periods, expressed in years, which is the capital cost of the project divided by the annual average return.
- Return on investment, which is the average annual return divided by the capital cost, expressed in per cent.

The discount rate chosen reflects the value of the lender places on the money and the risks that are anticipated in the project. In economic terms, the discount rate is an indication of the opportunity cost of the capital to the organization. Opportunity cost is the return on the next best investment and so is the rate below which it is not worthwhile to invest in the project.

In simple DCF analysis, general inflation is usually ignored and the discount rate is the so-called 'real' rate but it is possible to build in differential inflation if it is thought that some costs or income will rise at different rates over the life of the project. The financial model may be as sophisticated as required to reflect the concerns of the lender. It may, for example, include:

- Variations in electricity prices
- Tax
- Payments during constructions
- Residual values of the wind farm
- Operation and maintenance costs that vary over the life of the project.

There are generally considerable uncertainties with any financial modelling and it is often helpful to investigate the importance of the various assumptions and estimates in the model using a spider diagram illustrated in Figure 7.13. This is constructed by varying each of the important parameters one at a time from a base case. The slope of the resulting curves then indicates the sensitivity of the outcome to that particular parameter. Example 7.12 illustrates the concept.

EXAMPLE 7.12 Discounted cash flow analysis

Consider a 10 MW wind-farm project

Capital cost $60,00,000

Annual capacity factor 28 per cent

Annual operation and maintenance cost (O&M) $1,00,000

Sale price of electricity 0.04 $/kWh

Discount rate 10 per cent

Project lifetime 15 years

The annual capacity factor is defined as the energy generated during the year (MWh), divided by wind farm rated power (MW) and multiplied by the number of hours in the year. The determination of the capacity factor will, of course, be based on the best available information of wind speeds and turbine performance.

The annual energy yield is

$$0.28 \times 10 \times 8{,}760 = 24{,}528 \text{ MWh}$$

The gross annual income is: $24{,}528 \times 10^3 \times 0.04 = \$9{,}81{,}120$

Figure 7.13 shows the diagram to examine the sensitivity of the calculations while Table 7.3 shows the simple DCF analysis.

The example serves to illustrate a number of points:

1. The net present value (NPV) is positive and so the project is suitable for further consideration.
2. With a 10 per cent discount rate, the discount factor in year 15 is 0.239. This indicates that the present value of the income (and expenditure) stream in that year is only 24 per cent of its cash value.
3. The project is very sensitive to increase in capital cost or reductions in capacity factor. Any reduction in capacity factor could be caused either by low availability of the turbines or by low wind speeds. Steps need to be taken to control these risks by obtaining a firm price for the wind farm, with guarantees on the availability of the turbines, and by

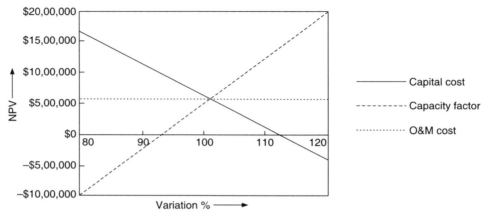

FIGURE 7.13 Sensitivity of NPV to parameter variations.

ensuring that the projections of the wind speeds at the site are based on the best data available.

TABLE 7.3 Simple discounted cash flow of a 10 MW wind farm

Year	Expenditure	Income	Net cash flow	Discount factor	Discounted cash flow
0	60,00,000		−60,00,000	1.000	−60,00,000
1	1,00,000	9,81,120	8,81,120	0.909	8,01,018
2	1,00,000	9,81,120	8,81,120	0.826	7,28,198
3	1,00,000	9,81,120	8,81,120	0.751	6,61,998
4	1,00,000	9,81,120	8,81,120	0.683	6,01,817
5	1,00,000	9,81,120	8,81,120	0.621	5,47,106
6	1,00,000	9,81,120	8,81,120	0.564	4,97,369
7	1,00,000	9,81,120	8,81,120	0.513	4,52,154
8	1,00,000	9,81,120	8,81,120	0.467	4,11,049
9	1,00,000	9,81,120	8,81,120	0.424	3,73,681
10	1,00,000	9,81,120	8,81,120	0.386	3,39,710
11	1,00,000	9,81,120	8,81,120	0.350	3,08,827
12	1,00,000	9,81,120	8,81,120	0.319	2,80,752
13	1,00,000	9,81,120	8,81,120	0.290	2,55,229
14	1,00,000	9,81,120	8,81,120	0.263	2,32,026
15	1,00,000	9,81,120	8,81,120	0.239	2,10,933
				NPV	4,90,936

4. The project is insensitive to operation and maintenance costs at the level included in the model.
5. In this simple model, the costs have been discounted to the commissioning date (year zero). The complete payment for the wind farm is assumed to be made at commissioning and the income and O&M flows are assumed to be at the end of each year's operation.
6. From trial and error calculations, the IPR is found to be 11.5 per cent.

7.7.1 Project Finance

Large companies (e.g. power utilities or major energy companies) may choose to develop small wind-farm projects using their own capital resources. The costs of the project are met from the general equity raised by the company, with the liabilities of the project secured against the main corporate assets. The cost of borrowing, which influences the required discount rate for the project, depends on the financial strength of the company and it is possible to avoid the considerable expense involved in raising external finance. Hence, this approach may allow cash rich companies to develop projects at low cost.

Larger projects, and those developed by entrepreneurs, tend to be based on project financing with a loan obtained from a bank or other financial institutions. This has the advantage of reducing the requirement for capital from the developer, but the loan repayment will have the first call on the income of the project. Project financing is also likely to offer tax advantages. If the loan is secured only on the project cash flow itself, this is referred to as *limited recourse* financing. Alternatively, if the developer has a large parent company willing and able to provide suitable guarantees, then the loan may be secured against the assets of the parent company, which then appears as a liability on its balance sheet. Clearly, the lenders of the debt will prefer to guarantee their loan against the assets of a large stable company rather than the project alone and this will be reflected in the repayment terms that are available.

Figure 7.14 shows a typical commercial structure of a wind-farm project. This is very similar to any other power project developed by independent power producers (IPPs), where the project company is merely the vehicle by which a large number of agreements are made. In order to limit risks, construction of the project does not start until all the agreements are in place and the so-called 'financial dose' is achieved.

The power purchase agreement is to sell the output electrical energy of the wind farm. To reduce risk, this should be at a defined price for the duration of the project.

The loan agreement is for the bank(s) to provide the debt finance for the project. An accurate and verifiable assessment of the wind resource is an essential prerequisite for this agreement, although there is also likely to be a 'due diligence' investigation of the whole project to ensure that all major risks are addressed.

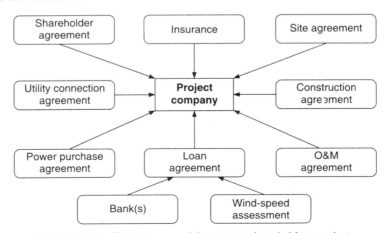

FIGURE 7.14 Typical commercial structure of a wind-farm project.

The construction agreement is to purchase the wind turbine and construct the wind farm. To reduce risk, this may be done on a 'turn-key' basis with the wind turbine manufacturer taking responsibility for the entire wind farm.

The O&M agreement is to operate and maintain the wind farm for the life of the project.

The site agreement is to define the relationship with the land owners and to ensure access to the site and the wind resource for the duration of the project.

The connection agreement is to allow the wind farm to be connected to the electrical power system and so export its output. In a deregulated power system, this is separate from the power purchase agreement.

The shareholder agreement is between the owners of the project company to define their rights and obligations.

It may be seen that a large number of agreements are required and it typically takes a year from project inception to financial close. During this time, the developer is operating at risk, expending considerable resources, and so full development of this administrative financial phase of the project is only commenced once feasibility has been demonstrated.

SOLVED EXAMPLES

1. Calculate the simple payback for a Bergey 1 kWh wind turbine. Go to www.bergey.com to get price and place it on a 20 m tower. It produces 2,000 kWh/year. Assume a value for O&M and fixed charge rate (FCR) = 0.

   Ans. Value of electricity for remote site is $0.50/kWh. From Bergey website, remote package, with batteries: $6,000. Installation adds another $500, so total cost is $6,500.

   $$\text{Value/year} = 2{,}000 \text{ kWh/year} * 0.50/\text{kWh}$$
   $$= \$1{,}000/\text{year}$$
   $$\text{Simple pay back} = SP = IC/(AKWH * \$/\text{kWh})$$

   where   IC = initial cost of installation, $
   AKWH = annual energy production, kWh/year
   $/kWh = price of energy displaced
   SP = 6,500/1,000 = 7 years

2. Calculate the cost of energy for a 400 W (Air X) wind turbine. Installed costs are $1,500, which include 10 m tower and battery. Annual energy production is 400 kWh/year. Assume FCR as 0.05 and AOM = 0.

   Ans. Cost of energy = COE = (IC * FCR + AOM)/AKWH
   where

   FCR = fixed charge rate, per year
   AOM = annual operation and maintenance cost, $/year
   For Air X,   IC = $1,200, AKWH = 400, FCR = 0.05, AOM = 0
   COE = ($1,200 * 0.05)/400 = $0.15/kWh

3. Calculate the cost of energy for a Bergey 10 kW wind turbine on a 30 m tower for a good wind regime.

   Ans.   COE = (IC * FCR + AOM)/AKWH

From www.bergey.com

Value package for 10 kW, grid intertie = $40,600 with shipping and installation

$$IC = \$46,000$$

Assume $\quad$ FCR = 0.08

$$AOM = \$100/\text{year}$$

Assume capacity factor = 25 per cent

$$AKWH = CF * kW * 8,760$$
$$= 0.25 * 10 * 8,760$$
$$= 21,900 \text{ kWh/year}$$
$$COE = (IC * FCR + AOM)/AKWH$$
$$= (\$46,000 * 0.08 + \$100)/21,900 \text{ kWh/year}$$
$$= \$0.17/\text{kWh}$$

4. Calculate the cost of energy for a 50 kW wind turbine, which produces 1,20,000 kWh/year. The installed cost is $1,50,000, fixed charge rate of 10 per cent, O&M is 1 per cent of installed cost, and levelized replacement costs are $4,000/year.

Ans. $\quad$ Cost of energy = COE = $\dfrac{(IC * FCR) + LRC + AOM}{AEP}$

where $\quad$ LRC = levelized replacement cost ($/year)

$\quad\quad\quad$ AEP = net annual energy production (kWh/year)

$\quad\quad\quad$ IC = $1,50,000, AKWH = 1,20,000, FCR = 0.10,

$\quad\quad$ AOM = 1,50,000 * 0.01 = $1,500, LCR = $4,000/year

Therefore COE = (1,50,000 * 0.10 + 1,500 + 4,000)/90,000

$\quad\quad\quad\quad\quad\quad$ = 20,500/1,20,000 = $0.17/kWh

5. Estimate the years to payback, IC = $1,50,000, r = 8 per cent, AKWH = 1,20,000 at $0.08/kWh. Assume a fuel escalation rate of 4 per cent.

Ans. $\quad\quad\quad\dfrac{f_0}{C} \geq \dfrac{(1+r)^L \, \alpha \, rL}{[(1+\alpha)^L][(1+r)^L - 1]}$

where $\quad f_0$ = value of energy saved per year, $

$\quad\quad\quad C$ = initial installed cost, $

$\quad\quad\quad L$ = years to pay back

$\quad\quad\quad \alpha$ = fuel inflation rate

$\quad\quad\quad r$ = interest rate

$\quad\quad\quad$ IC = $1,50,000, r = 0.08, AKWH = 1,20,000 at $0.08/kWh,

$\quad\quad\quad \alpha$ = 0.06

Value per year f_0 = 1,20,000 * $0.08/kWh

$\quad\quad\quad\quad\quad\quad\quad$ = $9,600 per year

$$\dfrac{f_0}{c} = \dfrac{\$9,600}{\$1,50,000} = 0.064$$

For values of α and r, guess at L and calculate right-hand side of equation. From first answer, estimate new L and calculate again. Let us choose $L = 10$ as a starting point. The answer is 12 years to pay back.

a	r			LHS	
0.04	0.08			0.64	
		Calculate RHS of Equation			
L	$(1+r)^L$	$(1+\alpha)^L$	Num.	Dem.	RHS
10	2.2	1.48	0.069	1.04	0.067
20	4.7	2.19	0.298	5.55	0.054
15	3.2	1.80	0.152	2.54	0.060
12	2.5	1.60	0.097	1.51	0.64

6. A 100 MW wind farm (100 wind turbines, 1 MW) is installed in the class 4 wind regime. The production is around 3,000 MWh/turbine/year. The utility company is paying an estimated $0.40/kWh for the electricity produced. Estimate the yearly income from the wind farm. If the landowners get 4 per cent royalty, how much money do they receive per year?

Ans.

$$\text{Total production} = 100 * 3{,}000 \text{ MWh/year}$$
$$= 3{,}00{,}000 \text{ MWh/year}$$
$$\text{Payment} = \$40/\text{MWh}$$
$$\text{Income} = \$40/\text{MWh} * 3{,}00{,}000 \text{ MWh/year}$$
$$= \$1{,}20{,}00{,}000/\text{year}$$
$$\text{Land owner's royalty} = 0.04 \times \$12 \times 10^6/\text{year}$$
$$= \$4{,}80{,}000/\text{year}$$

7. For Question 6, installed costs are $1,600/kW, FCR = 9 per cent, capacity factor = 35 per cent, AOM = 0.008/kWh. Calculate the COE. You will need to estimate the levelized replacement costs or calculate LCR. Compare your answer to the $0.05/kWh, which is the estimated price the wind farm is receiving. How can the wind farm make money?

Ans.

Calculate COE for one wind turbine:
$$IC = \$1{,}60{,}000, \quad FCR = 0.09$$
$$AKWH = 3 \times 10^6, \quad AOM = \$0.05/\text{kWh} \times 3 \times 10^6 \text{ kWh/year}$$
$$= \$15{,}000/\text{year}$$
$$\text{Estimated LCR} = \$15{,}000/\text{year}$$
$$COE = (16{,}00{,}000 \times 0.08 + 15{,}000 + 15{,}000)/(3 \times 10^6)$$
$$COE = \$0.053/\text{kWh}$$

The wind farm receives a production tax credit of $0.02/kWh for 10 years plus accelerated depreciation.

8. A power system consists of 10 kW wind turbine plus battery bank and inverter. IC = $4,500/kW, energy production = 50 kWh/day, FCR = 0.03, AOM = $0.01/kWh. Calculate COE.

Ans.   IC = $45,000, AKWH = 50 * 365 = 18,000
FCR = 0.03, AOM = $180
COE = (45,000 × 0.03 + 180)/18,000 = $0.085/kWh.

REVIEW QUESTIONS

1. A profit making industry wishes to set up a wind farm of 5 MW rating. The capital cost of the wind system is ₹ 24 crore. It is expected that the annual O&M cost will be ₹ 51 lakh. The company is permitted accelerated depreciation (80 per cent in the first year) and the company tax rate is 33 per cent. The industry purchases electricity from the grid at an average price of ₹ 5/kWh. The annual generation from the wind plant is expected to be 8,800 MWh. Compute the NPV and IRR for the company. The company has a hurdle rate (min expected rate of return) of 20 per cent. Should the industry invest on this?
2. A grid-connected wind farm is being set up—rated capacity 1 MW, cost ₹ 5 crore. The O&M cost is ₹ 0.4/kWh. Compute the LCC, ALCC, cost of generated electricity for a real discount rate of 30 per cent with life as 20 years and an annual capacity factor of 30 per cent. If the capacity factor is 20 per cent, what is the cost of generated electricity?
3. What type of incentives should there be for wind energy. Give a brief explanation for your choices.
4. What are ancillary costs?
5. Should governments provide incentives for wind energy? If yes, list your choices and explain why?
6. What incentives may be given from government for residential size wind system?
7. What is renewable energy portfolio standard? Explain with example.
8. Should there be a pollution tax for electricity produced by fossil fuels? If yes, how much per metric ton?
9. What are the two most important factors in the cost of energy from wind?
10. Explain the life cycle cost of a wind energy project.

Chapter 8

Environmental Impact

8.1 INTRODUCTION

The generation of electricity by wind turbines does not involve the release of carbon dioxide, acid rain, smog or radioactive pollutants. The use of wind energy reduces dependency on conventional fossil and nuclear fuels. Wind turbines also do not require water supplies, unlike many conventional and some renewable energy sources. Any development in the countryside will have an environmental impact of some kind, and wind energy is no exception. Wind energy development has both positive and negative environmental impacts. The scale of its future implementation will rely on successfully maximising the positive impacts while keeping the negative impacts to the minimum.

Dealing with uncertainties associated with siting wind power facilities is a challenge because it requires a management structure with high levels of social trust and credibility. Social learning is an approach with management seeking to enhance its capability to learn from experience and an expanding body of knowledge. Two models to decision making strategies are illustrated in Figure 8.1 to deal with environmental challanges of wind energy.

Model 1 assumes that sufficient research and assessment can be done before technology or management system is deployed. Model 2 is suited for wind energy, where experience and knowledge are still growing and where documented effects are strongly site-specific.

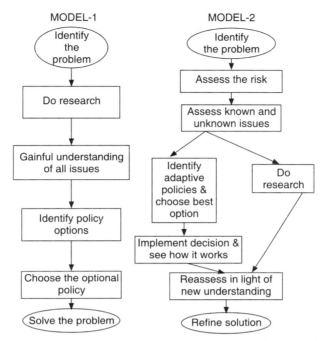

FIGURE 8.1 Linear decision strategy (command and control) and interative model with adaptive management principles. (*Source:* Morgan et al., 2007)

8.2 BIOLOGICAL IMPACT

The potential impact of wind energy project on wildlife is one of the primary factors to be considered in selecting sites for such facilities. At a wind energy project, the habitat loss includes the acres of habitat converted to permanent industrial facility, or the permanent project foot print. The permanent project foot print consists of all permanent facilities, including access roads, turbine locations, substations, O&M facilities, right-of-way under the transmission line, and any other ancillary facilities.

The potential for electrocution of large birds due to new transmission lines of the project can be minimised by adopting the following measures:

- Installation of bird flight diverters where the transmission line crosses riparian corridors
- Use of perch guards or insulated cover-ups
- Inspection and insulation of jumper/ground wires
- Construction of new transmission lines such that all transmission conductors are a minimum of 5 metre apart.

Environmental impacts of wind energy are minimised by siting wind projects in already disturbed landscapes that do not support native vegetation, such as agricultural cropland. In general, the best mitigation is to set turbines back from bird flight paths or to avoid habitat features attractive to relatively high number of birds. Examples of these habitat features include open water, wetlands, cliffs and caprocks that are used by nesting birds and where wind creates an updraft, or known migration or staging areas. Set backs should also be considered where species are present that are sensitive to habitat loss or show avoidance behaviour. A 60-metre set back can be used from the escarpement of a valley but there is no general rule; site-specific field studies recording flight paths of birds could be analyzed to identify such areas.

One method for minimising habitat impacts from wind energy facilities would be to site projects in lower-quality habitats when feasible. Examples of such habitat include:

- Active agriculture, row crops (e.g. wheat, corn, soybeans)
- Pastures
- Industrial sites (e.g. mines, landfills)
- Landscapes already fragmented in ways that reduce bird use
- Low quality, disturbed wastelands.

8.2.1 Ecological Assessment

Wind farms are often constructed in the areas of ecological importance, and the environmental statement will include a comprehensive assessment of the local ecology, its conservation importance, the impact of the wind farm and mitigation measures. When considering the ecological impact of renewable energy schemes, the following categories of effects should be considered:

- Immediate damage to wildlife habitats during construction
- Direct effect on individual species during operation
- Longer-term changes to wildlife habitats as a consequence of construction or because of changed land use management practices.

Thus, the scope of the ecological assessment is likely to include:

- Full botanical survey including identification and mapping of plant species on the site
- Desk and field survey of existing birds and non-avian fauna
- Assessment of how the site hydrological conditions relate to the ecology
- Evaluation of conservation importance of the ecology of the site
- Assessment of the potential impact of the wind farm.

The following are the mitigation measures that may be undertaken to protect important bird species while allowing wind energy development to continue:

1. Baseline studies should be undertaken at every wind farm site to determine which species are present and how the birds use the site. This should be a mandatory part of the environmental statement for all wind turbines.
2. Known bird migration corridors and areas of high bird concentration should be avoided unless site specific investigation indicates otherwise. Where there are significant migration routes, the turbines should be arranged to leave suitable gaps, e.g. by leaving large spaces between groups of wind turbines.
3. Particular care is necessary during construction and it is proposed that access to contractors should be limited to avoid general disturbance over the entire site. If possible, construction should take place outside the breeding season. If this is not possible, then construction should begin before the breeding season to avoid displacing nesting birds.
4. Tubular turbine towers are preferred to lattice structure. Consideration should be given to using unguyed meteorological masts.

5. Fewer large turbines are preferred to large numbers of small turbines. Larger turbines with lower rotational speeds are probably more readily visible to birds than smaller machines.
6. Within the wind farm, the electrical power collection system should be underground.
7. Turbines should be laid out so that adequate space is available to allow the birds to fly through them without encountering severe wake interaction. A minimum spacing of 120 m between rotor tips is tentatively suggested as having led to minimum collision mortalities. Turbines should be set back from ridges and avoid saddles and folds which are used by birds to traverse uplands.

8.2.2 Avian Issue

The bird mortality by wind energy's impact is quite small as illustrated in Figure 8.2. Currently, it is estimated that for every 10,000 birds killed by all human activities, less than one death is caused by wind turbines. Birds and bats may collide with wind turbines, as they do with any structure on their route. Careful planning can avoid the risk to bird population due to wind turbines.

Once the site for a wind farm is identified, the potential effects of the proposed development on the birds of concern should be systematically studied. If the available information is not sufficient to establish the biological suitability of the site, then conduct onsite monitoring and surveying using appropriate methods. All the bird groups of special concern, including the breeding, migrating and wintering species should be included in such studies. Apart from the mere number of birds and its distribution in the wind farm area, the functional behaviour of the species in the region should also be included in the studies.

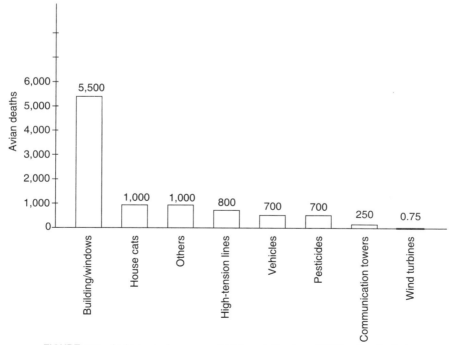

FIGURE 8.2 Anthropogenic causes of bird mortality. (per 10,000 avian deaths). (*Source:* Erickson et al., 2002)

8.3 SURFACE WATER AND WETLANDS

For conserving surface water and wetlands, typical best management practices include:

- Use of silt fences between construction area and waterbodies and/or wetlands.
- Installation of temporary water diversions at water channel crossings.
- Use of erosion control blankets or mats on slopes near waterbodies and/or wetlands.
- Construction of temporary bridges and culverts.
- Restoration of vegetative cover to the greatest extent practicable at the site.

For example, if one acre of a wetland is permanently impacted by the proposed project, one acre of replacement wetlands may be created.

8.4 VISUAL IMPACT

An adverse visual impact can be defined as an unwelcome visual intrusion that diminishes the visual quality of an existing landscape. Changes that can be perceived as visual intrusions generally result from the introduction of visual contrast to the existing scene based on differences in form, line, colour, and/or texture.

Wind turbines typically become the focal point of visual and aesthetic concerns on the basis of the visual patterns created by the wind turbines, such as their spacing and uniformity of appearance, as well as the physical markings or lighting on the turbines such as lighting required for aviation safety. In most cases, it is the simple size of the wind turbine that is the predominant source of visual contrast created by wind energy facility. As wind generating technology has advanced, so has the physical size of the structure. Due to its huge size, in a setting that is typically free of structures, trees, or intervening terrain and vegetation, the wind turbines will be visible.

Some options for key aesthetic design, construction and operation measures for consideration are as following:

- Employ turbine units (towers, nacelle and rotors) that are uniform and balanced in shape, colour and size. The colour should be used as off-white, earthy or in harmony with the background colour of landscape or skyline.
- The turbines will generally be seen against the sky and so an off-white or mid-grey ton is generally considered to be appropriate. Where the wind farm is seen against other backgrounds, a colour to blend in with the ground conditions may be more suitable. There is general agreement that the outer gel-coat of the blades should be a matt or semi-matt finish to minimise reflections.
- There are clearly engineering benefits associated with lattice towers, particularly in terms of the cost of the foundations, although the upper, tapered section of the tower must be such as to provide adequate clearance for the blades. From a distance and in certain light conditions, lattice towers can disappear leaving only the rotors visible and this effect is generally thought to be undesirable.
- Prohibit the use of commercial markings or messages on the turbines.
- Limit markings and lighting on the turbines to the minimum required for safety purposes, and synchronize flashing warning lights.

- Install power collection cables underground, wherever feasible.
- To the extent practicable, site substations, service buildings, and other project support facilities in locations where they will be less visible, and design the structures to harmonize with their visual setting.
- Locate project access roads to limit their visibility and potential to create erosion.
- Maintain project facilities regularly during operation (including repair or replacement of inoperable turbines or parts, regular painting and cleaning), minimize outside storage of material or equipment, promptly remove any debris or defective equipment, and keep the site in order.

8.4.1 Shadow Flicker

Shadow flicker is the term used to describe the effect caused by the shadows cast by moving wind turbine blades when the sun is visible. This can result in alternating changes in light intensity perceived by viewers. Since wind turbines are usually located relatively far from potential shadow receptors, shadow flicker typically occurs only at times and locations and low sun angles; this is most common just after sunrise and just before sunset, and in relatively higher latitudes. Shadow flicker does not occur when the sun is observed by clouds or fog or when the wind turbines are not operating, or when the blades are at a 90° angle to the rececptor.

The frequency that can cause disturbance is between 2.5–20 Hz. The effect on humans is similar to that caused by changes in intensity of an incandescent electric light due to variations in network voltage from a wind turbine. In the case of shadow flicker, the main concern is variations in light at frequencies of 2.5–3 Hz, which have been shown to cause anomalous EEG (electroencephalogram) reactions is some suffering from epilepsy. Higher frequencies (15–20 Hz) may even lead to epileptic convulsions. Large modern three-bladed wind turbines will rotate at under 35 rpm giving blade-passing frequencies of less than 1.75 Hz, which is below the critical frequency of 2.5˙Hz. A minimum spacing from the nearest turbine to a dwelling of 10 rotor diameter is recommended to reduce the duration of any nuisance due to light flicker.

While shadow flicker can be perceived outdoors, it tends to be more noticeable in rooms with windows oriented to the shadows. A wind turbine's shadow flicker impact area does not generally extend beyond 2 kilometres, and high impact durations are generally low located within approximately 300 metres of the turbine. A shadow flicker map is shown in Figure 8.3.

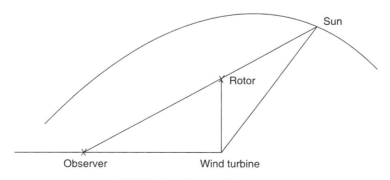

FIGURE 8.3 Shadow flicker map.

8.5 SOUND IMPACT

There are two main sources of wind turbine noise. One is that produced by mechanical or electrical equipment, such as the gearbox, bearings generator, yaw mechanism and pitch mechanism known as mechanical noise; the other is due to the interaction of the airflow with the blades, referred to as aerodynamic noise. Figure 8.4 shows the main sources emitting sound.

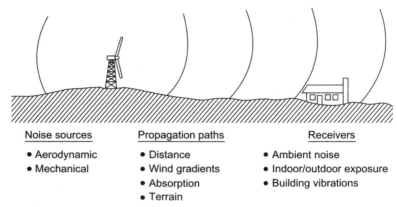

FIGURE 8.4 Factors contributing to wind turbine noise.

Sound waves are sequences of air pressure that can be detected by the human ear. The frequency that is audible to human hearing is from approximately 20 to 20,000 Hertz (Hz) (1 Hz = 1 cycle of vibration per second). Noise is unwanted sound. Sound is measured on a logarithmic scale in units known as decibels (dB). The threshold of hearing is defined as 0 dB. On this scale, a car passing 100 m away at 64 km/h has a sound level of approximately 55 dB. Doubling the sound power of the noise source increases the sound level by 3 dB.

The A-weighted (accoustically weighted) sound pressure curve, shown in Figure 8.5, describes the relative responsiveness of the ear to sound of different frequencies. Noise measurements are usually expressed not simply in decibels (dB) but on scale in which the sound levels are adjusted for different frequencies, in accordance with A-weighted sound pressure curve. The units used in this case are dB(A) (decibels, accoustic).

Sound level decreases further if you are away from the source of the noise: a simple rule of thumb is that noise levels reduce by 6 dB for every doubling of distance. The sound level at a given distance from the source tends to be greater downwind than at a similar distance upwind. Sound is absorbed as it passes through the air, but in a manner that varies according to frequency, humidity and temperature. It is also absorbed over long grass, vegetation and trees, but it is reflected, focussed or 'shaded' by hard surfaces such as roads, walls of buildings, etc.

The aerodynamic sound from wind turbines is caused by the interaction of the turbine blade with the turbulence produced both adjacent to it (turbulent boundary layer) and in its near wake as shown in Figure 8.6. Turbulence depends on how fast the blade is moving through the air. The main determinants of the turbulence are the speed of the blade and the shape and dimensions of its cross-section.

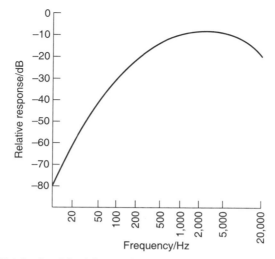

FIGURE 8.5 A-weighted (accoustically weighted) sound response curve.

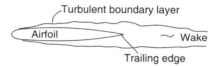

Turbulent boundary layer-trailing edge noise

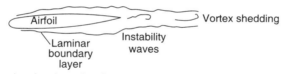

Laminar boundary layer-vortex-shedding noise

FIGURE 8.6 Sound produced by wind turbine flow.

The following points are to be noted for the flow conditions depicted in Figure 8.6.

- At high velocities for a given blade, turbulent boundary layers develop over much of the airfoil. Sound is produced when the turbulent layer passes over the trailing edge.
- At lower velocities, mainly laminar boundary layers develop, leading to vortex shedding at the trailing edge.

Other factors in the production of aerodynamic sound include the following:

- When the angle of attack is not zero in other words, the blade is tilted into the wind-flow separation can occur on the suction side near to the trailing edge, producing sound.
- At high angles of attack, large-scale separation may occur in a stall condition, leading to radiation of low frequency sound.
- A blunt trailing edge leads to vortex shedding and additional sound.
- The tip vortex contains highly turbulent flow.

The dominant sound is produced along the blade—nearer the tip end than to hub. Wind may also cause the sound level to be greater downwind of the turbine—that is, if the wind is blowing from the source towards a receiver—or lower, if the wind is blowing from the receiver to the source.

The sounds that most turbine emit are caused by the passage of the blades through the air—the aero-acoustic 'swoosh'. Aero-acoustic sound can occur in pulses corresponding to the passage of the turbine's blades as the rotor turns. It tends to be most noticeable in the middle range of turbine operating wind speeds, when the masking sounds of the wind are not at their highest level. Figure 8.7 illustrates sources of aerodynamics noise.

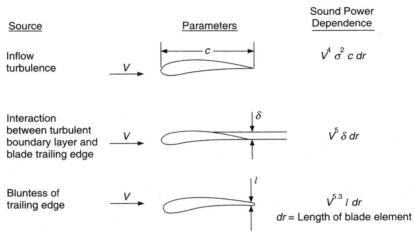

FIGURE 8.7 Sources of wind turbine broadband noise.

Some turbines also emit tonal sound, caused by mechanical components. While tonal sounds may not result in a higher overall level of sound emissions (in terms of loudness), they are more noticeable.

Most of the turbine manufacturers provide sound level data determined in accordance with the International Electrotechnical Commission (IEC) standard. The IEC method establishes that accoustic reference wind speed of 8 m/s at 10-metre height. Typically, the measurement is made using a microphone at ground level at a given distance from the turbine base, and then a series of mathematical calculations, defined by IEC and using the manufactures power curve for the turbine, are applied to the results to standardise them for the reference wind speed and location to allow comparison of different turbines and different hut heights. The sound power levels are then used to predict overall project sound emission level.

While human perception of sound level is substantially subjective, it is possible to accurately compare various sound levels with commonly experienced sounds as shown in Table 8.1.

Figure 8.8 shows a typical sound levels from a 1.5 MW wind turbine at different distances.

It is important to measure the background sounds at a particular location with the proper equipment typically prior to installation of the turbine(s). The relative increase in sound from the project is more important than the absolute sound levels of the project itself. Some of the mitigation techniques are as follows:

- Siting turbines beyond a minimum set back distance (typically 350 m) to all residential structures

Environmental Impact

TABLE 8.1 Noise of different activities compared with wind turbines

Source/Activity	Noise level in dB(A)*
Threshold of pain	140
Jet aircraft at 250 m	105
Pneumatic drill at 7 m	95
Truck at 48 km/h	65
Busy general office	60
Car at 64 km/h	55
Wind farm at 350 m	35–45
Quiet bed room	20
Rural night-time background	20–40
Threshold of hearing	0

*dB(A): decibels (accoustically weighted)

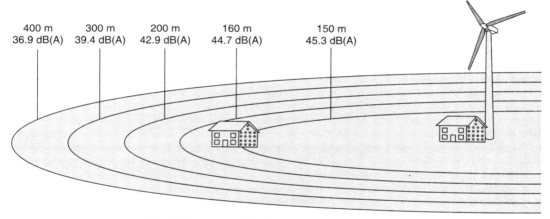

FIGURE 8.8 Sound level from a 1.5 MW wind turbine.

- Implementing best management practices for noise abatement during construction, including use of appropriate mufflers and limiting hours of construction
- Limiting the cutting/clearing of vegetation surrounding the proposed substation
- Adding landscape features to help screen specific receptors
- Keeping turbines in good running order throughout the operational life of the project
- Pursuing development agreements with neighbours whose residence is located within a certain distance of a project turbine
- Implementing a complaint resolution procedure to assure that any complaints regarding construction or operational noise are promptly and adequately investigated and resolved.

The following information is required for understanding the sound from a wind farm:

- Technical details of the turbine
- Hub height and rotor diameter
- Location of rotor: upwind or downward of the tower

- Speed(s) of rotation
- Cut-in wind speeds
- Predicted noise levels at specific properties closest to the wind turbine over the most critical range of wind speeds
- Measured background noise levels at the wind farm site and wind speed
- A scale map showing the proposed wind turbine(s), the prevailing wind conditions, and nearby existing development
- Results of independent measurements of noise emission from the proposed wind turbine, including the sound power level.

Noise is a sensitive issue and opposition to wind energy development is likely unless it is given careful consideration both at the wind turbine design and the project planning stages, taking into account the concerns of people who may be affected. Figure 8.9 shows the effect of wind induced refraction on accoustic rays radiating from an elevated point source. Figure 8.10 summarises factors and interactions to be considered in evaluating human response to wind turbine noise.

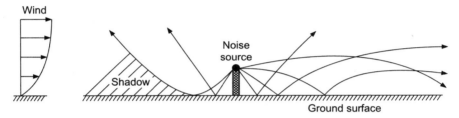

FIGURE 8.9 Effects of wind-induced refraction on accoustic rays radiating from an elevated point source.

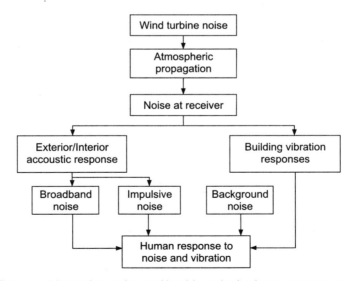

FIGURE 8.10 Factors and interactions to be considered in evaluating human response to wind turbine noise.

8.5.1 Measurement, Prediction and Assessment

The sound power level of a wind turbine is normally determined by field experiments. As per BS1999, JEC, 1998 outdoor experiments are necessary. The sound power level cannot be measured directly but is found from a series of measurements of sound pressure levels made around the turbine at various wind speeds from which the background sound pressure levels have been deducted. The method provides the apparent A-weighted sound power level at wind speed of 8 m/s, its relationship with wind speed and directivity of the noise source for a single wind turbine. It does not distinguish between aerodynamic and mechanical noise although there is method proposed to determine if tones are 'prominent'. In addition to A-weighted sound pressure levels. Octave or 1/3-octave and narrow-band spectra are measured.

The measurement are taken at a distance, R_0 from the base of the tower:

$$R_0 = H + \frac{D}{2} \quad (8.1)$$

where H is the hub height and D is the diameter of the rotor. The distance is a compromise to allow an adequate distance from the source but with minimum influence of the terrain, atmospheric conditions or wind-induced noise. The microphones are located on boards at ground level so that the effect of ground interference on tones may be evaluated. Four microphones are used as shown in Figure 8.11.

Simultaneous A-weighted sound pressure level measurements (more than 30 measurements, each of 1-2 min. duration) are taken with wind speed. All wind speeds are corrected to reference height of 10 m with a terrain roughnessc length of $Z_0 = 0.5$ m. The preferred method of determining wind speed, when the turbine is operating, is from the electrical power output of the turbine and the power curve. The main sound pressure level

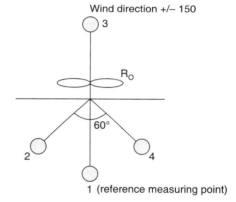

FIGURE 8.11 Recommended pattern for measuring points from IEA recommended practices for wind turbine testing (IEA, 1994 and BS EN61400-11).

measurement is that of the downwind position while the other three microphones are used for determining directivity. Measurements are taken with and without the turbine operating over a range of at least 4 m/s, which includes the 8 m/s reference. The sound pressure levels, with and without the turbine in operation, are then plotted against wind speed and linear regression used to find the 8 m/s values. The method of bins may be used to group the data. The sound pressure level of the turbine alone at the reference conditions, $L_{\text{Aeq,ref}}$ is then calculated using

$$L_{\text{Aeq,ref}} = 10 \log_{10} (10^{L_{S+N}/10} - 10^{L_N/10}) \quad (8.2)$$

where L_{S+N} is the sound pressure level of the wind turbine and the background sound at 8 m/s, and L_N is the sound pressure level of the background with the wind turbine parked at 8 m/s.

The apparent A-weighted sound power level of the turbine is then calculated from

$$L_{WA,\text{ref}} = L_{\text{Aeq,ref}} + 10 \log_{10} (4\pi R_i^2) - 6 \quad (8.3)$$

where R_i is the slant distance from the microphone to the wind turbine hub.

It may be seen that the calculation of sound power level assumes spherical radiation of the noise from the hub of the turbine. The subtraction of 6 dB is to determine the free field sound pressure level from the measurements and to correct for the approximate pressure doubling that occurs with the microphone located on a reflected board at ground level.

The directivity, DI, is the difference between the A-weighted sound pressure level at measurement points $i = 2–4$ and the reference point 1, downstream of the turbine. It is calculated from

$$DI_i = L_{\text{Aeq},i} - L_{\text{Aeq},1} + 20 \log_{10} \frac{R_i}{R_1} \quad (8.4)$$

To esimate the sound pressure level of a single turbine or group of turbines at a distance R provided that they are located in flat and open terrain. The calculation is based on hemispherical spreading,

$$L_P(R) = L_W - 10 \log_{10} (2\pi R^2) - \Delta L_a \quad (8.5)$$

The correction ΔL_a is for atmospheric absorption and can be calculated from $\Delta L_a = R\alpha$, where α is a coefficient for sound absorption in each octave band and R is the distance to the turbine hub.

If there are several wind turbines which influence the sound pressure level, the individual disturbances are calculated separately and summed using,

$$L_{1+2+\ldots} = 10 \log_{10} (10^{L_1/10} + 10^{L_2/10} + \cdots) \quad (8.6)$$

8.6 COMMUNICATION IMPACT

The wind turbines create physical obstructions that distort communication signals. The types of communications systems that may be affected include microwave systems, off-air television broadcast signals, land mobiles radio operations and mobile telephone services. Microware telecommunication systems are wireless point-to-point links that communicate between two sites (antennas) and require clear line-of-site conditions between antenna. Obstruction between transmitters reduces the reliability of the transmission.

Identification of communication systems near a wind project can avoid interference with signal transmission. Moving turbines outside of microwave paths and away from antennas would avoid obstructing microwave transmission and radio broadcasts. The following mitigation measures are available when turbines cannot be moved to accommodate communication systems:

- Adding transmitters and receivers to the communication system
- Installing satellite television service
- Installing cable television services when available
- Installing directive reception television antenna with amplifier
- Setting up wireless television distribution system for a cluster of homes affected by a wind project
- Repositioning land mobile radio repeaters or adding repeaters to the existing system.

The schematic plan view of the relative positions of a transmitter, receivers and a wind turbine are shown in Figure 8.12.

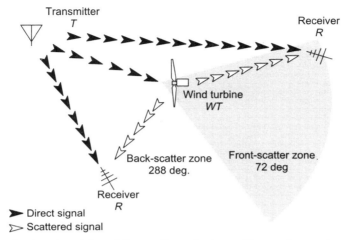

FIGURE 8.12 Schematic plan view of the relative positions of a transmitter, receivers and a wind turbine that may produce electromagnetic interference (EMI).

8.6.1 Electromagnetic Interference (EMI)

It is well known that any large structure, whether stationary or moving, in the vicinity of a receiver or transmitter of electromagnetic signals may interface with those signals and degrade the performance of the transmitter/receiver system. Under certain conditions, the rotor blades of an operating wind turbine may passively reflect a transmitted signal, so that both the transmitted signal and a delayed interference signal (varying periodically at the blade passage frequency) may exist simultaneously in a zone near the turbine. The nature and amount of electromagnetic interference (EMI) in this zone depend on a number of parameters, including location of the wind turbine relative to the transmitter and receiver, type of wind turbine, physical and electrical characteristics of the rotor blades, signal frequency and modulation scheme, receiver antenna characteristics, and the radio wave propagation characteristics in the local atmosphere. When the influence of these parameters on EMI is understood, wind turbine can usually be designed and sited so that any interference with communication signals may not exceed allowable levels.

Television interference (TVI) from wind turbines is characterized by video distortion that generally occurs in the form of a jittering of the picture that is synchronised with the blade passage frequency, i.e. the rotor speed times the number of blades. In the worst case, a strong interference signal may cause complete break-up of the picture. No audio distortion has been observed. Effects on frequency modulated (FM) broadcast reception appear as a background hissing noise pulsating in synchronism with the rotor blades. The interference to navigation systems has indicated that a stopped wind turbine may produce scalloping errors in the azimuthal bearings. However, when the wind turbine is operating, potential interference effects are much less than those under stopped conditions. Under normal siting conditions, i.e. when the wind turbine is not in close proximity to the transmitter or receiver, generally no degradation in communication performance is likely to occur. Theoretical studies of interference to microwave links indicate that the effects tend to smear out the modulation used in typical microwave transmission systems. Figure 8.13 shows interference mechanisms of wind turbines with radio systems.

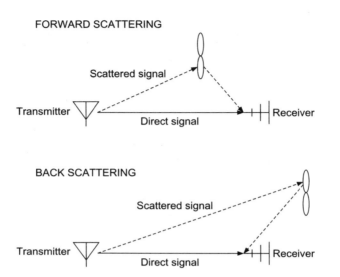

FIGURE 8.13 Interference mechanisms of wind turbines with radio systems.

REVIEW QUESTIONS

1. List major environmental issues as a case study for a large wind farm and mitigation strategies.
2. What are the major environmental concerns for off shore wind farm?
3. Calculate the carbon dioxide that wind displaced for the world. Use kilograms per kilowatt-hour from coal power electric plants for comparision. Use 25 per cent capacity factor for wind farms in estimating annual energy production.
4. Calculate the carbon dioxide that displaced for India, EU and for United States. Use kilograms per kilowatt-hour from coal power electric plants for comparison. Use 35 per cent capacity factor for wind farm in estimating annual energy production.
5. Compare the fatality rates for birds and bats from wind turbines.
6. What is the leading cause of death in wind industry? What are the leading causes of wind turbine accidents?
7. Is there a small wind turbine in your region? If yes, what is the visual impact from the neighbour's view and from the public view? Use Table 8.2 and Table 8.3 to estimate score?

TABLE 8.2 Criteria for points for visual impact of small wind turbines

	Neighbour's view				Public view	
	1	2	3	4	5	6
Point	View angle degree	Distance (m)	Prominent	Screened	Vista	Duration (s)
0	> 90	> 900	Below tree tops	Complete	Degraded	0
1	0–45	450–900	At horizon line	Multiple trees single tree	Common	< 15
2	50–60	150–450	Above horizon line	1/2–2/3	Scenic	< 30
3	60–90	< 150	Above tallest mountain	No screening	Highly scenic	> 60

TABLE 8.3 Rating of visual impact of small wind turbines

	Score	
	Neighbour	Public
Negligible	0–3	0–3
Minimal	3–6	3–9
Moderate	6–9	9–14
Significant	9–12	14–18

Chapter 9

Electrical and Control Systems

9.1 INTRODUCTION

Electricity and magnetism are related with charges and movement of charges. The following terms are related to electricity.

Current: Current is the rate of flow of charge (electrons). Charge is measured in coulomb. When the flow of charge is in one direction, it is direct current (DC) and flow changes direction it is called alternating current (AC). Frequency, that is, the number of cycles per second, is measured in terms of hertz (Hz).

$$\text{Current } I = \frac{\Delta q}{\Delta t} \tag{9.1}$$

where current I is measured in terms of ampere, Δq is charge and Δt is time. In India, the frequency of alternating current is 60 Hz.

Voltage: The potential energy (PE) to move charge divided by the charge is potential difference and measured in volts.

$$V = \frac{PE}{q} \text{ volts (V)} \tag{9.2}$$

Resistance: According to Ohm's law, current is linearly proportional to voltage, and constant of proportionality is called resistance.

$$V \propto I$$
$$V = RI \qquad (9.3)$$
$$R = \frac{V}{I} \; Ohm \; (\Omega)$$

In other words, there is resistance to the flow of charge across different elements in a circuit. In metals, resistance increases with temperature and more energy will be lost.

Power: The power in the circuit is the product of flowing current and voltage.

$$P = VI$$
$$P = I^2 R \qquad (9.4)$$

It means the conductor will be heated and power will be lost and it depends on the square of the current. Due to this, electric power is transmitted at higher voltages. Smaller diameter wire can be used for higher voltages for power transmission purpose. With large wind turbines, transformers are used to increase the voltage for transmission.

Capacitance: It is the amount of stored charge in capacitor. Capacitor is a device for storing charge for short duration of time between two metal plates separated by a small distance.

Inductance: Inductors are devices used for storing magnetic fields. A coil or wire is an example of inductor.

Electric field: An electric field E is around a charged particle. It may originate or terminate on it. If a charged particle is subjected to a force, it means it is in the electric field.

$$E = \frac{F}{q} \qquad (9.5)$$

Where E is the electric field, F is the force and q is the charge.

Magnetic field: Magnetic field B is due to moving charge. Few materials have a property to become magnets. Permanent magnets can be made of these materials, mostly rare earth materials and more expensive than iron, nickel and cobalt. Permanent magnet alternators are made by rare earth materials.

The theory of electromagnetism is formulated by Maxwell in the form of Maxwell's equations. Changing electric fields creates changing magnetic fields and vice-versa. If charged particles are placed in external electric fields and if moving charged particles are placed in external magnetic field, there will be a force on the charged particle. The amount of force depends on the strength of the electric and magnetic fields, the amount of charge and the velocity of the charge.

$$F = qE + q(V + B) \qquad (9.6)$$

Equation (9.6) is the basis of conversion of electric energy to mechanical energy, that is, the principle of working motor and the conversion of mechanical energy to electric energy, that is, the principle of working of generator.

Generator: A coil of wire is rotated by an external force (Figure 9.1). The power $P = Tw$, where T is torque and w is rotational speed in radians per second which may come from a rotor of wind turbine either directly in gearless machine or through a gearbox. If there is just one coil, it is a single phase generator. If there are three coils of wire, then it is three phase generator. The external magnetic field can be created by electromagnets or permanent magnets. The current in a coil creates a magnetic field in an iron core or magnetic field in the iron core will create electricity in the coil. The number of coils is referred as pole.

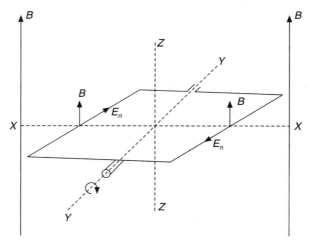

FIGURE 9.1 A coil rotated by an outside force with angular velocity ω in a uniform external magnetic field, a generator.

Motor: When current flows in the coil and an external magnetic field is applied, as shown in Figure 9.2, the force will be acted and gives rise to torque T on the coil to rotate about axis X–X. The torque T in the coil will depend on the current I, the area of coil A, and the strength of the magnetic field, B. The magnitude of torque T in the motor will be

$$T = I\,(A \times B) \tag{9.7}$$

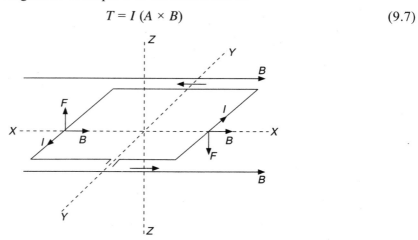

FIGURE 9.2 Forces on the sides of a current carrying coil in an external magnetic fields, a motor.

Electromagnetic induction: According to Faraday's law of electromagnetic induction, the magnetic flux ϕ_M is equal to the strength of the magnetic field B times the area of coil.

$$\phi_M = B \cdot A = BA \cos \theta \tag{9.8}$$

where θ is the angle between magnetic field and plane of coil. The electromotive force ϵ is then equal to the change in magnetic flux to time.

$$\epsilon = -\frac{\Delta \phi}{\Delta t} \tag{9.9}$$

In generators and motors, the magnetic field and area of coil can be kept constant and the angle between the two can be changed by rotating the coil of wire. It gives an alternating voltage and current, which varies like a sine wave as shown in Figure 9.3.

When two coils are placed near each other and changing magnetic flux in one coil causes a change in current in the other coil, it is called induction. A transformer works on the principle of induction.

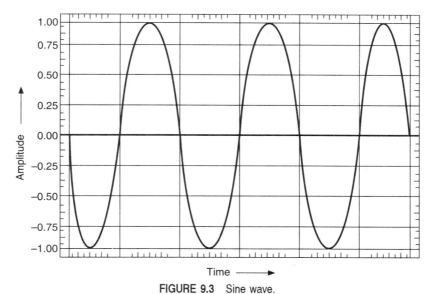

FIGURE 9.3 Sine wave.

Phase angle: For a load of pure resistance, the voltage is in phase with the current, for a capacitor the voltage lags the current by 90° and for an inductor the voltage leads the current by 90° as shown in Figure 9.4. The current is shown with reference to voltage.

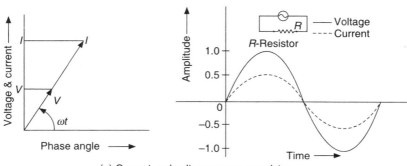

(a) Current and voltage across a resistor.

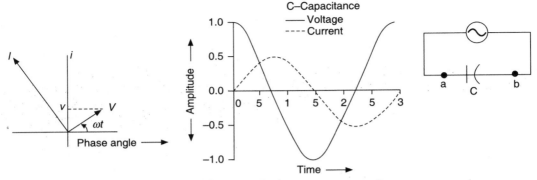

(b) Current and voltage across a capacitor.

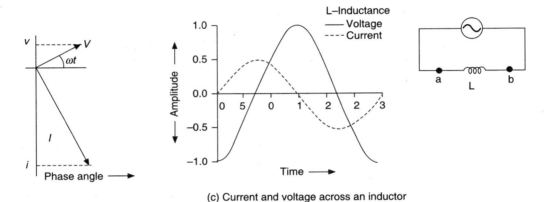

(c) Current and voltage across an inductor

FIGURE 9.4 Phase relationship between current and voltage across resistor, capacitor and inductor.

Power factor: Instantaneous current and voltage are given as,

$$I = I_p \sin(\omega t + \phi) \text{ and } V = V_p \sin(\omega t) \quad (9.10)$$

where I_p and V_p are the peak values

ω is angular velocity, $2\pi/t$ in rad/sec. t, time in seconds.

ϕ is the angle in degrees, phase difference between the instantaneous voltage and current.
For a resistor, voltage and current are in phase and the average power per cycle is

$$P = V * I = V_p \sin(\omega t) * I_p \sin(\omega t) = 0.5\, V_p I_p \quad (9.11)$$

For a capacitor, voltage and current are 90° out of phase and the average power per cycle is

$$P = V * I = V_p \sin(\omega t) * I_p \sin(\omega t + 90°) = 0$$

In real circuits, there is resistance, capacitance and inductance. Therefore, voltage and current will not be completely in phase. The instantaneous power to an arbitrary alternating current (AC) circuit oscillates because voltage and current will also oscillate. The instantaneous power is

$$P = VI = V_p \sin(\omega t) * I_p \sin(\omega t + \phi) \quad (9.12)$$

Average power is calculated by integrating over one cycle:

Average power, $$P_a = \frac{V_p I_p \cos\phi}{2} \tag{9.13}$$

The average values of current and voltage are measured in AC circuits as root mean square values for single phase.

Average voltage, $V_a = (V * V)^{0.5} = (V_p \sin(\omega t) * V_p \sin(\omega t))^{0.5}$

$$= V_p/2^{0.5} = 0.707 V_p \tag{9.14}$$

For three phases, the current is reduced by $\sqrt{3}$. For three-phase power transfer, each leg is transferring a current equal to coil current divided by $\sqrt{3}$. The real power generated or consumed is given as

Average power, $$P_a = VI \cos\phi \tag{9.15}$$

where $\cos\phi$ is the power factor. Adding a number of induction generators to the utility line can change the power factor and reduces the actual power delivered. Since the utility grid supplies the reactive power for the induction generators, capacitors are added to wind turbines or to the electric substation.

Wind power is easily convertible to electrical power due to the following reasons:

- Mechanical/electrical converters are robust, efficient and easily controllable. They could also be arranged to operate in a bi-directional way, i.e. they could be made to motor.
- Electrical transmission systems are very efficient and can transport energy over long distances economically and reliably.
- Interconnected electrical grids are spread geographically.
- It is therefore, in principle, easy to feed the converted wind energy into the grid.
- Electrical energy is a high grade energy.
- It is economically advantageous to offer wind energy in electrical form.
- Interconnected grids can easily accommodate the variability of electric energy from wind energy conversion systems.

Most wind-driven generators are connected to an existing utility grid that can supply electric power when wind speeds are low and can absorb wind power when speeds are high. The utility regulates the voltage within reasonable limits. The generators should be able to deliver power developed by the turbine over a wide range of wind speeds. It is not practical to design a generator to cover the highest possible wind speed. In general, the higher the peak power to be developed, the higher the cost. Since the highest wind speeds usually come in gusts, it is desirable to take advantage of the short-time overload capacity of a generator.

There are two general types of generators that should be considered: constant speed and variable speed. The simplest are the constant-speed generators either the induction generator with a squirrel-cage rotor (similar in construction to the simplest induction motor) or the synchronous generator (similar in construction to a synchronous motor). These generators run at, or near, a synchronous speed that is determined by the number of poles. For a 60 Hz power system, this synchronous speed in rpm is 7200, divided by the number of poles. If the synchronous speed is chosen to match some lower wind speed, it would then generate power only whenever the wind speed increased above this value.

With adjustable-pitch blades, the full advantage of the constant-speed generator can, in theory, be realised over a wide range of wind speeds. Even with fixed blades and constant speed, power can be developed over a considerable range of wind speeds, but with reduced efficiency except for a limited range. For smaller units, the low cost and simplicity of constant-speed generator may make it the best choice, even for a fixed-pitch blade. The choice is a matter of evaluating cost versus extra power generated. Two general types of variable-speed generators are available: the wound rotor induction generator and a synchronous generator with a frequency converter.

9.2 CLASSIFICATION OF GENERATORS

A classification of generating systems is shown in Figure 9.5. The principal division lies in the nature of the field patterns produced independently by the stator and the rotor. Machines with synchronous field patterns have in the steady-state a field pattern stationary with respect to one member, either the stator or rotor windings. The other member produces a pattern travelling with respect to itself, and electrical energy flows from (or to) its windings. The synchronous generator is the prime example of this type of machine. The stator produces a rotating field with respect to itself. The rotor producing a stationary field (with respect to itself), locks in with the field of the stator and travels with it.

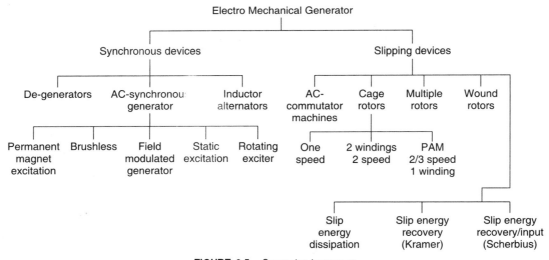

FIGURE 9.5 Generator taxonomy.

The normal DC-machine acts similarly. A stationary field is produced by the stator and interacts with that arising from the action of the rotor currents also stationary because of the commutator action. Another machine in this category is the inductor alternator in which the rotor produces field variations by reluctance variation, i.e. alternator rotor teeth and slots interacting with similar stator slots. Permanent magnet machines also are in this category, the field of one member, usually the rotor being obtained from permanent magnets. In summary, these machines as generator have:

- A field pattern stationary with respect to one member (say A).
- A field pattern moving with respect to one member (say B).

- Mechanical energy supplied to one member which is rotated with respect to the other.
- Energy supplied to member A, sufficient only to maintain the field current and supply the heating losses in the winding.
- Energy flowing from the winding of member B.
- No voltage induced in the windings of member A as it sees no field variation.

If the field in member A is excited with alternative current, the windings of B will have two frequencies: that arising from rotation so that any power flowing comes from mechanical rotation, and transformer-type voltages arising from the alternating character of member A current. Any energy arising at this frequency comes from the field supply of member A, not from the rotational motion.

The second category of machines is those that involve slip and in which the field pattern moves with respect to both members. A concomitant of the motion with respect to both members is that, energy is supplied to or from both members, or consumed in losses in one of them. This category includes all induction machines and double-fed machines. Whenever voltage is generated because of the relative motion of a field and a winding, some of the energy supplied to that winding is dissipated in the winding.

The slip, s, is the fraction of the field speed by which the rotor moves with respect to it. For normal generation with dissipation in the rotor or export of power, slip is negative and shaft speed exceeds that of the air-gap field. The performance in this mode is similar to that of a slipping clutch, energy entering in the shaft being imported to the output shaft by the ratio of their speeds, and the difference of energy being dissipated as frictional loss.

However, another mode of performance is possible in an induction device which is not feasible in a clutch. If the rotor is run below synchronous speed (slip is positive) and multiphase slip frequency power is supplied to the rotor to interact with the stator field, the stator output will be at the same frequency as before. However, the power is now flowing from the rotor mechanical input and the rotor electrical input in the ratio $\frac{1-s}{s}$.

Perhaps the best comparison is that of a differential gearbox in which the output speed and power is the sum or difference of the speeds applied to the other two shafts. Synchronous machine can be considered as the special case of the induction machine, which is rotor fed and for which $s = 0$ and all the power is generated in the windings of one member.

9.3 SYNCHRONOUS GENERATORS

Synchronous generators can be the salient-pole type with field windings or permanent-magnet machines. When connected to a public utility of fixed frequency, they have to run at constant speed. When they are directly connected to the turbine, their low RPM dictates a very short pole pitch since, the number of poles is equal to 7200, divided by the RPM. Although a permanent-magnet machine with many small poles can be used for such low speeds, it has the major disadvantage that the field cannot be removed in case of a fault to ground. Nevertheless, if the machine is grounded by a high resistance, it is possible to avoid serious burning if braking is provided to stop it quickly.

To synchronise the speed and the angle, both must be very close to correct values. This may be difficult to attain with the turbine torque changing rapidly with wind gusts. Typically, the speed should be matched within 1 per cent and the angle within 30 electrical degrees. To do so will require some form of brake to hold the speed nearly steady. With sudden torque changes of 50 per cent, as might well occur with high wind gusts, the speed might change 1.0 per cent in a little over 1 second, and the chance of catching the angle within the required limits is low. Some time might be required to match speed and angle correctly, and one might even have to wait for the wind to steady. Thus, an incentive exists for either using an automatic synchroniser or building this capability into a programmable controller.

A salient-pole synchronous motor with field coils on the pole is identical to a synchronous generator, except that the power is reversed. Typical synchronous motors have an 80 per cent power factor and a 150 to 250 per cent pull out torque. The rating chosen should thermally match the maximum rms load over a 10-min period. The pull-out power should at least match the maximum load it expects to carry without tripping off the line.

9.4 INDUCTION GENERATOR

The squirrel-cage induction generator (SCIG) is the simplest and usually the cheapest type of generator. It has a limited range of speed. When it runs at synchronous speed, the fact that the magnetic flux developed by the stator winding is travelling at the rotor speed means that no current is induced in the rotor and no torque is produced. At a higher wind speed, the turbine tries to run at a higher speed. This increase represents a 'slip speed' that causes the flux to cut through the rotor bars, thereby inducing a voltage that circulates a current, producing power and a counter torque.

Over a limited range, the power and the torque are proportional to the slip, but the current produces a reactive voltage (and flux) drop that is initially out of phase and has little effect. At larger values, the drop rapidly reduces the flux, requiring more current for a given torque until a 'pull-out slip' is reached. Beyond this point, the torque and the power decrease. It is not practical to operate in the range beyond 'pull-out slip' not only because the torque is dropping and speed may increase excessively, but also because the power loss in the stator windings, rotor bars, and end rings is excessive and might over heat these parts very rapidly. The most limiting part of this rotor heating is usually the end ring, which expands radially and tends to bend and fatigue the rotor bars or the brazed connections. Extra rotor heat also causes a higher air temperature, which raises stator-coil temperatures. Stator windings roast out at these high loadings.

The other most important characteristics is the per cent pull-out torque. The cost of a generator is nearly proportional to the continuous rating and about proportional to the square root of the pull-out torque. The problem of picking the most economical generator to do the job requires knowing the turbine power needed over the range of wind speeds and the power speed characteristics of the chosen generator. Since there will be rare cases of excessive wind speed with turbine torque that exceeds the pull-out value of the generator, it is necessary to establish this pull-out value and provide for tripping the generator off the line if the pull-out speed or power is exceeded for more than about 10 seconds. The generator can be restored to the line at any time the turbine speed falls below the pull-out speed of the generator.

At low wind speeds, the turbine speed will drop below the synchronous speed, and the generator torque will reverse and drive the turbine at just below synchronous speed. The only problem in this range is the loss of power, and if this persists for appreciable time, the machine should be removed from the line.

The curves of Figure 9.6 show the slip and rotor windings loss for a typical generator as a function of torque. The per cent rotor loss that can be tolerated depends on the ventilation and the construction. Typically any rotor loss greater than 25 per cent of rating that exists for more than 10 minutes at a time would give excessive temperature and stress. Short-time overloads in excess of this value can be judged by assuming the heat to go up with the product of slip and torque and using a thermal time constant of 10 minutes.

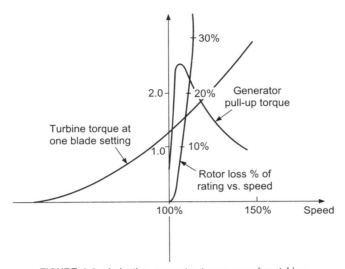

FIGURE 9.6   Induction generator torque-speed matching.

Another characteristic of the induction generator is its power factor, which varies with the load. The kW load divided by the power factor is the total kVA (the product of the line voltage and current multiplied by the square root of 3, divided by 1000) for three-phase generators. The current has two components—real and reactive—which add as vectors at right angles. The reactive current is out of phase with the voltage and represents no average power.

For small users of power, since the utility does not normally measure power factor or reactive kVA, there is no penalty. For larger connected loads, there is often a demand charge based on maximum kVA for any 15-min period. In these cases, a bank of shunt capacitors can be used to compensate for the reactive current at some load. This bank, however, should have its own switch and a voltage relay set to trip it off for any voltage more than 115 per cent of the normal. This breaker should open in 15 cycles or less since the capacitor current may exceed the magnetizing current if the main breaker opens, and the voltage may rise to excessive values.

9.5   VARIABLE-SPEED GENERATORS

Variable-speed generators can be used to take best advantage of turbine characteristics and generate the maximum power available at any wind speed up to some peak limit set by the rating of the generator and the frequency converter. Two general types of variable-speed generators are practical, but both are more expensive than constant-speed generators. The additional cost must be evaluated against the additional power generated. The simplest, and usually the cheapest, variable-speed generator uses a synchronous machine with a solid-state DC-link frequency changer as shown in Figure 9.7. The thyristor bridge connected to the machine terminals is a controlled rectifier, and the DC-current loop feeds a second thyristor bridge that functions as an inverter to feed power back into the power system.

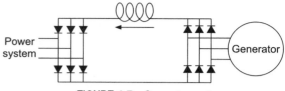

FIGURE 9.7 Generator system.

The rating of the generator and the solid-state converters should be chosen to match, not the highest possible wind speed, but rather some value that will be reached a sufficient percentage of the time to justify the incremental expense of the increased ratings.

9.6 CONTROL SYSTEMS

Controls are used for the following functions:

- To enable automatic operation
- Keep the turbine aligned with the wind
- To engage and disengage the generator
- To govern the rotor speed
- To protect the turbine from overspeed or damage caused by very strong winds
- To sense malfunctions and warn operators of the need to perform maintenance of repair, etc.

Controls for an individual machine may be integrated into a data-gathering system for a wind farm, and a master computer can act as a supervisory control and diagnostic tool for the entire farm. For a small machine, the controls must be simple and passive. For a large machine, the cost of controls represents but a small percentage of the total cost, and very elegent and precise controls can often be justified. Although passive controls do their own sensing and use natural forces for their actuation, active controls may use electrical, mechanical, hydraulic, pneumatic, or other means in any combination to suit the desired purpose.

Active control systems depend upon transducers to sense many variables—rotation, speed, power produced, generator temperature, voltage, current, wind speed, pitch angle, etc.—that will determine the control action needed. Transducers used for control must be of high quality—rugged and extremely reliable. They must not only accurately measure the variable they are supposed to sense, but also refrain from responding to erroneous inputs. Another important point is that the machine should not be encumbered with so many interacting protective controls that it is hardly ever allowed to run.

9.6.1 RPM Control (Mechanical)

Few wind systems have controls that constantly change the rotor-blade angle of attack (pitch angle) to match the instantaneous wind velocity. Even though this feature might produce more power, it has not been considered practical. Most rotors can be considered fixed pitch over most of the operating range of wind velocities. Many designers have found it necessary to control the blade pitch angle to accomplish the following:

- Start the rotor turning
- Set the pitch angle to a 'run' position
- Control RPM to prevent the rotor from overpowering the generator
- Protect the rotor and system from high wind-velocity damage.

9.6.2 Electronic Controls

A small computer determines the system's operational status and commands the appropriate actions. It receives input from sensors monitoring wind speed, RPM, vibration, power and blade pitch angle. From this input, the desired blade-pitch angle can be obtained by a command to the pitch actuators. This technique controls rotor RPM throughout the entire range of operation. If any problems are detected, such as exceeding the preset operational limits, the system commands the blade-pitch servos to put the rotor in a non-operational position (pitch angle of 90 degrees) and an operator is required to reset the system prior to restarting. The designer of small wind turbines should adhere to the adage, 'keep it simple' (KIS), and attempt to build in passive, reliable controls that will manage to keep the energy-gathering performance as high as possible. The designer of medium to large wind turbines can utilize more complex controls to relieve loads and maximise performance and machine life.

If the primary control systems fail, for any reason, it is important to have a back-up system that can prevent machine damage. A back-up overspeed control system can take many forms including tip brakes or spoilers, blade pitch changes to cause stall, and others. It is important to note that a brake located on the rotor shaft will not necessarily stop the rotor in very high winds. Since power is obtained largely by the outer third of the blade, this is also the most effective place to destroy it. Very small control surfaces that ruin the smooth flow of air over the airfoil near the tips or greatly increase the blade profile drag will have a very large effect on output power. Likewise, a small change in angle that results in stall of this section of the blade will reduce rotor power output considerably.

Many of the upwind and downwind rotors of DC power-generating systems are fixed pitch. Among additional RPM control techniques, these systems use the following:

- A tail vane to rotate the rotor edgewise to the wind direction.
- Blade tip plates that extend to cause braking of the rotor when overspeed is experienced.
- In VAWT, an axis rotational control system spring that assists the rotation of the rotor and generator vertically. The effect is similar to that produced by a tail vane except that the rotor moves from in front of the generator to a position above it and the tail remains oriented in line with the wind direction.

Many of these systems try not to shut down completely but maintain some power production during these high wind speed conditions. This allows them to return to operation without operator assistance or the use of a sophisticated electronic control system. Such wind systems were designed for a particular mode of operation, and this may or may not be applicable to all systems.

Mechanical methods have been tried, unsuccessfully, to initiate the electrical connection of AC generators to the utility. A governor-type device that closed a microswitch, which in turn caused the electrical contactor to connect with the generator. This type of device was found to be inaccurate (i.e. off-synch), unreliable (different cut-in RPM's) and hard to adjust. It thus created severe loadings on the drive train and rotor. The immediate consequence was a rotor-shaft coupling failure that indicated the severity of the loadings. High currents were also experienced across the power slip rings.

9.6.3 Electrical Cut-in

The electronic sensor to connect the generator correctly will differ for AC and DC systems. DC generators have a very low power output at low RPM. For them, the correct correction is made when the rotor RPM is adequate to generate some power but remains low enough for it not to cause a sudden jolt to the rotor system when electrical connection is made. These control systems are usually generator current or voltage activated. They are very common, economical, and no further sophistication is required. AC generators, on the other hand, represent a design condition that is more difficult to meet. First of all, the connection to the utility must be made as close to synchronous frequency as possible to help eliminate rotor-shaft and generator torque spikes.

The RPM sensing devices are usually accurate to some small per cent of the synchronous RPM (for example ± 4 per cent of 1800 = ± 72 RPM). If the torque curve for the generator in the synchronized mode indicates that the torque at 1728 or 1872 RPM is high, the sensing device is not the proper one, and a more accurate device must be found. Total loss of power is an additional problem for an electronically controlled system. In this case, when the power to the computer is gone (unless there is a battery backup), the power to operate the blade-pitch servo actuators is also gone. This is potentially a very dangerous situation, for the rotor will probably remain at fixed pitch and we are back to the old rotor run-away situation again.

9.7 POWER COLLECTION SYSTEM

The main power circuit shown in Figure 9.8 is from the generator, via three flexible pendant cables to a moulded case circuit breaker (MCCB). The MCCB is filled with an overcurrent protection trip, with an instantaneous settings against faults and a delayed (thermal) function which operates at a lower current level and is intended to detect overload of the generator. There are a number of additional circuits including those for the power factor correction capacitors (PFC) and the auxiliary supplies. The voltage level chosen for the main power circuit from the generator to the tower base is usually less than 1000 V and is selected to be one of the internationally standard voltages. The low generator voltage leads to the requirement of a

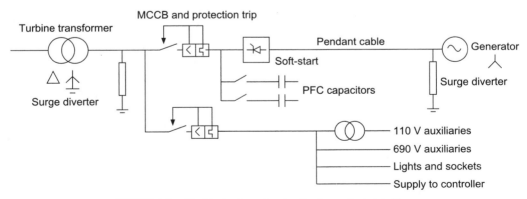

FIGURE 9.8 Electrical schematic of a fixed-speed wind turbine.

transformer located in the tower or immediately adjacent to it. In some early wind farms, a number of small wind turbines were connected to one transformer as shown in Figure 9.9, but as turbine ratings have increased, individual turbines have their own transformer.

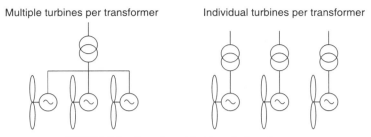

FIGURE 9.9 Grouping of turbines to transformers.

The design of the power collection is very similar to that of any medium-voltage power network. It is generally not necessary to provide any redundancy in the circuits to take account of medium voltage equipment failures. Wind farm power collection networks tend to consist of simple radial circuits with limited switchgear for isolation and switching. The wind turbine transformers are connected directly to the radial circuits.

Any wind farm project which reaches the stage of requiring a detailed design of the electrical system should also have a good estimate of the wind farm output. Thus, it is straightforward to use these data to calculate the losses in the electrical equipment at various wind turbine output powers using a load-flow program, sum these over the life of the project, and using discounted cash flow techniques, choose the optimum cable size and transformer rating. In countries where wind-generated electricity attracts a high price, then it is likely to be cost-effective to install cables and transformers with a thermal rating considerably in excess of that required at full wind-farm output in order to reduce losses. It may also be worthwhile to consider 'low-loss' transformer designs.

9.8 EARTHING (GROUNDING) OF WIND FARMS

All electrical plants require a connection to the general mass of the earth in order to:

- Minimise shock hazards to personnel
- Establish a low-impedance path for earth-fault currents and hence satisfactory operation protection
- Improve protection from lightning and retain voltages within reasonable limits
- Prevent large potential differences being established which are potentially hazardous to both personnel and equipment.

For earthing (grounding) of AC substations, the US standard (ANSI/IEEE, 1986) is widely applied. Wind farms, however, have rather unusual requirements for earthing. They are often very extensive, stretching over several kilometres, subject to frequent lightning strikes because of the height of modern wind turbines, and are often on high resistivity ground being located on the tops of hills. Thus, normal earthing practice tends not to be easily applicable and special consideration is required. The IEEE Recommended Practice (1991) says 'that the entire wind farm installation should have a continuous metallic ground system connecting all equipment. This should include, but not be limited to, the substation, transformers, towers, wind turbine generators and electronic equipment. This practice is generally followed with bare conductor being laid in the power-collection cable trenches to provide both bonding of all parts of the wind farm and a long horizontal electrode to reduce the impedance of the earthing system. The performance of a wind-farm earthing system may be understood qualitatively by considering Figure 9.10.

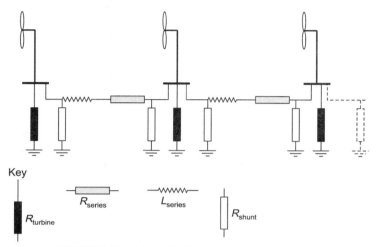

FIGURE 9.10 Schematic of wind farm earthing system.

At each turbine, a local earth is provided by placing a ring of conductor around the foundation at a depth of 1 m and by driving vertical rods into the ground. It is common to bind the steel reinforcing of the wind-turbine foundation into this local earthing network. The purpose of this local earth is to provide equipotential bonding against the effect of both 'lightning and power frequency fault currents and provide one element of the overall wind-farm earthing system'.

9.8.1 Lightning Protection

It was thought that as the blades of wind turbines are made from non-conducting material, i.e. glass reinforced plastic (GRP) or wood-epoxy, it was not necessary to provide explicit protection for these types of blades provided they did not include metallic elements for the operation of devices such as tip brakes. However, there is now a large body of site experience to show that lightning will attach to blades made from these materials and can cause catastrophic damage if suitable protection systems have not been fitted. Of course, if carbon fibre (which is conducting) is used to reinforce blades, then additional precautions are necessary.

Lightning is a complex natural phenomenon, often consisting of a series of discharges of current. The term 'lightning flash' is used to describe the sequence of discharges which use the same ionized path and may last upto 1s. The individual components of a flash are called *strokes*. Lightning flashes are usually divided into four main categories:

- Downward inception, negative and positive polarity
- Upward inception, negative and positive polarity.

Generally, flashes which start with a stepped leader from the thunder cloud and transfer negative charge to earth (downward inception–negative polarity) are the most common. Downward negative flashes typically consist of a high-amplitude burst of current lasting for a few microseconds, followed by continuing current of several hundred ampere (A). Then, following the extinction of the initial current transfer between cloud and earth, there may be a number of restrikes. Although the maximum peak current of this form of lightning is low (some 15 kA), the charge transfer can be very high and hence has a significant potential for damage. The top of wind turbine blades can now be over 100 m above ground and so there is a growing concern over the effect of upward negative hightning flashes. Upward positive flashes are rare.

Table 9.1 shows the parameters normally used to characterize lightning and some aspects of their potential for damage in wind turbines. The peak current of a single lightning stroke is over 200 kA, but the medium value is only approximately 30 kA. The corresponding values for charge transfer are 400 C (peak), 5 C (median) and specific energy 20 MJ/Ω (peak) and 500 MJ/Ω (median). The very large range of these parameters implies that the initial step in any consideration of lightning protection of a wind farm or wind turbine is to undertake a risk assessment (IEC, 2000a). The risk assessment includes consideration of the location of the turbine as the frequency and intensity of lightning varies considerably with geography and topography.

TABLE 9.1 Effects of various aspects of lightning on a wind turbine

Parameter	*Effects on wind turbine*
Peak current (A)	Heating of conductors, shock effects, electromagnetic forces
Specific energy (J/Ω)	Heating of conductors, shock effects
Rate of rise of current (A/s)	Induced voltages on wiring, flashovers, shock effects
Long duration charge transfer (C)	Damage at arc attachments points or other arc sites (e.g. bearing damage)

The techniques commonly used to protect blades against lightning are shown in Figure 9.11. The main distinction is whether a limited number of receptors are used to intercept the lightning (Type A and B), or if an attempt is made to protect the entire blade (Type C and D). The main

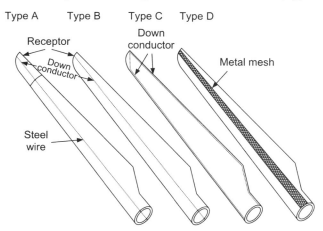

FIGURE 9.11 Blade lightning-protection methods.

mechanism of damage is when the lightning current forms an arc in air inside the blade. The pressure shock wave caused by the arc may explode the blade or, less dramatically, cause cracks in the blade structure. Thus, for effective protection, it is essential that the lightning is attached directly to the protection system and then is conducted safely down the length of the blade in metallic conductor of adequate cross section.

9.9 WIND POWER INTEGRATION INTO GRID

The impact of wind power becomes significant as the penetration level of amount of electricity from wind turbines is increasing into the grid. It is becoming necessary to upgrade the existing grid network. The upgradation exercise is very complex and requires short-term and long-term measures. Short-term measures are optimum use of grid network by grid management procedures and with addition of some tie lines. Whereas long-term measures require expansion of existing grid nework operating at high and extra high voltage levels in order to transmit power over long distances.

The variability of wind speed affects the electricity generation. The control of generation or supply and demand load are difficult to match. The large percentage of wind power affects the stability of grid network. The variability of wind power can be classified as short-term variability and long-term variability. Short-term variability affects the scheduling of generation units, and it determines the magnitude of reserve power. Long-term variability is determined by seasonal meteorological patterns and inter-annual variations of wind patterns. The impact of variability can be tackled in three different time farmes:

- Frequency regulation by automatic generation control by an action of governor can take place in a period of few seconds to a minute.
- Load follow up by economic dispatch and generation forecasts are guided by unit-commitment schedule can take place in a period of few minutes to a couple of hours.
- Unit commitment based on load profile prediction and characteristics of available generating units can take place in a period of several hours to few days.

An energy system operates with optimum reserve capacities for economy, which is based on balancing demand and supply continuously. To maintain quality of electricity, the two parameters to be maintained are voltage and frequency under close limits. To maintain a close balance between demand and supply is the main objective of management of intermittency of wind power output from individual as well as from cluster of wind turbines and not the steady output of power. Some of the following strategies are adopted:

- Combining ancillary power plants like combined cycle gas turbine or oil-fired power plant with wind farm can provide short-term operation reserves as well as long-term capacity reserves.
- Using hydro storage reserve offers large storage capacity and quick response for balancing high demand and supply.
- Interconnection with other grids of the region will improve the reliability of intermittancy management.
- Charging consumers at different rates at different time slots of the day. During peak demand time, extra charge is levied. Less charge is levied during off peak demand hours.

- During overloading of the distribution line the wind farm of that region is disconnected from the network to provide stability in the system.

9.9.1 Power System Stability

It is defined as the ability of a power system to operate in the state of equilibrium under normal conditions and regain the state of equilibrium after a disturbance. Once the penetration of wind power in the power system network becomes substantial, it will impact distribution system as well as the transmission system. The grid stability will depend upon the following features:

- In a wind turbine technology where standard squirrel-cage induction generator (SCGI) is connected directly to grid, the rotation speed of rotor is matched with generator through gearbox. The operating speed range of turbine is governed by torque–speed characteristics of generator. The generator draws reactive power from the grid to maintain voltage stability. It is based on reactive power compensation from the grid.
- The wound rotor generator with variable resistance technology offers limited range of a variable speed to wind turbine. In a wound rotor induction generator with variable resistance, blade pitch control upto 10 per cent of speed variation is allowed.
- For doubly fed induction generator (DFIG) in which wound rotor generator with four quadrant power converter is used. It offers real and reactive power flow in either direction that is from generator to grid and vice-versa. A gearbox is used between turbine rotor and generator with active pitch control of blades for maximum power production. It is a complex technology with ± 30 per cent variable speed. The stability of the system is achieved by reactive power control.
- In permanent magnet synchronous generator (PMSG) and permanent magnet induction generator (PMIG), the generator is coupled to the grid through AC/DC/AC power convertor. The substantial decoupling of electrical generator dynamics from the grid is ensured by power convertor. Due to complexity of technology and control, it is costlier than other systems.

9.9.2 Economics of Grid Network

The integration of wind power in the grid network requires the assessment of additional cost due to four main parameters:

- Balancing energy generation and demand
- Operation reserves, reinforcement and management
- Capacity reserves, impact and losses
- Extension of transmission and distribution lines

The economic viability of wind power generation system is evaluated for the following parameters: cost of energy (COE), life cycle cost (LCC), pay-back period, internal rate of return (IRR), net present cost (NPC) and benefit to cost ratio.

9.9.3 Codes and Standards for Grid Integration

Small capacity wind farms are normally connected to distribution networks, whereas large ones are connected to transmission networks. There are specific set of requirement known as grid codes and standards for integrating large wind farms. The codes and standards ensure that the wind farms do not adversely affect in terms of security of supply, reliability and power quality of operation of power system. The grid codes and standards enumerate the requirements of frequency, voltage and behaviour of wind turbine in the event of grid faults. Following are the main requirements:

- Active power and frequency control in power system to control amount on injected power and frequency to prevent over-loading of transmission lines.
- Voltage control requires for reactive power compensation. The power factor is to be maintained from 0.925 (leading) to 0.85 (lagging).
- Faulty ride-through (FRT) requirement ensures the connection of wind power plant to power system network during fault. The above requirement varies from country to country. Figure 9.12 shows the low voltage ride-through (LVRT) requirements envisaged by American Wind Energy Association (AWES) for minimum required wind power plant response to emergency low voltage.

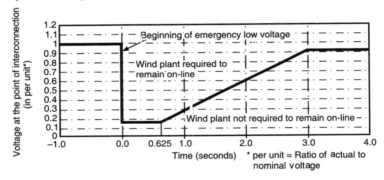

FIGURE 9.12 Low voltage ride-through (LVRT) standard.
(www.ferc.gov/whats-new/comm-meet/052505/E-1.pdf)

Two key features of this regulation are:

A wind power plant must have low voltage ride-through capability down to 15 per cent of rated line voltage for 0.625 seconds;

A wind power plant must be able to operate continuously at 90 per cent of rated line voltage, measured at high voltage side of the wind power plant sunstation transformers. Following important points are related to power regulation in the wind farm:

- Supervisory control and data acquisition (SCADA) is required in wind power plant to remotely control the power and important parameters for scheduling and forecasting. It is a versatile management information system for wind farms.
- Voltage and current ram rates limitations requirement to suppress frequency fluctuations caused by start-up and shutdown of wind power plant. It varies from country to country, for example in Germany it is 10 per cent of rated power per minute for start-up.
- Power quality requirements are specified for active and reactive power injected in the power system network. The parameters are harmonic component in current, voltage fluctuation or flicker, number of switching operation to connect to grid, etc. For wind power plant, the power quality requirements are described in IEC 61400-21:2001 standard.

- Wind farm protection system is required against disturbances in the network due to over-current, under-volatge, over-volatge, etc.
- Wind power modelling and simulation provides the model to investigate connection and the integration of wind farm with the electrical grid.

9.10 EMBEDDED (DISPERSED) WIND GENERATION

The wind is a diffused source of energy with wind farms and individual turbines often distributed over wide geographical areas, and so the public electricity distribution networks, which were originally constructed to supply customer loads, are usually used to collect the electrical energy. Thus, wind generation is said to be embedded in the distribution network or the generation is described as being dispersed. The terms 'embedded generation' and 'dispersed generation' can be considered to be synonymous. Figure 9.13 is a diagrammatic representation of a typical modern electric power system. The electric power is generated by large central generating sets and is then fed into an interconnected high-voltage transmission system. The generating units may be fossil-fuel, nuclear or hydro sets and will have capacities of upto 1000 MW. The generating voltage is rather low (typically around 20 kV) to reduce the insulation requirements of the machine windings and so each generator has its own transformer to increase the voltage to that of the transmission system. The transmission network is interconnected, or meshed, and so there can be many paths for the electrical power to flow from the generator to the bulk supply transformers. The bulk supply transformers are used to extract power from the transmission network and to provide it to the distribution networks at lower voltage. Practice varies from country to country, but primary distribution voltages can be as high as 150 kV. Most wind farms are connected to rural, overhead distribution lines. The design of these circuits tends to be limited by consideration of voltage drop rather than the thermal constraints and this severely limits their ability to accept wind generation.

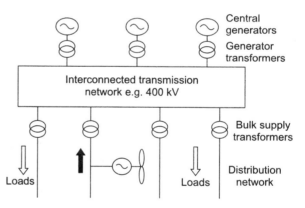

FIGURE 9.13 A typical large utility power system.

Primary distribution circuits can accept power injections of up to 100 MW and so most on-shore wind generation is likely to be embedded in a distribution system. There is also considerable interest in much smaller-scale renewable energy schemes, e.g. individual wind turbines or dispersed wind farms.

9.10.1 Electrical Distribution Networks

The conventional function of an electrical distribution network is to transport electrical energy from a transmission system to customers' loads. This is to be done with minimal electrical losses and with the quality of the electrical power maintained. The voltage drop is directly proportional to the current, while the series loss in an electrical circuit is proportional to the square of the current. Therefore the currents must be kept low which, for constant power transmitted, implies that the network voltage level must be high. However, high-voltage plants (e.g. lines, cables and transformers) are expensive due to the cost of insulation, and so the selection of appropriate distribution network voltage level is an economic choice.

A schematic representation of a typical distribution system is shown in Figure 9.14. Power is extracted from the interconnected transmission grid and then transformed down to the primary distribution voltage (132 kV in this case). The electrical energy is then transported via a series of underground–cable and overhead lines circuits to the customers.

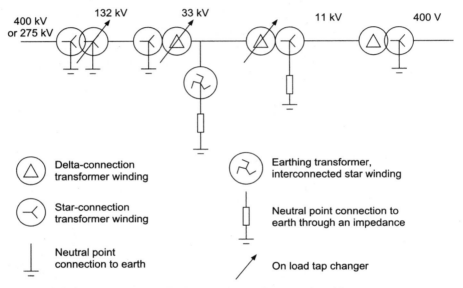

FIGURE 9.14 Typical distribution voltage transformer and earthing arrangements.

Current passing through a circuit leads to a change in voltage and this is compensated for by altering the ratio (or taps) of the transformer. 11 kV/400 V transformers have fixed taps, which can only be changed manually when no current flows. However, higher-voltage transformer have on-load tap changers that can be operated automatically when load current is passing. The simplest control strategy is to use an automatic voltage controller (AVC) to maintain the lower voltage terminals of the transformer close to a set-point as shown in Figure 9.15. The AVC operates by measuring the voltage on the busbar, comparing it with set-point value and then issuing an instruction

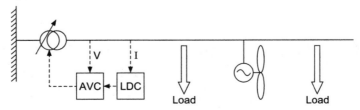

FIGURE 9.15 Typical voltage control of a distribution circuit.

to the on-load tap changer to alter the ratio of the transformer. Control systems of this type are unaffected by the presence of wind generators on the network.

The voltage change permitted in 11 kV circuits is small (typically ±1 per cent to ±2 per cent) as any voltage variation is passed directly through the fixed tap transformers to customer's supply. In contract, 33 kV and 132 kV circuits are allowed to operate over a wider voltage range (up to ±6 per cent) as the automatic on-load tap changes can compensate for variations in network voltage. Table 9.2 gives some indications of the maximum capacities which, experience has indicated, may be connected. Table 9.2 assumes that the wind farms are made up of a number of turbines and so the connection assessment is driven by voltage rise effects and not by power quality issues due to individual large machines.

TABLE 9.2 Indication of possible connection of wind farms

Location of connection	Maximum capacity of wind farm (MW)
Out on 11 kV network	1–2
At 11 kV busbars	8–10
Out on 33 kV network	12–15
At 33 kV busbar	25–30
On 132 kV network	30–60

9.10.2 Power Flows, Slow Voltage Variations and Network Losses

If the wind generator operates at unity power factor (i.e. reactive power $Q = 0$), then the voltage rise in a lightly-loaded radial circuit shown in Figure 9.16 is given approximately by:

$$\Delta V = V_1 - V_0 = \frac{PR}{V_0} \qquad (9.16)$$

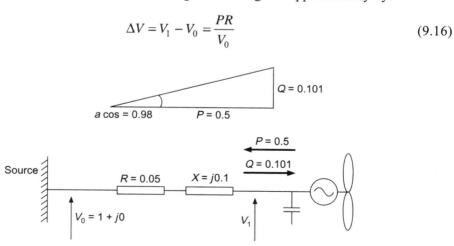

FIGURE 9.16 Fixed-speed wind turbine on a radial circuit.

Operating a generator at a leading power factor (absorbing reactive power) acts to reduce the voltage rise but at the expense of the increased network losses. In this case, the voltage rise is given by:

$$\Delta V = V_1 - V_0 = \frac{PR - XQ}{V_0} \tag{9.17}$$

The source impedance of an overhead 11 kV distributor circuit may, typically have a ratio of inductive reactance to resistance (X/R ratio) of 2. An uncompensated induction generator at rated output, typically, has a power factor of 0.89 leading, i.e. $P = -2Q$. Thus, under these conditions, there is no apparent voltage rise in the circuit at full power. However, the real power loss (W) in the circuit is given approximately by:

$$W = \frac{(P^2 + Q^2)R}{V_0^2} \tag{9.18}$$

The reactive power drawn by the generator acts to limit the voltage rise, but higher real power losses are incurred in the circuit.

A simple but precise calculation for voltage rise in any radial circuit may be carried out. Using an iterative technique as follows:

At the generator busbar of Figure 9.16, the apparent power (sometimes known as the complex power) S_1 is given by:

$$S_1 = P - jQ \tag{9.19}$$

(for a generator operating at lagging power factor, exporting VArs, S_1 would be given by $P + jQ$).

By definition $S = VI^*$, where * indicates the complex conjugate. Therefore, the current flowing in the circuit is given by:

$$I = \frac{S_1^*}{V_1^*} = \frac{P + jQ}{V_1^*} \tag{9.20}$$

The voltage rise in the circuit is given by IZ and so:

$$V_1 = V_0 + IZ = V_0 + \frac{(R + jX)(P + jQ)}{V_1^*} \tag{9.21}$$

It is common for the network voltage V_0 to be defined and the generator busbar voltage V_1 to be required. V_1 can be obtained using the simple iterative expression:

$$V_1^{(n+1)} = V_0 + \frac{(R + jX)(P + jQ)}{V_1^{*(n)}} \tag{9.22}$$

where n is the iteration number. This is a very simple form of the conventional Gauss–Seidel load flow algorithm.

EXAMPLE 9.1 Calculation of voltage rise in a radial circuit (Figure 9.15).

Consider a 5 MW wind farm operating at a leading power factor of 0.98. The network voltage (V_0) is $(1 + j0)$ per unit and the circuit impedance (Z) is $(0.05 + j0.1)$ per unit on a 10 MVA base.

Solution: Thus,

$$V_1^{(n+1)} = 1 + \frac{(0.05 + j0.1)(0.5 + j0.101)}{V_1^{*(n)}}$$

For the first iteration ($n = 0$), assume $V_1^{*(0)} = 1 + j0$; then $V_1^{(1)}$ may be calculated as

$$V_1^{(1)} = 1.0149 + j0.0551$$

For the 2nd iteration ($n = 1$),

$$V_1^{*(1)} = 1.0149 - j0.0551$$

then

$$V_1^{(2)} = 1.0117 + j0.0549$$

For the 3rd iteration ($n = 2$),

$$V_1^{*(2)} = 1.0117 - j0.0549$$

then

$$V_1^{(3)} = 1.0117 + j0.0551$$

and the procedure has converged.

Therefore, $V_1 = 1.013$ per unit at an angle of 3°; i.e. the voltage at the generator terminal is 1.3 per cent above that at the source. The angle between the two voltage vector is small (3°).

The approximation calculation ($V_1 = V_0 + PR - XQ$) indicates a voltage V_1 of 1.015 per unit (i.e. a rise of 1.5 per cent).

The current (I) in the circuit may be calculated from:

$$I = \frac{S_1^*}{V_1^*} = \frac{0.5 + j0.101}{1.0117 - j0.0551}$$

$$= 0.4873 + j0.1264 \text{ per unit}$$

$$|I| = 0.503 \text{ per unit}$$

With a connection voltage of 33 kV, the base current is given by

$$I_{base} = \frac{VA_{base}}{\sqrt{3} \times V_{base}}$$

$$= \frac{5 \times 10^6}{1.732 \times 33 \times 10^5} = 87.5 \text{ A}$$

and so the magnitude of the current flowing in the 33 kV circuit is 44 A.

The real power loss in the circuit (W) is

$$W = I^2 R$$

$$= 0.0127 \text{ per unit, or } 127 \text{ kW}.$$

The symmetrical short-circuit level at the generator, before connection of the generator, is simply:

$$S'' = \frac{1}{|z|} = \frac{1}{(0.05^2 + 0.1^2)^{\frac{1}{2}}}$$

$$= 8.94 \text{ per unit or } 89.4 \text{ MVA}.$$

9.10.3 Power Quality

Power quality is the term used to describe how closely the electrical power delivered to customers corresponds to the appropriate standards and so operates their end-use equipment correctly.

The particular importance of the influence of wind turbines on power quality has been recognised in the creation of an international standard (IEC, 2000b). It lists the following data as being relevant for characterizing the power quality of a wind turbine:

- Maximum output power (10 min average, 60 s average and 200 ms average values)
- Reactive power (10 min average) as a function of active power
- Flicker coefficient for continuous operation as a function of network source impedance phase angle and annual average wind speed
- Maximum number of wind turbine starts within 10 min and 120 min periods
- Flicker step factor and voltage step factor at start-up as a function of network source impedance phase angle
- Harmonic currents during continuous operation as 10 min averages for each harmonic current up to fiftieth.

There are a number of difficulties when assessing the power quality of wind turbines as their performance will depend on:

- Design of the entire wind turbine (including the aerodynamic rotor and control system)
- The conditions of the electrical network to which it is connected
- The wind conditions in which it operates.

9.10.4 Wind Farm and Generation Protection

A typical protection arrangement is shown in Figure 9.17 for a wind farm of fixed-speed wind turbines with generator voltages of 690 V and with a collection circuit voltage of 11 kV.

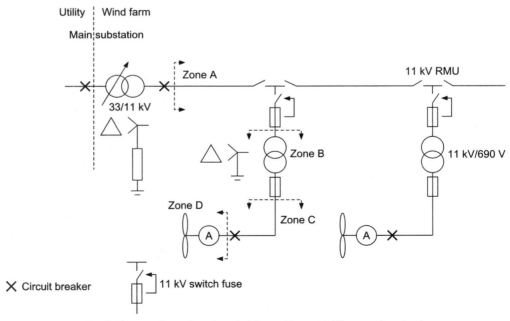

FIGURE 9.17 Protection of a wind farm with an 11 kV connection circuit.

The 11 kV circuit is fed from a 33/11 kV Delta/Star wound transformer, with the 11 kV neutral grounded either directly or through a resistor. The 11/0.69 kV transformers are also wound Delta/Star and so the 690 V neutral points of each circuit may be directly grounded. The neutral point of the generator is not connected to ground. There are a number of zones of protection. At the base of the wind turbine tower, a 690 V circuit breaker fitted to protect the pendant cables and the generator. This is shown as zone D. Zone C is the 690 V cables running from the turbine transformer to the tower-base cabinet. Fuses or another moulded case circuit breaker may be fitted to the 690 V side of the turbine transformer to provide protection of the cables and also a point of isolation so that all the electrical circuits of the turbine may be isolated without switching at 11 kV. In some designs of wind turbine, the main incoming busbar at the bottom of the electrical cabinet at the tower base consists of exposed conductor.

As wind turbine ratings increase and large wind farms are constructed, 11 kV circuits cease to be cost-effective and so a number of large wind farms have been constructed with a collection voltage of 33 kV. Figure 9.18 shows a typical arrangement where a 33 kV wind farm is connected directly to a 33 kV public utility network. This arrangement poses a number of additional difficulties for the electrical protection. 33 kV switch fuses are not readily available and so it can be difficult to provide comprehensive protection of the 33 kV/690 V transformer for the full range of prospective fault currents. Effective protection for a single phase to ground fault on the low-voltage terminals is particularly difficult.

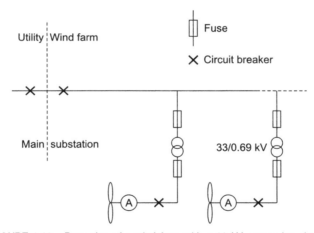

FIGURE 9.18 Protection of a wind farm with a 33 kV connection circuit.

9.10.5 Interface Protection

The protection is required to ensure that the wind farm does not feed into faults on the distribution network or attempt to supply an isolated section of network. The problem is illustrated in Figure 9.19. For the faults on the network, the difficulty is that wind turbines are not a reliable source of fault current and so circuit breaker B cannot be operated by over-current protection. Thus, for the fault shown, the current-operated protection on the network is used to open circuit breaker A. This, then, isolates the wind turbine which begins to speed up as the wind input remains, but it is no longer possible to export power to the network.

FIGURE 9.19 Protection of the distribution network from a wind turbine.

Most distribution utilities are extremely sensitive to possible islanded operation for a number of reasons:

- The possibility that customers may receive supply outside the required limits of frequency and voltage
- The possibility that part of the network may be operated without adequate neutral earthing
- The danger associated with out-of-phase reclosing
- It is against the regulations governing operation of the distribution network
- The potential danger to staff operating the distribution network.

9.10.6 Losses in Generation

Wind generation embedded in the distribution network alters the flows of real and reactive power and so changes the electrical losses within the network. This is illustrated in Figure 9.20. A load of real power, P, and reactive power Q, are fed from a distribution feeder of resistance R. Now, the electrical power losses in a circuit are by definition:

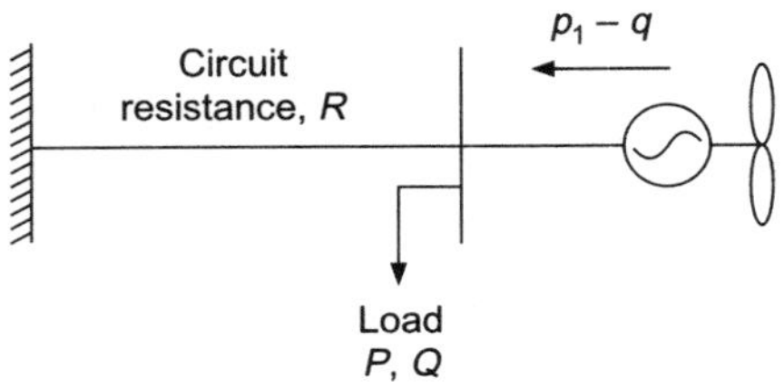

$$W = I^2 R$$

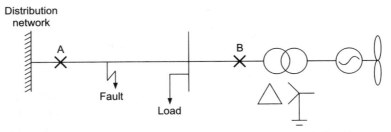

FIGURE 9.20 Illustration of effect of wind generation on losses in a distribution network.

which is approximately equal to:

$$W = \frac{(P^2 + Q^2)R}{V_0^2} \quad \text{[see Eq. (9.18)]}$$

and so the losses in the feeder may be estimated from Eq. (9.18). If a wind turbine is then connected to the busbar and exports real power output p and imports reactive power q, the losses will now be approximately:

$$W = \frac{[(P-p)^2 + (Q+q)^2]R}{V_0^2} \quad (9.23)$$

REVIEW QUESTIONS

1. Explain the electrical design of a typical wind farm, say, of 2 MW wind turbines and such 15 turbines are installed in a wind farm.

2. What are the control policies of output electrical power of a wind turbine in particular and of a wind farm in general?

3. Explain the effect of wind turbulence and wind farm geometry on system voltage flicker.

4. How power electronics interfaces is used for variable speed operation of wind turbine?

5. Explain the following: (i) induction generators and (ii) synchronous generators.

6. Explain why teetering is necessary for 2-bladed machines. Is it necessary for 3-bladed machine? What is the excitation frequency for this machine due to vertical wind profile?

7. What is stall regulation? Explain stall with the help of lift vs. angle of incident characteristics.

8. What is the voltage drop across a 100-ohm resistance if the current is 2 amp? How much power is lost as heat through that resistance?

9. The maximum power rating of a wind turbine is 30 kW, and it has a single-phase, 240 V generator. What is the maximum current produced? Remember the difference between root mean square values and peak values. What are the peak voltages for 110, 240 and 480 VAC?

10. If the phase angle in a 240 VAC, 20-amp circuit is 20°, how much is the power reduced from maximum power?

11. What does a three-phase generator mean?

12. What is the angular velocity for 60 Hz frequency?

13. The synchronous point on an induction generator is 1200 RPM. If the generator is rated at 500 kW, what is the shaft torque into that generator?

14. Look at Figure 9.21. At what slip is the efficiency maximised for generator operation?

15. If a 25 kW rated wind turbine has a three phase, 480 V generator, what maximum size wire will be needed for each phase to connect the wind turbine to a load that is 50 m away?

[**Hint:** You need to count the length of wire down the tower, 25 m tall. Peak power can be 30 kW. Calculate maximum current and reduce it by a factor of 1.7 since it is a three-phase system. Each leg (wire) of the three-phase system carries 1/3 of the current. Use Table 9.3.]

16. A 100 kW, 480 V generator, three-phase, is connected to a transformer within 10 m of the base of a wind turbine. Peak power can be 120 kW. Remember, you need to count the length of wire down the tower, 30 m tall. What minimum size wire is needed for each phase? Calculate current and reduce it by a factor of 1.7. Since it is a three-phase system, it carries part of the current. Use Table 9.3.

318 Wind Energy: Theory and Practice

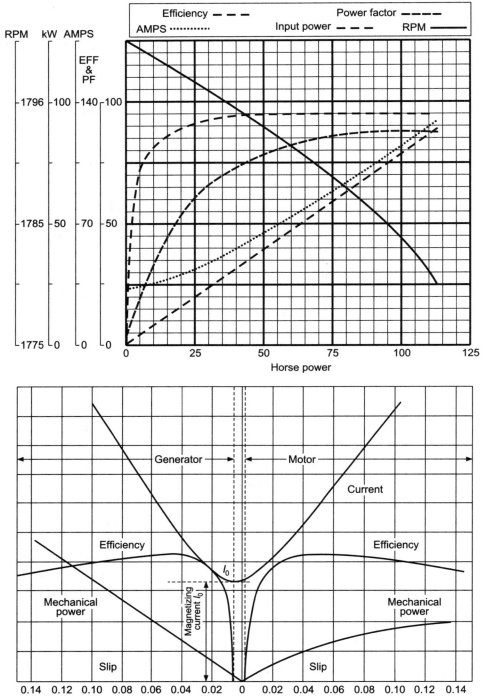

FIGURE 9.21 Operating characteristics of induction motor (420 V, 75 kW). The curves for the induction generator are essentially a mirror image, as shown by the bottom graph.

TABLE 9.3 Wire size, copper, 480 V, three-phase, 2% voltage drop

Load (amps)	Type insulated	Overhead (Bare, covered)	30	46	60	76	91	107	122	137	152	168	183
5	12	10	12	12	12	12	12	12	12	12	12	12	12
7	12	10	12	12	12	12	12	12	12	12	10	10	10
10	12	10	12	12	12	12	12	10	10	10	10	8	8
15	12	10	12	12	12	10	10	10	8	8	8	8	6
20	12	10	12	12	10	10	8	8	8	6	6	6	6
25	10	10	12	10	10	8	8	6	6	6	6	4	4
30	10	10	12	10	8	8	6	6	6	6	4	4	4
35	8	10	10	10	8	6	6	6	4	4	4	4	4
40	8	10	10	8	8	6	6	4	4	4	4	3	3
45	6	10	10	8	6	6	6	4	4	4	3	3	2
50	6	10	10	8	6	6	4	4	4	3	3	2	2
60	4	8	8	6	6	4	4	4	3	2	2	2	1
70	4	8	8	6	4	4	4	3	2	2	1	1	1
80	4	6	8	6	4	4	3	2	2	1	1	0	0
90	3	6	6	6	4	3	2	2	1	1	0	0	0
100	3	6	6	4	4	3	2	1	1	0	0	00	00
115	2	4	6	4	3	2	1	1	0	0	00	00	000
130	1	4	6	4	3	2	1	0	0	00	00	000	000
150	0	2	4	3	2	1	0	0	00	00	000	000	4/0
175	00	0	4	3	1	0	0	00	000	000	4/0	4/0	4/0
200	000	00	4	2	1	0	00	000	000	4/0	4/0	250	250
250	250	00	3	1	0	00	000	4/0	4/0	250	250	300	300

Source: From *Agriculture Wiring Handbook*, 3rd ed., 1993. National Food and Energy Council, Colombia, Missouri.
Note: First column is the current; the next two columns give minimum size of wire to use for type of insulation and for bare wire. With longer wire runs, a longer-diameter wire is needed. The length of the total wire run is in bold, m. Other numbers are size of wire. 4/0 means 0000 size. A smaller number means a larger-diameter wire, and more zeros mean larger-diameter wire. After 4/0, the number is a thousand, circular mills.

17. What is power factor? What affects the value of the power factor for a wind farm?
18. List two advantages and two disadvantages with induction generator, constant RPM, and stall control.
19. List two advantages and two disadvantages with doubly fed induction generators, variable RPM, and pitch control.
20. What happens at the wind farm when there are faults on the utility line?
21. Why are supervisory control and data acquisitions (SCADA) used in wind farms?
22. What type of wind turbines use power electronics? Why?
23. List three functions for a wind turbine controller.
24. What is the function of an inverter? What type and size of wind turbines use inverters?

Appendix A

Units and Conversion

The International System of units (SI) is an absolute system. The base units are the metre for length, the kilogram for mass and the second for time, and so on. With SI units, standard gravity is 9.806 m/s². Table A.1 lists the seven base units.

TABLE A.1 Base units

Quantity	Name	Symbol
Length	meter	m
Mass	kilogram	kg
Time	second	s
Electric current	Ampere	A
Temperature	Kelvin	K
Amount of substance	mole	mol
Luminous intensity	candela	cd

The radian (symbol rad) is a supplemental unit in SI for a plane angle.

A second class of SI units comprises the derived units, that is, units derived from the base units. Table A.2 is a list of most useful derived units.

TABLE A.2 Some useful derived units

Quantity	Unit	SI symbol	Formula
Acceleration	metre per second squared		ms^{-2}
Angular acceleration	radian per second squared		$rad.s^{-2}$
Angular velocity	radian per second		$rad.s^{-1}$
Area	square metre		m^2
Circular frequency	radian per second	ω	$rad.s^{-1}$
Density	kilogram per cubic metre		$kg.m^{-3}$
Energy	Joule	J	N.m
Force	Newton	N	$kg.m.s^{-2}$
Force couple, moment	Newton metre		N.m
Frequency	Hertz	Hz	s^{-1}
Power	Watt	W	Js^{-1}
Pressure	Pascal	Pa	$N.m^{-2}$
Quantity of heat	Joule	J	N.m
Speed	revolution per second		s^{-1}
Stress	Pascal	Pa	$N.m^{-2}$
Torque	Newton metre		N.m
Velocity	metre per second		$m.s^{-1}$
Volume	cubic metre		m^3
Work	Joule	J	N.m

CONVERSION FORMULA

Linear

To convert	Multiply by
inch into millimetres	2.54
metre into inches	2.54 × 10
feet into centimetres	30.48
feet into metres	0.3048
yard into metres	0.9144
miles into metres	1609.344
miles into kilometres	1.609344

Area

To convert	Multiply by
Sq. inches into sq. centimetres	6.4516
Sq. feet into sq. centimetres	929.03
Sq. feet into sq. metres	0.092903
Sq. yards into sq. metres	0.8361

Sq. miles into sq. kilometres	2.58999
Sq. miles into hectares	258.999
Acres into sq. metres	4046.856
Acres into hectares	0.4046856

Volume and Capacity

To convert	Multiply by
Cu. inches into cu. centimetres	16.3871
Cu. inches into litres	0.0163871
Cu. feet into cu. metres	0.028317
Cu. feet into cu. litres	28.32
Cu. yards into cu. metres	0.7646
pints into litres	0.56826
quarts into litres	1.13652
UK gallon into litres	4.54609
US gallon into litres	3.7854

1 US beer barrel = 117.347765 litres

Mass

To convert	Multiply by
ounces into grams	28.3495
pounds into grams	453.6
pounds into kilograms	0.4536
ton into kilograms	1016.047
tahils into grams	37.799
kati into kilograms	0.60479
grains into grams	0.0648

Force

To convert	Multiply by
pounds force into Newtons	4.44822
poundals into Newtons	0.138255

Power

To convert	Multiply by
Horsepower into kilowatt	0.7457
Horsepower into metric horsepower	1.01387
Foot pound force per second into kilowatt	0.001356

Celcius Temperature Conversion

	From celcius	*To celcius*
Fahrenheit	$[°F] = [°C] \times \dfrac{9}{5} + 32$	$[°C] = \{[°F] - 32\} \times \dfrac{5}{9}$
Kelvin	$[K] = [°C] + 273.15$	$[°C] = [K] - 273.15$
Rankine Conversion	$[R] = \{[°C] + 273.15\} \times \dfrac{9}{5}$	$[°C] = \{[R] - 491.67\} \times \dfrac{5}{9}$

CONVERSIONS

Speed
1 m/s = 3.6 km/s
1 mph = 0.446 m/s
1 knot = 1.15 mph

Length
1 metre = 3.28 feet
1 foot = 0.305 metre
1 kilometre = 0.602 mile
1 mile = 1.61 kilometres

Area
1 square kilometre = 0.386 square mile
1 square kilometre = 1,000,000 square metres
1 square kilometre = 100 hectares
1 hectare = 10,000 square metres
1 hectare = 2.47 acres
1 acre = 0.405 hectare
1 acre = 4049 square metres

Volume
1 litre = 0.264 gallons
1 gallon = 3.78 litres
1 cubic metre = 1000 litres
1 cubic metre = 264 gallons

Flow Rate
1 litre/second = 15.8510 US gallons/minutre
1 gallon/minute = 0.063 litre/second
1 cubic metre/minute = 264 US gallons/minute

Weight

1 metric ton = 1.10 tons
1 kilogram = 2.20 pounds
1 pound = 0.454 kilogram

Energy Equivalence of Common Fuels

1 kWh = 3413 Btu
     = 3.14 cubic foot of natural gas
     = 0.034 gallon of oil
     = 0.00017 cord of wood

1 therm = 1E + 05 Btu
       = 100 cubic foot of natural gas
       = 1 gallon of oil
       = 29.3 kWh
       = 0.005 cord of wood

1 gallon of oil = 1E + 05 Btu
1 cord of wood = 2E + 07 Btu
1000 cubic feet of natural gas = 1E + 06 Btu

TABLE A.3 Scale of equivalent power (Instantaneous Power)

Typical wind turbine rating by rotor diameter (m)	Power (kW)	Equivalent
1.5	0.25	1/3 horse power electric motor
2	0.50	2/3 horse power electric motor
3	1	Hair dryer, electric space heater
7	10	Garden tractor
10	25	
18	100	Passenger car engine
25	250	
40	500	Heavy truck engine
50	1,000	Small diesel locomotive, race car engine
100	3,000	Diesel locomotive
	5,00,000	Coal-fired generator, small nuclear reactor
	10,00,000	Large nuclear reactor

TABLE A.4 Scale of equivalent energy

Energy	Approximate size wind turbine to provide the same annual energy output by rotor diameter * (m)	Equivalent
1		Auto battery, 100-watt light for 10 hours
10		Electric space heater for 10 hours
50	0.25	Average per capita consumption in India
1,000	1.5	
3,500	2.5	Average residential consumption in northern Europe
4,000	3	
6,450	4	Average residential consumption in California
12,000	5	Average residential consumption in Texas
15,000	6	
20,000	7	Average residential consumption in Oklahoma
75,000	10	
2,50,000	18	
5,00,000	25	
7,50,000	33	
10,00,000	36	

* Class 4 wind resource 7 m/s, 30 m hub height, 30 per cent conversion efficiency.

Air Density

$$\rho = \frac{p}{RT}$$

where ρ is air density, p is air pressure, T is temperature and R is universal gas constant. If relative humidity ϕ is also considered for calculation of density, then

$$\rho = \frac{1}{T}\left[\frac{p}{R_o} + \phi p_w \left(\frac{1}{R_o} - \frac{1}{R_w}\right)\right]$$

where

ϕ is relative humidity of moist air,
R_o is gas constant of air (dry) 287.05 J/kg K
R_w is gas constant of water vapour 461.5 J/kg K
p_w is vapour pressure in Pascal and given as $p_w = 0.0000205 \, e^{0.0631848T}$

Appendix B

Wind Characteristics of MANIT, Bhopal, India

Latitude: 23° 12′ 25″ N
Longitude: 77° 24′ 29″ E
AMSL: 537 m
Duration: 2004–2005
Instrument: NRG Systems Anemometer Direction Sensor and Data Logger on 25 m Tower

The hourly time series wind data of speed and direction of the year 2004–2005, measured and recorded (under the sponsored project 'Development of Wind Energy Laboratory' of MHRD, GOI, scheme of thrust areas in Technical Education, grant No. F. 27/2003) at 10 m and 25 m above ground level at MPSTEP, MANIT, Bhopal, India is plotted in Figure B.1. The plots illustrate the summary of wind characteristics as wind rose, frequency distribution of speed, Weibull scale and shape parameters, average wind speed and wind power density, monthwise. The location of wind monitoring station is shown in Figure B.2.

Appendix B—Wind Characteristics of MANIT, Bhopal, India

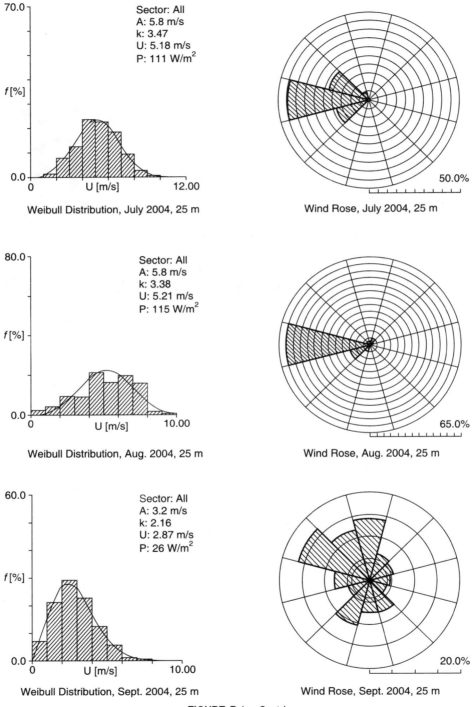

FIGURE B.1 Contd.

Appendix B—Wind Characteristics of MANIT, Bhopal, India **329**

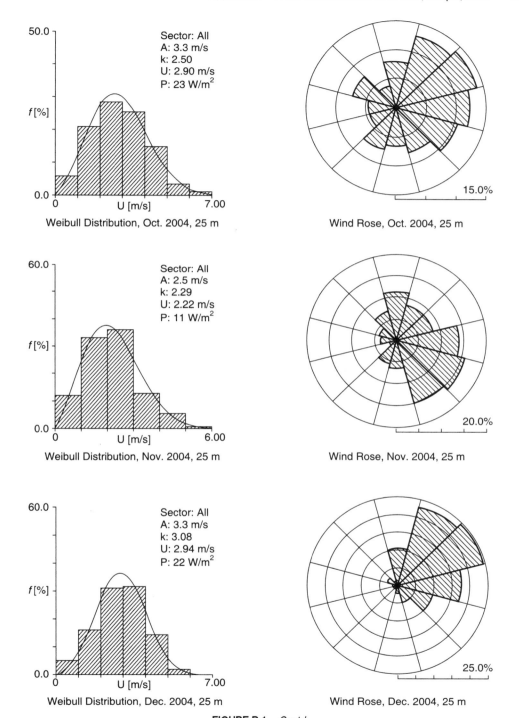

FIGURE B.1 *Contd.*

Appendix B—Wind Characteristics of MANIT, Bhopal, India

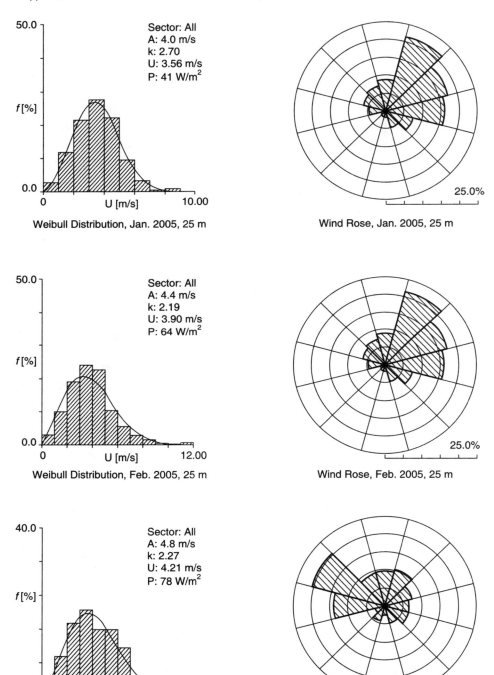

FIGURE B.1 *Contd.*

Appendix B—Wind Characteristics of MANIT, Bhopal, India

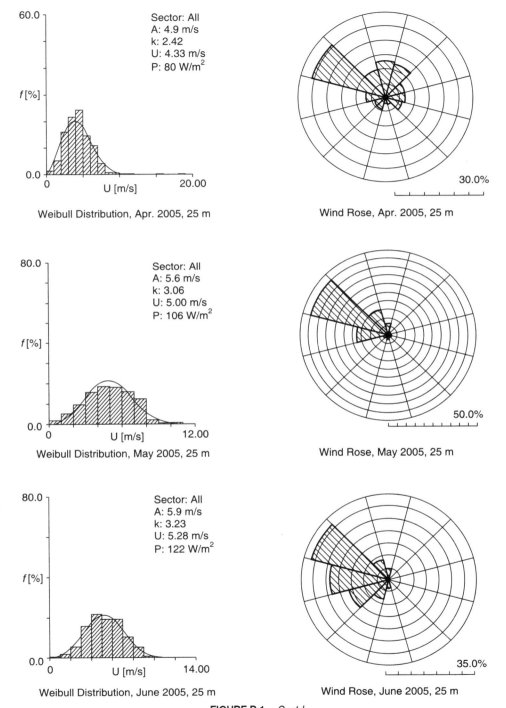

FIGURE B.1 Contd.

Appendix B—Wind Characteristics of MANIT, Bhopal, India

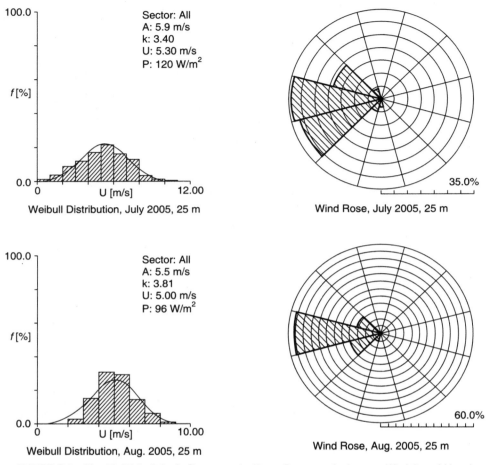

FIGURE B.1 Monthly Weibull Scale Parameter A, Shape Parameter k, Average Wind Speed U and Wind Power Density P with accompanying Wind Rose at MANIT-MPSTEP-Site.

Appendix B—Wind Characteristics of MANIT, Bhopal, India 333

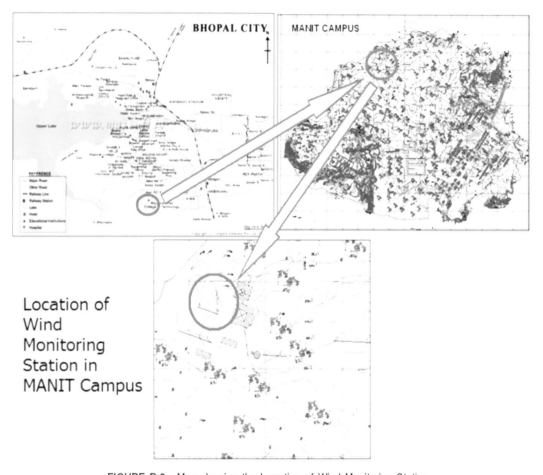

FIGURE B.2 Map showing the loacation of Wind Monitoring Station.

Appendix C

Newton–Raphson Method

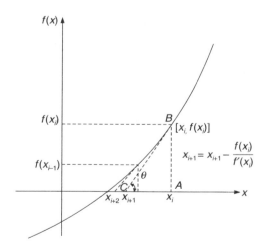

FIGURE C.1 Geometric illustration of the Newton–Raphson method.

$$\tan\theta = \frac{AB}{AC}$$

$$f'(x_i) = \frac{f(x_i)}{x_i - x_{i+1}}$$

$$x_{i+1} = x_i - \frac{f(x_i)}{f'(x_i)}$$

Steps

I Evaluate $f'(x)$.

II Use an initial guess of the root, x_i, to estimate the new value of the root, x_{i+1} as

$$x_{i+1} = x_i - \frac{f(x_i)}{f'(x_i)}$$

III Find the absolute relative approximate error $|\epsilon|$ as

$$|\epsilon| = \left|\frac{x_{i+1} - x_i}{x_{i+1}}\right| \times 100$$

IV Compare the absolute relative approximate error with pre-specified relative error tolerance ϵ_s. Also, check if the number of iterations has exceeded the maximum number of iterations allowed.

Glossary

Aerodynamic noise: It is a 'swishing' sound, affect by the shape of the blades; the interaction of the airflow with the blades and the tower; the shape of the blade trailing edge; tip shape; whether or not the blade is operating in stall conditions; and turbulent wind conditions, which can cause unsteady forces on the blades, causing them to radiate noise.

Aerofoil section: A wind turbine blade has a distinctive curved cross-sectional shape, which is rounded at one end and sharp at the other. The shape of the blade's cross section is the key to how modern wind turbines extract energy from the wind. This special profile shape is known as an *aerofoil section*.

Ambient noise level: The composite of noise from all sources near and far. The normal or existing level of environmental noise at a given location.

Anemometer: One of the components of a meteorological tower, the anemometer is a sensor that measures wind speed.

Angle of attack: The angle that an object makes with the direction of an airflow (or fluid flow), measured against a reference line in the object is called the *angle of attack*. The reference line on an aerofoil section is usually referred to as the chord line.

Asymmetrical airfoil: The shape of the airfoil is asymmetrical about its central axis (or chord), it is optimised to produce most life when the underside of the aerofoil is closest to the direction from which the air is flowing.

Availability: Availability or availability factor is measurement of reliability of a wind turbine or other power plant. It refers to the percentage of time that a plant is ready to generate that is not out of service for maintenance or repair.

A-weighted sound pressure level (dBA): The sound pressure level in decibels as measured on a sound level meter using the A-weighted filter network. The A-weighted filter de-emphasizes the very low and very high frequency components of the sound in a manner similar to the frequency response of the human ear and correlates well with subjective reactions to noise.

Base: The structure below the generator of a wind turbine that supports the turbine, houses the metres and wires, and keeps the turbine high above the ground level to protect the surrounding area and people from the force of the blades. The base also serves to elevate the turbine above surrounding obstacles that could otherwise block the wind.

Bus: An electrical conductor that serves as a common connection for two or more electrical circuits.

Bus bar: The point at which power is available for transmission.

Camber: Arching a flat plate to cause it to induce higher lift force for a given angle of attack.

Capability: The maximum load that a generating unit, generating station, or other electrical apparatus can carry under specific conditions for a given period of time without exceeding approved limits of temperature and stress.

Capacity: The amount of electrical power delivered or required for which manufacturers rate a generator, turbine, transformer, transmission circuit, station or system.

Capacity factor (CF): A measure of the productivity of a power plant, calculated as the amount of energy that the power plant produces over a set time period, divided by the amount of energy that would have been produced if the plant had been running at full capacity during that same time interval.

Capacity penetration: The ratio of nameplate rating of the wind plant (wind-farm) capacity to the peak load.

Capital cost: The total investment cost for a windfarm, including auxiliary costs.

Circuit: An interconnected system of devices through which electrical current can flow in a closed loop.

Complex-terrain: Terrain surrounding the test site that features significant variations in topography and terrain obstacles that may cause flow distortion.

Concentrating structures: Funnel shapes and deflectors fixed statically around the turbine to draw the wind into the rotor.

Conductor: The material through which electricity is transmitted, such as an electrical wire.

Constant speed turbine: Wind turbine that operates at a constant rotor revolutions per minute (rpm) and is optimised for energy capture at a given rotor diameter at a particular speed in the wind power curve.

Conventional fuel: Coal, oil and natural gas (fossil fuels), also nuclear fuel.

Cup anemometer: A device rotates by drag force. The shape of this cup produces a nearly linear relationship between rotational frequency and wind speed.

Cut-in speed: The lowest wind speed at which a wind turbine begins producing usable power.

Cut-out speed: The highest wind speed at which a wind turbine stops producing power.

Cut-in wind speed: The lowest wind speed at which a wind turbine will operate.

Cut-out wind speed: The maximum wind speed at which a wind turbine is designed to shutdown. It is also known as 'shutdown wind speed'.

Cycle: In AC electricity, the current flows in one direction from zero to a maximum voltage, then back down to zero, then to a maximum voltage in the opposite direction. This comprises one cycle. The number of complete cycles per second determines the frequency of the current in Hertz (Hz). The standard frequency of AC electricity in India is 50 cycles/sec. and in US, it is 60 cycles/sec.

Darrius rotor: This rotor has two or three thin curved blades with an airfoil section. The driving forces are lift, with maximum torque occurring when a blade is moving across the wind at a speed much greater than wind speed.

Data logger: One of the components of a meteorological tower, the data logger records the measurements.

Data-set: Collection of data that was sampled over a continuous period.

Decibel (dB): A unit describing the amplitude of sound, equal to 20 times the logarithm to the base 10 of the ratio of the measured pressure to the preference pressure, which is 20 micropascals (μ Pa).

Dispatch: The physical inclusion of a generator's output onto the transmission grid by an authorized scheduling utility.

Distribution: The process of distributing electricity. Distribution usually refers to the series of power poles, wires and transformers that run between a high-voltage transmission substation and a customer's point of connection.

Drag force: Is the force component in line with the wind relative velocity.

Due diligence: Concept involving either an investigation of a business or person prior to signing a contract or an act with a certain standard of care. It can be a legal obligation but the term will more commonly applied to voluntary investigation.

Electricity generation: The process of producing electricity by transforming other forms or source of energy into electrical energy. Electricity is measured in kilowatt-hours.

Energy: The capacity for work. Energy can be converted into different forms, but the total amount of energy remains the same.

Energy penetration: The ratio of the amount of energy delivered from one type of resource to the total energy delivered.

Evans rotor: The vertical blades twist about a vertical axis for control and fail safe shut down.

Fatal flow analysis: Any problem, lack or conflict (real or perceived) and its analysis that will destroy a solution process. A negative effect that can not be offset by any degree of benefits from other factors.

Feather: In a wind energy conversion system, to pitch the turbine blades so as to reduce their lift capacity as a method of shutting down the turbine during high wind speeds.

Feed-in law: A legal obligation on utilities to purchase electricity from renewable sources. Feed-in laws can also dictate the price that renewable facilities receive for their electricity.

Flow distortion: Change in airflow caused by obstacles, topograpical variations or other wind turbine that results in deviation of the measured wind speed from that free stream speed and in a significant uncertainty.

Frequency: The number of cycles through which an alternating current passes per second, measured in Hertz.

Furling: The process of forcing, either manually or automatically, the blades of a wind turbine out of the direction of the wind in order to stop the blades from turning. Furling works by decreasing the angle of attack, which reduces the induced drag from the lift of the rotor, as well as the cross-section. One major problem in designing wind turbines is getting the blades to stall or furl quickly enough should a gust of wind cause sudden acceleration. A fully furled turbine blade, when stopped, has the edge of the blade facing into the wind.

Gearbox: A system of gears in a protective casing used to increase or decrease shaft rotation speed.

Generator: A device for converting mechanical energy to electrical energy.

Gigawatt (GW): A unit of power, which is instantaneous capability, equal to one million kilowatts.

Gigawatt-hour (GWh): A unit or measure of electricity supply or consumption of one million kilowatts over a period of one hour.

GIS (Geographical information system): A system of hardware and software used for storage, retrieval, mapping, and analysis of geographic data.

Global warming: A term used to describe the increase in average global temperature caused by greenhouse effect.

Greenhouse gases (GHGs): Gases such as water vapour CO_2, methane, and low level ozone that are transparent to solar radiation, but opaque to long-wave radiation. These gases contribute to the greenhouse effect.

Greenpower: A popular term for energy produced from renewable energy sources.

Grid: A common term that refers to an electricity transmission and distribution system.

Grid codes: Regulations that govern the performance characteristics of different aspects of the power system, including the behaviour of wind farms during steady-state and dynamic conditions. These fundamentally technical documents contain the rules governing the operations, maintenance and development of the transmission system and the coordination of the actions of all users of the transmission systems.

Gust: More than five times of speed of mean wind speed.

Heat rate: A measure of thermal efficiency of generating station. Commonly stated as British thermal units (Btu) per kilowatt-hour. Heat rates can be expressed as either gross or net heat rates, depending whether the electricity output is gross or net generation. Heat rates are typically expressed as net heat rates.

Hertz (Hz): A unit of measurement of frequency; the number of cycles per second of a periodic waveform.

Hub height: In a horizontal-axis wind turbine, the distance from the ground to the centre-line of the turbine rotor. See also wind turbine tower height.

Incidental take: Unintentional removal that may occur during otherwise lawful activities. If a project results in 'incidental take' of a listed species, an incidental take permit is required in USA.

Instantaneous penetration: The ratio of wind farm (wind plant) output to load at a specific point in time, or over a short period of time.

Investment tax credit (ITC): A tax credit that can be applied for the purchase of equipment such as renewable energy system.

Kilowatt (kW): A standard unit of electrical power which is instantaneous capability equal to 1000 watts.

Kilowatt-hour (kWh): A unit of measure of electricity supply or consumption of 1000 watts over a period of one hour.

Leading edge: The surface part of wind turbine blade that first comes into contact with wind.

Lift: The force that pulls a wind turbine blade.

Lift force: Is the force component perpendicular to drag force.

Load (electricity): The amount of electrical power delivered or required at any specific point or points on a system. The requirement originates at the consumer's energy-consuming equipment.

Load factor: The ratio of average load to peak load during a specific time.

Load following: A utility's practice in which more generation is added to available energy supplies to meet moment-to-moment demand in the utility's distribution system, or in which generating facilities are kept informed of load requirements. The goal of the practice is to ensure that generators are producing neither too little nor too much energy to supply the utility's customers.

Megawatt (MW): The standard measure of electricity power plant generating capacity. One megawatt is equal to 1000 kilowatt or 1 million watts.

Megawatt-hour (MWh): A unit of energy or work equal to 1000 kilowatt-hour or 1 million watt-hour.

Met tower: A meteorological tower erected to verify the wind resource found within a certain area of land.

Molinology: Study of wind energy conversion systems.

Musgrove rotor: The blades of this form of rotor are vertical for normal power generation, but tip or turn about a horizontal point for control or shut down.

Nacelle: The cover for the gear box, drive train, generator, and other components of a wind turbine.

Nameplate rating: The maximum continuous output or consumption in MW of an item of equipment as specified by the manufacturer.

Noise: Implies the presence of sound but also implies a response to sound noise is often defined as unwanted sound.

Non dispatchable: The timing and level of power plant output generally cannot be closely controlled by the power system operator. Other factors beyond human control, such as weather variations, play a strong roll in determining plant output.

Orographic features: Orography refers to the terrain and elevation of an area, e.g. mountain ranges, hills, valleys, cliffs, escarpments, ridges, etc. that influences the wind flow.

Pitch angle: Angle between the chord line at a defined blade radial location (usually 100 per cent of the blade radius) and the rotor plane of rotation.

Pitch control: A method of controlling the speed of a wind turbine by varying the orientation, or pitch, of the blades, and thereby altering its aerodynamics and efficiency.

Pitching: The blade of the wind turbine is rotated few degrees about its longitudinal or 'spanwise' axis (called the pitch axis) until it reaches the optimum 'pitch angle', at which it will produce the maximum power at that wind speed.

Power: The rate of production or consumption of energy.

Power electronics: Converts the variable-frequency, voltage and current from the generator to the utility grid constant-frequency, voltage and current within the range set by the utility to ensure quality of power delivered to the grid.

Power grid: A common term that refers to an electricity transmission and distribution system.

Power marketers: Business entities engaged in buying and selling electricity. Power marketers do not usually own generating or transmission facilities, but take ownership of the electricity and are involved in its trade.

Power purchase agreement (PPA): A long-term agreement to buy power from a company that produces electricity.

Power quality: Stability or frequency and voltage and lack of electrical noise on the power grid.

Public utility commission: A governing body that regulates the rates and services of a utility.

Ramp rate: The rate at which load on a power plant is increased or decreased. The rate of change in output from a power plant.

Rated wind speed: The wind speed at which wind turbine produces maximum output power.

Renewable energy: Energy derived from resources that are regenerative or that cannot be depleted. Types of renewable energy sources include wind, solar, biomass, geothermal and moving water.

Repowering: Repowering is the process of replacing older, smaller wind turbines with modern and more powerful machines, which would reap considerably more power from the same site.

Restructuring: The process of changing the structure of the electric power industry from a regulated guaranteed monopoly to an open competition among power suppliers.

Rime icing: Occurs when the structure is at a sub-zero temperature and is subject to incident flow with significant velocity and liquid water content.

Rotor: The blades and other rotating components of a wind turbine.

Savonius rotor (turbo machine): A device with a complicated motion of the wind through and around the two curved sheet airfoils. The driving force is principally drag. The high solidity produces high starting torque, so savonius rotors are used for water pumping.

Sea breezes: Are generated in coastal areas as a result of the different heat capacities of sea and land, which give rise to different rates of heating and cooling. The land has a lower heat capacity than the sea and heats up quickly during the day, but at night, it cools more quickly, than the sea. During the day, the sea is therefore cooler than the land and this causes the cooler air to flow shorewards to replace the rising warm air on the land. During the night, the direction of airflow is reversed.

Shadow flicker: The effect caused by the sun's casting shadows from moving wind turbine blades.

Siting: To assess and predict relative desirability of potential sites for wind farms.

Sound: Describes wave-like variations in air pressure that occur at frequencies that can stimulate receptors in the inner ear and, if sufficiently powerful, be appreciated at a concious level.

Solar energy: Electromagnetic energy transmitted from the sun, i.e. solar radiation.

Solidity: This is the ratio of the total area of the blades at any one moment in the direction of the airstream to the swept area across the airstream. Thus, with identical blades, a four-bladed turbine presents twice the solidity of a two-bladed turbine.

Stall control: Let us assume that a wind turbine is rotating at a constant rotation speed, regardless of wind speed, and that the blade pitch angle is fixed. As the wind speed increases, the tip speed ratio decreases. At the same time, the relative wind angle increases, causing an increase in the angle of attack. It is possible to take advantage of this characteristic to control a turbine in high winds, if the rotor blades are designed so that above the rated wind speed they become less efficient because the angle of attack is approaching the stall angle. This results in a loss of lift, and thus torque, on the regions of the blade that are in stall.

Strobe effect: Long shadow during sunset and sunrise.

Swept area: For a horizontal axis wind turbine, the projected area of the moving rotor upon a plane normal to the axis of rotation.

Symmetrical airfoil: The shape of the airfoil is symmetrical about its central axis (or chord), and able to induce lift equally well (although in opposite directions) when the airflow is coming from either side of them.

Topographical features: Variation of surface elevation of earth on a small scale, e.g. roughness, etc. and measured by contour values.

Trade wind: The consistent system of prevailing winds occupying most of the tropics. Trade winds, which constitute the major component of the general circulation of the atmosphere, blow northeasterly in the northern hemisphere and southeasterly in the southern hemisphere. The trades, as they are sometimes called, are the most persistent wind system on earth.

Trailing edge: The part of a wind energy conversion device blade, or airfoil, that is the last to contact the wind.

Turbine: A term used for a wind energy conversion device that produces electricity.

Turbulence: A swirling motion of the atmosphere that interrupts the flow of the wind.

Upwind turbine: A type of wind turbine in which the rotor faces the wind. The basic advantage of upwind designs is that one avoids the wind shade behind the tower. By far the vast majority of wind turbines have this design. On the other hand, there is also some wind shade in front of the tower, i.e. the wind starts bending away from the tower before it reaches the tower itself, even if the tower is round and smooth. Therefore, each time the rotor passes the tower, the power from the wind turbine drops slightly. The basic drawback of upwind designs is that the rotor needs to be made rather inflexible, and placed at some distance from the tower. In addition, an upwind machine needs a yaw mechanism to keep the rotor facing the wind.

Utility grid: A common term that refers to an electricity transmission and distribution system.

Variable-speed wind turbines: Turbines in which the rotor speed increases and decreases with changing wind speeds. Sophisticated power control systems are required on variable-speed turbines to ensure that their power maintains a constant frequency compatible with the grid.

Viewshed: The landscape or topography visible from a geographic point, especially those that have an aesthetic value.

Volt (V): A unit of electrical force.

Voltage: The amount of electromotive force, measured in volts, between two points.

Watt (W): A unit of power.

Watt-hour (Wh): A unit of electricity consumption of one watt over the period of one hour.

Whirlpools of air: Rotational movement of the air occurs as the airstream flows around the blade. This may be apparent as distinct vortexes and eddies created near the surface. Vortex shedding occurs as these rotating masses of air break free from the surface and move away, still rotating, with this airstream. Such three-dimensional rotations do not occur off airplane wings.

Wind: Moving air. The wind's movement is caused by the sun's heat, the earth and the oceans, which force air to rise and fall in cycles.

Wind energy: Energy generated by using a wind turbine to convert the mechanical energy of wind into electrical energy.

Wind generator: A wind energy conversion system designed to produce electricity.

Windmill: A wind energy conversion system that is used primarily to grind grain. Windmill is commonly used to refer to all types of wind energy conversion systems.

Wind power: Power generated by using a wind turbine to convert the mechanical power of the wind into electrical power.

Wind power class: A scale for classifying wind power density. There are seven wind power classes, ranging from 1 (lowest wind power density) to 7 (highest wind power density). In general, sites with a wind power class rating of 4 or higher are now preferred for large-scale wind farms.

Wind power density: A useful way to evaluate the wind resource available at a potential site. The wind power density, measured in watts per square metre, indicates the amount of energy available at the site for conversion by a wind turbine.

Wind power plant: A group of wind turbines interconnected to a common utility system.

Wind project: Wind projects vary in size, from a small project of one to a few turbines (known as *behind the meter* or *distributed wind systems*) serving individual customer, to large projects ('utility' or 'commercial scale' or 'wind forms') designed to provide wholesale electricity to utilities or an electricity market.

Wind resource assessment: The process of characterizing the wind resource and its energy potential for a specific site or geographical area.

Wind rose: A diagram that indicates the average percentage of time that the wind blows from different directions, on a monthly or annual basis.

Wind speed: The rate of flow of wind when it blows undisturbed by obstacles.

Wind speed duration curve: A graph that indicates the distribution of wind speeds as a function of the cumulative number of hours that the wind speed exceeds a given wind speed in a year.

Wind speed frequency curve: A curve that indicates the number of hours per year that specific wind speeds occur.

Wind speed profile: A profile of how wind speed changes at different heights above the surface of the ground or water.

Wind shear profile: The profile of increase in wind speed with height from ground is called wind shear profile.

Wind turbine: A term used for a device that converts wind energy to electricity.

Wind turbine braking: Overspeed control of a wind turbine is exerted in two main ways: aerodynamic stalling or furling, and mechanical braking. Furling is the preferred method of slowing wind turbines. Braking of a wind turbine can also be done by dumping energy from the generator into a resistor bank, thereby converting the kinetic energy of the turbine rotation into heat. This method is useful if the connectic load on the generator is suddenly reduced or is too small to keep the turbine speed within its allowed limit. Cyclically braking causes the blades to slow down, which increases the stalling effect, reducing the efficiency of the blades. This way, the turbine's rotation can be kept at a safe speed in faster winds while maintaining (nominal) power output. A mechanical drum brake or disk brake is used to hold the turbine at rest for maintenance. Such brakes are usually applied only after blade furling and electromagnetic braking have reduced the turbine speed, as the mechanical brakes would wear quickly if used to stop the turbine from full speed.

Wind turbine rated capacity: The amount of power a wind turbine can produce at its rated wind speed.

Yaw: The rotation of a horizontal-axis wind turbine around its tower or vertical axis.

Zone of visual influence (ZVI): Area from which a wind farm is theoretically visible. It is usually represented as a map using colour to indicate visibility. The use of computational design tools allows ZVI or visible footprint, to be calculated to identify from where the wind farm will be visible.

Bibliography

Abbot, I.H. and A.E.V. Doenhoff, *Theory of Wing Sections*, Dover Publications, New York, 1959.

Ahmed, Siraj, M.M. Pandey and A.R. Ansari, Computer Aided Aerodynamic Modeling of Horizontal Axis Wind Turbine Rotor, Proc. of 17th National Convention of Mechanical Engineers at the Advent of Millennium, pp. 443–453, Indore, 2001.

Ahmed, Siraj, Lecture Notes, MHRD-AICTE, Staff Development Programme on Solar and Wind Energy: Theory and Practice, MANIT, Bhopal, 2009.

Ahmed, Siraj, Lecture Notes, TEQIP, Short Term Training Programme on Recent Developments in Solar and Wind Energy Systems, MANIT, Bhopal, 2008.

Ahmed, Siraj, Lecture Notes, TEQIP, Short Term Training Programme on Advances in Wind and Solar Energy Technology, MANIT, Bhopal, 2006.

Ahmed, Siraj, Training Report, 'Wind Energy and its Utilisation', Indo-UK, RECs' Project, Energy Theme, University of Salford, UK, 1996.

Ahmed, Siraj, Wind Energy Generation and its Utilization–Lecture Notes, vol. I and II, AICTE, ISTE Winter School, MANIT, Bhopal, India, 2000.

Alfredo Pena et al., Remote Sensing for Wind Energy, DTU Wind Energy–E–Report–0029(EN), ISBN 978-87-92896-41-4, Technical University of Denmark, 2013.

Article on Best Practice for Accurate Wind Measurement, www.wind-energy-the-facts.org.

ASME, Performance Test Code 42, United Engineering Centre, New York, 1998.

Burton, Tony, David Sharpe, Nick Jenkins and Ervin Bossanyi, *Wind Energy Handbook*, John Wiley & Sons, New York, 2001.

Dado, V.J. et al., Assessment of Wind Energy Potential of Trombay, Mumbai (19.1° N, 72.8° E), India, *Int. J. of Energy Convers. and Mgmt*, vol. 39, No. 13, pp. 1351–1356, 1998.

David, S. Renne, A Practical and Economic Method for Estimating Wind Characteristics at Potential Wind Energy Conversion Sites, *Int. J. of Solar Energy*, vol. 25, pp. 55–65, 1980.

Directory Indian Wind Power, Consolidated Energy Consultants Ltd., Bhopal, 2010.

Dunn, P.D., *Renewable Energies—Sources, Conversion and Application*, Peter Peregrinus, UK, 1986.

Earnest Joshua and Tore Wizelius, *Wind Power Plants and Project Development,* PHI Learning, New Delhi, 2010.

Eggleston, D.M. and F.S. Stoddard, *Wind Turbine Engineering Design*, Van Nostrand Reinhold Co., New York, 1987.

Elisa Wood, Offshore Awakening, *Renewable Energy World Magazine*, 2010.

Freris, L.L., A Plain Man's Guide to Electrical Aspects of Wind Energy Conversion System, Proceedings of Conference on Research Needs for the Effective Integration of New Technologies into the Electrical Utility, Maryland, USA, 1982.

Frost, Walter, B.H. Long and R.E. Turner, *Engineering Handbook on the Atmospheric Environmental Guidelines for Use in Wind Turbine Generator Development*, NASA Technical Paper 1359, 1978.

Gipe, Paul, *Wind Energy—How To Use It*, Stackpole Books, USA, 1983.

_____ *Wind Energy Comes of Age*, John Wiley & Sons, New York, 1995.

Golding, E.W., *The Generation of Electricity by Wind Power*, E. and F. Spon, London, 1976.

Gourieres, D. Le, *Wind Power Plants—Theory and Design*, Pergamon Press, Oxford, 1982.

Heister, T.R. and W.T. Pennell, *The Siting Handbook for Large Wind Energy Systems*, Windbooks Inc., St. Johannesburg, USA, 1981.

Hunt, V. Daniel, *Wind Power: A Handbook on Wind Energy Conversion Systems*, Van Nostrand Reinhold Company, N.Y., 1981,

Jamil, M., Wind Power Statistics and Evaluation of Wind Energy Density, *Int. J. of Wind Engineering*, vol. 18, No. 5, pp. 227–240, 1994.

Jangamshetty, Suresh H. and Rau Guruprasada V., Site Matching of Wind Turbine Generators: A Case Study, *Int. J. of IEEE on Energy Conversion*, vol. 14, No. 4, pp. 1537–1543, 1999.

_____ Optimum Siting of Wind Turbine Generator, *Int. J. of IEEE Transactions on Energy Conservation*, vol. 16, No. 1, pp. 8–13, 2001.

Johnson, Gary L., *Wind Energy Systems,* Prentice Hall, NJ, 1985.

Krauze, Richard, Mitigating Variability: Advances in Wind Resource Assessment and Modellin, *Renewable Energy World Magazine*, 2009.

Mani, Anna and S. Rangarajan, *Wind Energy Resource Survey in India*–IV, Allied Publisher, New Delhi, 1996.

Mathew, Satyajith, *Wind Energy: Fundamentals, Resource Analysis and Economics*, Springer, 2006.

Mathew, Sathyajith, et al., Analysis of Wind Regimes for Energy Estimation, *Int. J. of Renewable Energy*, vol. 25, No. 3, pp. 381–399, 2002.

Martin, O.L. Hansen, *Aerodynamics of Wind Turbines,* 2nd ed., Earthscan, 2008.

McVeigh, J., *Energy Around the World*, Pergamon Press, Oxford, 1985.

Miller, Rene H., J. Dugundji, et al., Wind Energy Conversion MIT Aeroelastic and Structures Research Lab, TR-184-7 through TR-184-16. DOE Control No. COO-4131-T 1, distribution category UC-60, Sept. 1978.

Mohamed, Kotb and Hassan Soliman, Design Chart for HAWT's, *Int. J. of Engineering*, vol. 15, No. 3, pp. 155–162, 1991.

Mortensen, Niels G., Anthony J. Bowen and Ioannis Antoniou, Improving WAsP Prediction in (Too) Complex Terrain, www. risoe. dk, 2004.

Nivedh B.S. et al., Repowering of Wind Farms: A Case study. www. risoe.dk/rispubl/NET-DK-5841.pdf.

Pandey, M.M., Studies on design and performance aspects of low speed horizontal axis wind turbine rotor, *Ph.D. Thesis*, IIT Kharagpur, 1988.

Pandey, M.M. and Chandra Pitam, Determination of Optimum Rated Speeds of Wind Machines for a Particular Site, Technical Note, *Int. J. of Solar and Wind Technology*, vol. 3, No. 2, pp. 135–140, 1986.

Pandey, M.M., K.P. Pandey and T.P. Ojha, An Analytical Approach to Optimum Design and Peak Performance Prediction for Horizontal Axis Wind Turbines, *Int. J. of Wind Engineering and Industrial Aerodynamics*, pp. 247–262, 1989.

_____Aerodynamic Characteristics of Cambered Steel Plates in Relation to their Use Wind Energy Conversion System, *Int. J. of Wind Engineering*, vol. 12, No. 2, pp. 90–104, 1988.

Park, J., *The Wind Power Book*, Cheshire Books, 1981.

Performance Standards for Wind Turbines, American Wind Energy Association, VA, 1995.

Proceedings of CEP Course on Wind Energy Technology, IIT Mumbai, 2007.

Proceedings of CEP Course on Wind Energy Technology, IIT Mumbai, 2008.

Proceedings of National Seminar on Wind India—2006, Pune.

Proceedings of Wind Power, Shanghai, 2007.

Rohtagi J.S. and V. Nelson, *Wind Characteristics: An Analysis for the Generation of Wind Power'*, West Texas, A and M University, TX. 1994.

Salmach, Ziyad M. and Irianto Safari, Optimum Wind-mill-site Matching, *Int. J. of IEEE Transactions on Energy Conversion*, vol. 7, No. 4, pp. 669–675, 1992.

Sasi, K.K. and Sujay, Basu, On the Prediction of Capacity Factor and Selection of Size of Wind Electric Generators—A Study on Indian Sites, *Int. J. of Wind Engineering*, vol. 21, No. 2, pp. 73–87, 1997.

Sastry, S.S., *Engineering Mathematics*, vol. 2, Prentice Hall of India, New Delhi, 2003.

Seguro, J.V. and T.W. Lambert Modern Estimation of the Parameters of the Weibull Wind Speed Distribution for Wind Energy Analysis, *Int. J. of Wind Engineering and Industrial Aerodynamics*, vol. 85, Issue 3, pp. 309–324, 2000.

Seguro, J.V. and T.W. Lambart, Modern estimation of the parameters of the Weibull wind speed distribution for wind energy analysis, *Journal of Wind Engineering and Industrial Aerodynamics*, vol. 85(1), 2000.

Seguro, J.V. and T.W. Lambart, Modern estimation of the parameters of the Weibull wind speed distribution for wind energy analysis, *Journal of Wind Engineering and Industrial Aerodynamics*, vol. 85(1), 2000.

10 Steps in Building a Wind Farm, AWEA, Wind Energy Fact Sheet.

Singh, Shikha, T.S. Bhatti and D.P. Kothari, The Power Coefficient of Wind Mills in Ideal Conditions, *Int. J. of Global Energy Issues*, vol. 21, No. 3, 2004.

Spears, D.A., *Wind Turbine Technology*, ASME USA, 1988.

_____*Wind Turbine Technology*, ASME Press, New York, 1994.

Stevens, M.T.M. and P.T. Smulders, The Estimation of the Parameters of the Weibull Wind Speed Distribution for Wind Energy Utilization Purposes, *Int. J. of Wind Engineering*, vol. 3, No. 2, pp. 132–140, 1999.

Stewart, W.H., H.J. Stewart and D.L. Staples, Optimal Design Techniques for Horizontal Axis Wind Turbines, Report WER-17, Wichita State University, Centre for Energy Studies, Feb. 1984.

10 Steps in Building a Wind Farm, AWEA, Wind Energy Fact Sheet, 2011.

T521 Renewable Energy, A Resource Pack For Tertiary Education, The Open University Teaching Package, UK, 1994.

TAPS-2000, Provisional Type Certification Scheme for Wind Generator Systems in India, 2003.

The Economics of Wind Energy, www.awea.org.

The Law of the Wind: A Guide to Business and Legal Issues, www.stoel.com, 2007.

Twidell, J.W. and Weir, A.D., *Renewable Energy Resources*, E. and F. Spon Ltd., London, 1986.

US Department of Energy Report-2008: 20% Wind Energy by 2030.

US Department of Energy, Home Windpower, Garden Way Publishing, Charlotte, Vermont, USA, 1981.

Ushiyama, Izumi and Hiroshi Nagai, Optimum Design Configurations and Performance of Savonious Rotor, *Int. J. of Wind Engineering*, vol. 12, No.1, pp. 59–75, 1988.

Wegley, H.L., J.V., Rasdell, M.M., Orgill, and R.L., Drake, *The Siting Handbook for Small Wind Energy Conversion Systems*, Windbooks Inc., St. Johnsburg, USA, 1989.

Wind Energy Applications Guide, AWEA, January, 2001.

Wind Energy: Siting Handbook, American Wind Energy Association, 2008.

Wind Resource Assessment Handbook, National Renewable Energy Laboratory, Colorado, 1997.

Wind Resource Survey in India–VII, Centre for Wind Energy Technology, India, 2005.

Wind Turbine Sound and Health Effect—An Expert Panel Review, AWEA and CWEA, December 2009.

Wolfe, B. and Brabant, E.J., *Your Wind Driven Generator*, Van Nostrand Reinhold Co. Inc, New York, 1984.

www.awea.org

www.bwea.com

www.ireda.nic.in

www.leonardoenergy.org

www.mnes.nic.in

www.windpower.dk

www.windpower.org

www.windpowerindia.com

www.windpoweronline.com

www.wpm.co.nz

www.niwe.res.in

www.windpower.sandia.gov/topical.htm

IEC 16400–1 Standard for Turbulence in Turbine.

ISO 9631 Standard on Noise Propagation.

IEC 61400–2 Standard for Design Requirement of Wind Turbine.

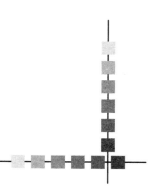

Index

Accumulated net present value, 253
Accumulated present value, 244
Active pitch control, 164
    braking, 165
Active stall control, 164
Actuator disc, 81
    model, 80
Adaptive management principles, 275
Aero-acoustic sound, 282
Aerodynamic centre, 149
Aerodynamic characteristics, 146
Aerodynamic forces, 66, 68, 76
Aerodynamic lift, 113
Aerodynamic sound, 280
Aerodynamics, 63
    noise, 282
Aerofoil, 64
    section, 65
Aerogenerator, horizontal-axis wind turbine, 1
Aerosol, 48
Ainslie, 226
Airfoil, 69

Airfoil characteristics, 68, 70
Airfoil designation, 70
Albert Betz, 83
Analytical method, 96
Anemometer, 36
Angle of attack, 64, 75, 126
    distribution, 146
Angular speed interference factor, 88
Annual cost of operation, 253
Annual energy output, 241
Annular element, 88
Anthropogenic causes, 277
Apparent A-weighted sound power level, 286
Apparent power, 312
Array, 219
Asymmetrical aerofoil, 149
Atmospheric boundary layer, 17, 30
Average electrical power, 192
Average power, 295
Average wind power density, 24
Avian Issue, 277
A-weighted sound power level, 285

A-weighted sound pressure curve, 280
Axial flow devices, 74
Axial force variation, 147
Axial interference factor, 82
Axial momentum theory, 80

Basins, 212
Benefit/cost, 248
    ratio, 248, 257
Bernoulli effect, 64
Bernoulli's equation, 81, 87
Betz criterion, 84
Betz limit, 2, 87, 129
Betz model, 85
Betz's coefficient, 83
Biological impact, 275
Bird species, 276
Bi-static continuous wave, 48
Blade angle, 154
Blade design, 146, 166
Blade element, 89, 127
    parameters, 133
    theory, 89, 90
Blade feathering, 164
Blade geometry, 153
    parameter, 142
Blade length, 120
Blade lightning-protection, 305
Blade line load, 157
Blade loading coefficient, 132
Blade loads, 156
Blade materials, 168
Blade pitch, 111
    angle, 75
Blade solidity, 128
Blade span, 77
Blade structure, 167
Blade twist distribution, 144
Blade velocity, 76
Blades, 119
Boundary layer, 73
Boundary layer thickness, 72
Brag scattering, 53
Braking by pitching blade tips, 165
Braking systems, 165
Breezes, 15
Broadband noise, 282

Brushing, 214

Cambered, 69
    aerofoils, 148
Capacitance, 291
Capacitor, 294
Capacity factor, 189, 191, 192, 194, 241, 252
Capacity penetration, 6
Capital recovery factor, 241, 248
Carbon-fibre composites, 168
Carpeting, 214
Carrier to noise ratio, 48
Characteristic aerodynamic load, 153
Chord, 65
    length, 69, 99
    line, 64
Circulation reduction factor, 91
Classes of wind power density, 33
Classification of wind turbines, 115, 116
Cliff section, 213
Cliffs, 212
Clipping, 214
Codes and standards, 159, 307
Coefficient of
    moment, 67
    performance, 86, 99
    power, 83
    thrust, 82
Communication impact, 286
Complex terrain, 205
Concentrator, 117, 118
Connection circuit, 315
Constraint map, 198, 199
Consumer price index, 246
Continuity equation, 85
Continuous wave (CW) lidar, 50
Control systems, 300
Convective-scale, 15
Coordinate system, 153
Coriolis forces, 11
Cost model, 161, 163
Cost of wind energy, 251
Cost per unit of electricity, 240
Cross flow devices, 74
Cumulative distribution fnction, 27
Cup anemometer, 40, 115
Current, 290

Cyclonic storms, 16

Danish standard, 151, 159
Darrieus rotor, 115
Data transfer, 43
Declining balance depreciation, 261
Deflation, 246
Degree of certainty, 34
Depreciation, 249, 261
Design angle, 101
Design loads, 159
Design tasks, 125
Designation of NACA, 149
Deterministic loads, 156
Digit depreciation, 262
Direct-drive, 123
    systems, 122
Directivity, 286
Discount rate, 243
Discrete gust model, 22
Distortion energy, 175
Distribution circuit, 310
Distribution network, 309, 316
Distribution voltage transformer, 310
Diurnal variations, 17
Doldrums, 16
Doppler shift, 47, 48, 51
Doubly fed induction generator, 307
Downstream wind speed, 224
Drag component, 72
Drag force, 63, 66, 127
Drag translator device, 99, 100
Dual optimum, 138
Dynamic viscosity, 72

Earthing, 303
    system, 304
Easterlies, 15
Ecological assessment, 276
Ecological indicators, 214
Economic appraisal, 239
Economic merit, 256
Economics worksheet, 264
Eddy viscosity model, 224, 225
Effect of dirt, 172

Effect of spacing, 218
Effects of wind-induced refraction, 284
E-glass, 170
Electric field, 291
Electrical cut-in, 302
Electromagnetic induction, 292
Electromagnetic interference, 287
Electronic controls, 301
Elemental coefficient of performance, 95
Elementary actuator disc theory, 131
Embedded (Dispersed) wind generation, 309
Energy content, 32
Energy pattern factor, 32
Energy penetration, 6
Energy spectrum, 14
Environmental challanges, 274
Environmental impact, 274
Environmental issues, 198
Equivalent neutral wind, 54
Equivalent wind speed, 52
Escalation rate, 246
Estimation of capacity factor, 190
Euler's equation, 85
European Wind Energy Association, 241
Exceedance statistics, 22
External gust, 22
Extrapolation of wind speed, 201

Faraday's law, 292
Fatigue loading regime, 160
Fibreglass, 119
Fixed and variable costs, 253
Fixed-pitch, 74
Fixed-speed wind turbine, 302
Flagging, 214
Flat terrain, 34
Flat-topped ridge, 208
Flicker step factor, 314
Flow velocity, 68
Four-digit designation, 149
Fractional probability, 27
Free-stream, 80
Frequency, 20
    distribution function, 24
Froude efficiency, 136, 137
Furling wind speed, 186

Gamma function, 25, 27, 191
Gaps and gorges, 210
Gaussian profile, 227
Gaussian wind speed, 227
Gauss–Seidel load flow algorithm, 312
Generation protection, 314
Generator, 291
    system, 300
    taxonomy, 296
Geographical positioning system, 44
Geostrophic wind, 12
Germanischer Lloyd's regulation, 159
Gin-pole, 44
GL rules, 159
Glass reinforced plastic, 304
Glass/epoxy, 170
Glass/polyester, 170
Glass-fibre blade, 170
Glass-fibre construction, 168
Glass-fibre-reinforced plastic, 6
Glauert momentum vortex theory, 130
Global positioning system, 38
Global wind pattern, 12
Goodman relationship, 175
Grid network, 307
Grid stability, 307
Griggs-Putnam classification, 214
Grontmij, 231

HAWT, 74, 117
Hemispherical spreading, 286
History of wind machines, 111
Horizontal axis machines, 115
Hub loads, 157

Idealized ridge, 207
IEA recommended practices, 285
IEC 61400-1, 159, 160
IEC 61400-12-1 standard, 47
IEC 61400-21:2001 standard, 308
IEEE Recommended Practice (1991), 303
Inductance, 291
Induction generator, 298, 299
Inflation, 246
Inflow angle, 143
Instantaneous penetration, 6

Instantaneous power, 294
Instrumentation, 40
Integration into grid, 306
Intensity of wakes, 187
Inter tropical convergence zone, 16
Interface protection, 315
Interference factor, 82, 83, 98
Interference mechanisms, 288
Interference parameter, 86
Internal rate of return, 248, 258
International Electrotechnical Commission (IEC), 282
Isolated hill, 209

Kite anemometer, 187

Lagrange multipliers, 135
Land–sea breeze system, 19
Large eddy simulation, 31
Lattice structure, 276
Law of conservation of energy, 85
Layout of wind farm, 217
Layout of wind turbines, 219
Leading edge, 65
Least squares method, 26
Levelized equivalent amount, 263
Lidar, 47, 48, 52
Life cycle costing, 250
Lift and drag, 67
    coefficients, 66
Lift coefficient, 144, 153
Lift force, 63, 66
Lift-to-drag ratio, 70, 114
Lightning, 305
    flash, 304
    protection, 304
Linear decision strategy, 275
Linear taper blade, 145
Load calculation, 151
Load matching, 103
Local inflow angle, 142
Local power coefficient, 136
Local solidity, 132
Local speed ratio, 90
Local tip speed ratio, 88
Logarithmic wind profile, 163

Losses in generation, 316
Low plant load factor, 229
Low sharp-crested ridge, 208
Low voltage ride-through (LVRT) standard, 308
Lowest resonance frequency, 157

Mach number, 68
Magnetic field, 291
Maintenance cost, 241
Maximum elemental power, 93
Maximum likelihood method, 27
Mean wind speed, 190
Measure-Correlate-Predict (MCP) technique, 222
Measured line of sight, 50
Measurement parameters, 39
Measurement plan, 34
Median and quartile, 26
Mesas and buttes, 214
Mesoscale wind systems, 15
Micro-meteorological range, 20
Micrositing, 33, 222
Modern wind turbines, 111
Modified maximum likelihood method, 28
Momentum theorem, 81
Monitoring duration, 34
Monitoring tower, 38
Monostatic lidar, 49
Monsoons, 15
Mountain pass, 210

NACA aerofoil, 149
Nacelle, 122, 123
Navier-Stokes equations, 225
Negative pitch control, 164
Net present value, 248, 256, 266
New wind profile, 201
Newton–Raphson method, 98, 249, 258
Noise of different activities, 283
Non-flat or complex, 204
Non-uniform wind field, 21
Normalized power, 242
Number of blades, 90
Numerical flow modelling, 188
Numerical weather prediction, 188

Obstruction effects, 39
Off-design performance prediction, 99
Off-optimum operation, 142
Offshore prospects, 8
Offshore wind energy, 8
Offshore wind potential, 36
One seventh exponent, 31
Optimal rotors, 134
Optimum design, 95, 96, 100, 139
Optimum performance, 101
Optimum turbine size, 162
Overall eddy viscosity, 226

Park model, 224
Partial safety factors, 160
Passes and saddles, 209
Passive control, 166
Passive pitch control, 164
Passive stall control, 163
Pay back period, 257
Peak performance, 95
    prediction, 96
Permanent magnet, 122
    induction generator, 307
    synchronous generator, 307
Persistence, 20
Phase angle, 293
Physical modelling, 187
Pitching moment coefficient, 149
Planar gust front, 22
Power, 131, 291
    coefficient, 2, 88, 93, 129
    collection system, 302
    curve, 152
    flux, 86
    law, 36
    quality, 313
    system stability, 307
    train, 120, 121
Prandtl correction for finite, 134
Prandtl correction function, 134
Prandtl tip loss factor, 92
Prandtl's approximation, 92
Prandtl's model, 91
Preliminary site characterization, 194
Present value of annual costs, 253

Present worth approach, 243
Pressure and velocity relationships, 76
Pressure gradient, 72
Pressure vessel code, 175
Probability density function, 190
Project appraisal, 266
Project finance, 269
Project viability, 240
Propeller anemometer, 40
Pulsed LIDAR, 51

Quasi-static, 20, 21

Radial circuit, 311
Radio frequency, 44
Radio systems, 288
Rankine–Fraude momentum theory, 80
Ranking of ridge shape, 206
Rated electrical power, 192
Rayleigh distribution, 25
Rayleigh probability density function, 191
Real power loss, 312
Reference annual wind speed, 25
Reference wind speed, 159
Referenced data, 36
Relative wind angle, 73, 75
Relative wind velocity, 73, 76
Remote sensing technique, 47
Re-powering, 228, 231
Resistance, 291
Response amplification, 158
Resulting turbine cost, 162
Reynolds number, 68, 143, 148
Reynolds stresses, 225
Ridge, 205
Root-mean-square average, 20
Rotation of wake, 87
Rotor, 118
   angular velocity, 75
   blade, 75
   frequency, 154
   hub, 173
   loads, 157
   sizing, 150
   thrust loading, 173

Roughness, 202
   lengths, 163
RPM control (Mechanical), 300

Sailing boat, 73
Salvage value, 249
Savonious rotor, 115
Scale factor, 28
Scale parameter and shape parameter, 155
Scattered light, 49
Scatterometry, 52
Scatterring geometry, 53
Sensor configuration, 46
SERI blade, 172
   sections, 171
Shadow flicker map, 279
Shape factor, 28
Shape parameter, 190
Shear exponent, 30
Shelterbelts, 204
Silicon carbide, 123
Simplified cost model, 161
Simpson's rule, 99
Sine wave, 293
Site analysis, 216
Site in flat terrain, 199
Site ranking, 38
Site selection, 220
Site suitability, 207
Siting, 185
   effort, 34
Skin friction, 72
Slant distance, 286
Slip speed, 298
Slip-stream, 80
Slip stream boundary, 91
Slope of the ridge, 207
Small HAWT, 121
Sodar, 47
Solidity, 3
   ratio, 90, 96, 98
Sound impact, 280
Sound level, 283
Sound response curve, 281
Source assessment, 33
Spherical hub, 174

Squirrel-cage induction generator, 298
Stall behaviour, 71
Stall control, 77
Stall region, 148
Stand-alone system, 7
Standard approach or first order approach, 93
Standard deviation, 23, 40, 190
Steady wind, 14, 19
Stiffness modulus, 171
Stochastic, 21
  load, 155
  models, 22
Straight line depreciation, 261
Streamline, 66, 71
Streamlined shapes, 64
Strength-to-stiffness ratios, 170
Strip theory, 90, 93
Strokes, 304
Structural properties, 169
Sub-polar flows, 15
Supervisory control and data acquisition, 308
Swept area, 3
Swoosh, 282
Synchronous generators, 297
Synchronous speed, 297
Synoptic fluctuations, 13
Synoptic-scale, 15
Synthetic aperture radar, 47, 52

Techno-economic analysis, 230
Teetering, 165
Television interference, 287
Terrain, 200
Throwing, 214
Thrust, 131
Thumb rules, 223
Time value of money, 243
Tip loss, 91
Tip loss correction, 93
Tip speed ratio, 3, 132
Topographic indicators, 37
Topographic maps, 38
Torque, 130
Torque coefficients, 134
Total power coefficient, 99
Tower, 120
  shadow, 21

Trade wind, 14
Trailing edge, 65
Transitional process, 73, 80
Transmitter, 51
Travelling-wave, 14
  systems, 16
Triangular-shaped ridge, 206
Tri-cylindrical hub, 174
True north, 44
Tubular turbine towers, 276
Turbine cost, 162
Turbines configuration, 218
Turbulence, 20, 37
  intensity, 22, 159, 227, 228
  models, 21
Turbulent viscosity, 225
Twist-flap coupled, 120
Typical commercial structure, 269
Typical power output, 113
Typical spacing, 217
Typical wind monitoring station., 43

Ultrasound anemometers, 42
Uniform cash flow, 244
Uniform present value factor, 248
Uniform roughness, 200
Unsteady flow processes, 79
Urban heat island effect, 19
US standard (ANSI/IEEE, 1986), 303
Utility power system, 309
Utility scale wind energy, 239

Vacuum infusion process, 119
Valley or canyon, 210, 211
Variable-speed generators, 299
Variable speed machines, 103
Variance, 26, 191
Variation of blade geometry, 181
VAWT, 74, 76, 117, 301
Vector diagram, 75
Velocity diagram, 130
Velocity profile, 72
Vertical axis machines, 115
Vertical profile, 30
Vertical-axis wind turbine, 1
Viscous boundary layer, 72

Visual impact, 278
Voltage, 290
Voltage step factor, 314
von Karman constant, 226
Vortex theory, 89

Wake behaviour, 203
Wake decay constant, 225
Wake effects, 217
Wake model, 224, 226
Wake profile, 226
Wake velocity, 82
Wake width, 227
Water pumping, 8
Weibull equation, 24
Weibull frequency distribution function, 25
Weibull k parameter, 152
Weibull parameters, 27, 31
Weibull scale, 190
   factor, 24
Weibull shape factor, 24
Westerlies, 12, 15
Wholesale price index, 246
Wind characteristics, 79
Wind data, 36
Wind electrical system, 192
Wind energy conversion system, 1
Wind farm project, 196
Wind flow modelling, 187

Wind flow models, 77
Wind flow pattern, 79
Wind machine characteristics, 101
Wind monitoring station, 45
Wind power density, 23, 36
Wind profiler, 51
Wind resource, 3, 195
Wind resource atlas, 35
Wind rose, 32
Wind shear, 79, 114, 163
Wind shear power law, 30
Wind speed profiles, 202
Wind turbine, 1, 110, 112, 118
   classes, 159
   generator, 1
   noise, 280
   size, 114
Wind vanes, 42
Windbreaks, 204
Wood/epoxy blade, 167, 168

Yaw, 111
Yaw control, 165
Yearly cost of operation, 254

Zero set pitch, 76
Zone of disturbed flow, 203